Cryocoolers

Part 1: Fundamentals

THE INTERNATIONAL CRYOGENICS MONOGRAPH SERIES

H. J. Goldsmid
Thermoelectric Refrigeration

G. T. Meaden
Electrical Resistance of Metals

E. S. R. Gopal
Specific Heats at Low Temperatures

M. G. Zabetakis
Safety with Cryogenic Fluids

D. H. Parkinson and B. E. Mulhall
The Generation of High Magnetic Fields

W. E. Keller
Helium-3 and Helium-4

A. J. Croft
Cryogenic Laboratory Equipment

A. U. Smith
Current Trends in Cryobiology

C. A. Bailey
Advanced Cryogenics

D. A. Wigley
Mechanical Properties of Materials at Low Temperatures

C. M. Hurd
The Hall Effect in Metals and Alloys

E. M. Savitskii, V. V. Baron, Yu. V. Efimov, M. I. Bychkova, and L. F. Myzenkova
Superconducting Materials

W. Frost
Heat Transfer at Low Temperature

I. Dietrich
Superconducting Electron-Optic Devices

V. A. Al'tov, V. B. Zenkevich, M. G. Kremlev, and V. V. Sychev
Stabilization of Superconducting Magnetic Systems

G. Walker
Cryocoolers, Part 1: Fundamentals
Cryocoolers, Part 2: Applications

Cryocoolers

Part 1: Fundamentals

Graham Walker
The University of Calgary
Calgary, Alberta, Canada

PLENUM PRESS • NEW YORK AND LONDON

Library of Congress Cataloging in Publication Data

Walker, G. (Graham), 1930–
Cryocoolers.

(The International cryogenics monograph series)
Includes bibliographical references and indexes.
Contents: pt. 1. Fundamentals—pt. 2. Applications. 1. Low temperature engineering. I. Title. II. Series.
TP482.W34 1983 621.5′9 83-2166
ISBN 0-306-40715-9 (v. 1)
ISBN 0-306-41219-5 (v. 2)

A Division of Plenum Publishing Corporation
233 Spring Street, New York, N.Y. 10013

Printed in the United States of America

High-capacity, three-stage Vuilleumier cycle cryocooler.
(Courtesy of Hughes Aircraft Co., Los Angeles.)

Foreword

The rapidly expanding use of very low temperatures in research and high technology during the last several decades and the concurrent high degree of activity in cryogenic engineering have mutually supported each other, each improvement in refrigeration technique making possible wider opportunities for research and each new scientific discovery creating a need for a refrigerator with special features. In this book, Professor Walker has provided us with an excellent exposition of the achievements of this period, the fundamental principles involved, and a critical examination of the many different cryogenic systems which have led to a new era of low-level refrigeration.

I feel fortunate to have had a part in the developments discussed in this book. During the early 1930s I constructed several rotary engines using leather vanes. Their performance was not good, but I was able to liquefy air. I had been impressed by the usefulness of leather cups in tire pumps and in Claude-type engines for air liquefaction. I was trying to find a way to avoid that part of the friction generated by a leather cup as a result of the radial force of the working gas on the cylindrical part of the cup. During the 1950s I built two efficient helium liquefiers in which essentially leather pistons were used. A steel core was encased in a stack of leather rings separated by thin steel rings of slightly smaller diameter, the stack of alternate leather and steel rings being compressed to provide a rigid piston which could be machined to fit the cylinder. The wearing quality was excellent. One of these liquefiers provided thousands of liters of liquid helium during about two years of use. Instead of becoming smaller in diameter because of wear, the diameter of the pistons actually increased by absorption of water. The necessity of controlling the water content was a disadvantage.

I was also intrigued by Heylandt's crowned piston concept during the 1930s but ruled it out because of the sealing problem at the warm end of the piston. I had seen some of them in action in oxygen plants. A slight air leakage was of no importance but a helium leak of such magnitude could not be tolerated. Little did I realize then that by 1960 I would find an adequate solution of the leakage problem by the use of rubber O-rings. During the late 1930s I experimented with free piston expanders and compressors and with diaphragm engines. When the war began in 1941 I was able to secure substantial funds for the development of oxygen generators and decided to see if Kapitza's engine could be useful. At least two attempts had been made to reproduce Kapitza's liquefier in the U.S. One was a complete failure and the other almost so.

I had tried several combinations of metals for the piston and cylinder. The problem was to produce a fit between piston and cylinder close enough to reduce leakage to a reasonable value but not close enough to encourage seizure. Of the four combinations I tried, Kapitza's choice of bronze and stainless steel was the least desirable. Kapitza found it necessary to employ a relatively wide annular gap and contrived a way for the piston to travel very rapidly during the power stroke so as to minimize leakage. By making both piston and cylinder of nitrided alloy steel (hardness = 90 RC) I was able to reduce the radial clearance by an order of magnitude and thus reduce leakage to a negligibly small value. It became possible, therefore, to use conventional gear for controlling the motion of piston and valves.

Several dozen engines with nitrided pistons and cylinders were manufactured for oxygen production during the war. After the war this type of engine was incorporated in helium liquefiers and was manufactured by Arthur D. Little and its successor, CTI-Cryogenics. Between 1947 and 1970, 365 units were marketed.

I started using oil-lubricated rubber O-rings to seal piston rods and valve stems in nitrogen and oxygen plants in 1950 and in helium liquefiers a few years later. Finally, I returned to the Heylandt type of piston, first in nitrogen plants, and in 1960 in a helium liquefier. The piston was made from a rod of laminated phenolic plastic (Micarta) about 2 ft long machined to fit loosely in a stainless steel cylinder sealed at the room-temperature end by a single O-ring. The advantages over the earlier engine proved to be enormous. The pistons and valves could be removed easily for inspection without disturbing the insulating vacuum or breaking any piping. There was no friction between piston and cylinder except when massive amounts of air or water gained access to the working fluid. The reliability was very high.

The loss of efficiency from the flow of gas to and fro in the annular space between the hot and cold ends was very slight and mostly from the mismatch of temperature between cylinder wall and piston surface. The thin (0.002–0.006-in.) annular passageway became an effective regenerator.

The inefficiency, as in other reciprocating engines, was mostly the result of irreversible heat transfer between the working gas and the walls of the expansion chamber in response to the relatively great temperature change the gas undergoes. Engines working at lower temperature levels tended to be more efficient because the drop in temperature of adiabatic expansion was smaller. In fact, the efficiency of the two-phase engine was apparently above 90%. Not only was the change in temperature very small, but also the vanishingly small heat capacity of cylinder and piston made the expansion truly adiabatic.

When I retired from MIT in 1964, I joined the engineering staff of A. D. Little and soon thereafter put together an experimental model of their standard liquefier equipped with Micarta pistons sealed by O-rings. A considerable number were produced over the next few years. Concurrently, I developed a slightly larger machine, again with Micarta pistons, which led to the 1400 series, completely replacing the earlier series. There are 200 of these units now in use.

A two-phase engine has replaced the Joule–Thomson valve in all of the recent machines I have built and in some that CTI has manufactured. The size of plant on which this change has been made ranges from 5 to 1000 l/hr. The gain in liquid production has been 25% to 33%.

One of the Naval Research Laboratory liquefiers has an inverted engine, that is the cold end of the cylinders is at the top. It was astonishing that the two-phase engine behaved normally and gave the expected improvement over the Joule-Thomson. It seemed strange to have liquid helium formed on top of the piston without appreciable loss by flowing down the annular gap toward the hot end of the piston. Apparently convection is very ineffective in thin layers of the order of 0.004 in. when so little time (a period of one stroke) is available.

There is a wealth of information in this book. The comprehensive Bibliography and Guide to Cryogenic Engineering Literature (Appendix II of Part 2) is indicative of the prodigious effort which Professor Walker has expended in preparation for writing this book.

Professor Walker and those who contributed certain chapters are to be congratulated on this scholarly and timely treatise on a very interesting subject.

S. C. Collins

Preface

The kindest thing to be said about this book is that it is like the curate's egg—good in parts. Many times in its composition I have felt I would be better engaged in *reading* rather than *writing* a book on cryocoolers. Perhaps my modest effort will stimulate others better qualified than I to do the job.

Completion is due entirely to Klaus Timmerhaus, Associate Dean of Engineering at the University of Colorado, and to Dr. Alan Clark of the National Bureau of Standards, Boulder, Colorado, joint editors of the International Cryogenics Monograph Series published by Plenum Press. They saw a draft of my first book on Stirling-cycle machines and flattered me with an invitation to write another on cryocoolers. With the starlight in my eyes I signed the contract that has been their rod to beat me with since. Now at the end I am most grateful for their unquenchable interest and enthusiasm despite the discouragements with which I have confronted them. They have proved the burr in my saddle that only a completed manuscript will remove.

Luster has been added to my effort by the substantial contributions of others. I am indebted first to Samuel Collins for the Foreword. Ray Radebaugh of the Thermophysical Properties Division of the National Bureau of Standards, Boulder, Colorado, contributed Chapter 11, dealing with his interest and expertise in the fundamentals of cryocooling and Chapter 12, dealing with the systems and techniques for achieving very low temperatures.

Fred Chellis, senior Applications Engineer of Cryogenic Technology Inc., Waltham, Massachusetts, has contributed Chapter 10, dealing with the practical problems of cryocooler design and operation. Fred was highly qualified to do this, having been around cryocoolers longer than he cared to remember and really should have been the author of this book. He died unexpectedly during the production of this volume; his death was a great loss to the cryocooler community.

Evgeny Mikulin, a Professor of Cryogenic Engineering at the Baumann Institute in Moscow, contributed Chapter 13. This review of the state of development in the USSR of cryogenic cooling engines was a most valuable contribution to offset my predominantly North American view. Mikulin spent two months visiting the University of Calgary in 1977 and we came to know him well. In a similar way Yoshihiro Ishizaki of the University of Tokyo contributed Chapter 14, a review of cryocooler development in Japan.

In Chapter 4, I have reproduced, with permission, a great deal of an unclassified report on Vuilleumier cooling engines by Ronald White of the U.S. Air Force Flight Dynamics Laboratory of the Wright-Patterson Air Force Base, Ohio.

Professor Kurti of the University of Oxford convinced me to use the descriptor *cryocoolers* rather than the more prosaic *cryogenic cooling engines* that came naturally to me.

I am most grateful to all these men.

The book is dedicated to a trio who in different ways, contributed much to my interest in this field. Aubrey Burstall, Professor Emeritus of the University of Newcastle, perceived with great foresight, the future significance of Stirling engines and initiated research at Newcastle on the subject soon after World War II. With much persistence and charm he secured delivery of the first Philips Stirling engine liquefier in Great Britain. It was my good fortune to install and research this machine under Burstall's guidance.

Jan Köhler led the team from Philips in their brilliant development of a commercial Stirling cryocooler. He is recognized as a principal figure in modern regenerative cryocooler development. I am most grateful to Köhler for the insight and inspiration I gained from his stimulating public presentations and technical writings.

Samuel Collins completes the trio to whom this book is dedicated. While serving as professor at the Massachusetts Institute of Technology, Collins invented a machine permitting the virtual routine production of liquid helium and thus unlocked the way to widespread superconducting research and development. When the ramifications are worked through and superconducting systems are commonplace, Collins' place in technology will surely be recognized as the equal of James Watt and the condenser: the development that prefaced the industrial revolution.

My own debt to Collins lies principally in his book *Expansion Machines for Low Temperatures*. By good fortune this became available simultaneously with the Philips engine and my need of it at Newcastle. I gained much insight and encouragement from his book.

Since those days, over twenty years ago, many have helped me to understand things better. My old friends Ted Finkelstein and William Beale come quickly to mind. More recently I have benefited much from mutual exchanges with Bill Martini, Costa Rallis, and E. H. Cooke-Yarborough.

My colleagues at the University of Calgary withstand, good-humoredly, my preoccupation with regenerative machines and my insistence on discussing with them matters I suspect are not their principal interest. In particular, John Kentfield is a worthy foil off which I have bounced many ideas. The Head of the Department, Dr. Peter Glockner, has helped in his encouragement and advice.

I wrote this text as my principal activity on a six-month sabbatical leave from the University of Calgary. I am most grateful to the University for the opportunity to devote myself to the task unhampered by my normal academic duties.

Many people helped in the production of the book. Bert Unterberger and his assistants worked tirelessly on the diagrams and illustrations contained herein. Karen Undseth, Edie Schulz, and Pamela Appleton did a good job transforming my chicken-scratching into a readable text. Marlene Stewart and Karen Undseth labored indefatigably in their customary fashion on the index, corrections to the original text, and the galley proofs thereby relieving me entirely of this onerous last lap. I am most grateful for all their help.

My children Josephine and Christopher have been deprived of their rightful allocation of my time but seem to have survived in good order.

Finally I owe my greatest thanks to my wife Ann. For over twenty years now she has heard *all* about Stirling and other engines and still manages a credible show of interest. With unfailing good humor she somehow, in the end, makes it all worthwhile.

G. Walker
Calgary, Alberta

Contents of Part 1

Contents of Part 2

Chapter 8 **Heat Exchangers in Cryocoolers**

Chapter 1

Introduction

DEFINITION

A cryocooler is a device or ensemble of equipment for producing refrigeration at temperatures less than 120 K (216 R). The "quality" or worth of a unit of such refrigeration depends on the temperature at which the refrigeration is available. This is illustrated in Fig. 1.1. The figure shows the theoretical ideal work required to generate a unit of refrigeration as a function of the refrigeration temperature.

At 0.5 K, 600 W of work are necessary to generate 1 W of cooling. At 4 K this has declined to 74 W of work input. At 10 K, 29 W are necessary, and at 120 K, only 1.5 W are required theoretically per watt of refrigeration. These values were calculated for an ideal concept based on the Carnot thermodynamic cycle with a "sink" temperature of 300 K. Unfortunately, these ideal values cannot be attained in practice. Historically, the actual work input has never been less than twice the Carnot value. In most cases, it is many times the ideal, perhaps ten or even a hundred times. Nevertheless, the example shown in Fig. 1.1 illustrates the point that the refrigeration temperature is as important a criterion as the actual quantity of refrigeration.

CLASSIFICATION OF CRYOCOOLERS

Temperature Level

For classification purposes it is expedient to divide the temperature range, 0–120 K, into six levels:

i. Level 1: 60–120 K
ii. Level 2: 20–60 K
iii. Level 3: 10–20 K
iv. Level 4: 4–10 K
v. Level 5: 1–4 K
vi. Level 6: less than 1 K

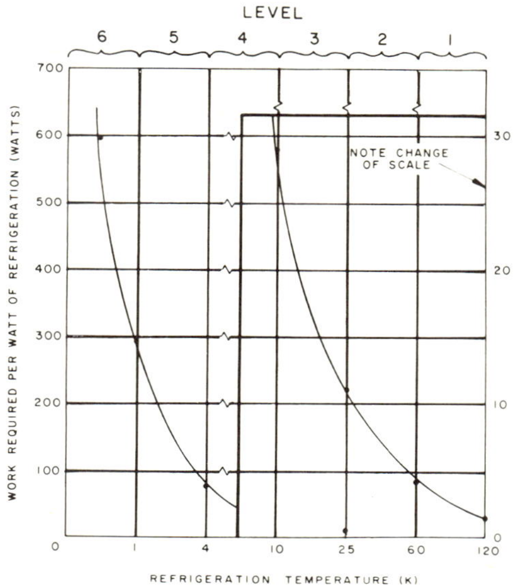

Fig. 1.1. Theoretical ideal work to generate a unit of refrigeration as a function of the refrigeration temperature (assumed to be the Carnot cycle with a sink temperature of 300 K).

Refrigeration Capacity

For the purposes of classification, another important parameter is refrigeration capacity. It is necessary to specify not only the refrigeration capacity, but also the temperature at which it is associated. None of the cryocoolers known at present can operate effectively over the whole cryogenic temperature range. Therefore, for classification and reference, benchmark temperature levels of 1 K, 4 K, 20 K, 80 K, and 120 K are proposed. These correspond approximately to the liquefaction temperatures at normal pressure of: helium (4 K), hydrogen (20 K), nitrogen (80 K), and methane (120 K).

A possible classification scheme for cryocoolers is shown in Table 1.1. In this classification five ranges of cooling capacity are prescribed: microminiature, miniature, small, intermediate, and large.

The principal means of achieving cryogenic refrigeration in the level 1, 2, 3, 4 temperature range (120–4 K) is the expansion of compressed gas in some form of engine. There are many varieties of cooling engines. They may be broadly classified as

i. Rotary machines.
ii. Reciprocating machines.

Table 1.1. Proposed Classification of Cryogenic Cooling Systems Based on Cooling Capacity and Refrigeration Temperature.

Designation	Cooling capacity				
	1 K	4 K	20 K	80 K	120 K
(a) Microminiature (very low capacity for electronic applications)	—	—	<0.25 W	<1 W	<1.5 W
(b) Miniature (low capacity electronic and instrument applications)	—	<0.5 W	<2 W	<8 W	<12 W
(c) Small (general purpose laboratory and commercial systems)	<1 W	<10 W	<100 W	<0.8 kW	<1.2 kW
(d) Intermediate (small-scale gas liquefaction and industrial superconducting systems)	<25 W	<100 W	<1 kW	<15 kW	<25 kW
(e) Large (large-scale gas liquefaction systems, LNG production, tonnage oxygen, helium recovery, etc.)	>25 W	>100 W	>1 kW	>15 kW	>25 kW

Rotary Machines

Rotary machines, turbines, and other forms of rotary expander, are characterized by small size and weight, high speed, and high fluid flows at relatively low compression ratios (maximum pressure/minimum pressure). Rotary machines are relatively inefficient in small sizes but improve with increases in size and speed. They are relatively inflexible for operation away from the design condition and so are best suited for continuous, steady-state operation. There are few bearings and no rubbing surfaces in contact so that reliability and life expectancy can be very high. Rotary machines are, inherently, in near perfect dynamic balance.

Reciprocating Machines

With reciprocating machines the effect of size is much less dramatic. To a first approximation they are as efficient in small sizes as in large. Of course the relative significance of the mechanical and thermofluid losses tends to decrease as the size of the machine increases.

Reciprocating engines can operate over a very wide pressure range and operate quite well even when conditions are reasonably removed from the design condition. They are relatively large, heavy machines and, to minimize friction and wear, must operate at low speeds, often an order of magnitude less than that of rotary systems. The increased number of moving parts and various rubbing seals and sliding bearings involved decrease the prospects for high reliability and long life so that relatively high maintenance costs might be anticipated for reciprocating systems. The out-of-balance reciprocating forces must be balanced in some way or absorbed in relatively complicated foundation or mounting installations.

Mixed Units

A cryocooler generally comprises a compression unit and an expansion unit. Both can be rotary or reciprocating or can be mixed. It is not uncommon to find in intermediate and the smaller large-scale cooling systems a combination of reciprocating compressor and turbine expander. This is because the process of *expanding* a gas in a turbine is immeasurably easier to achieve than compressing one. In a turbine expander the gas readily decreases to a lower energy level. In a rotary compressor the gas must be ‘coaxed'' by ingenious aerodynamic design to ascend in pressure against all natural inclination. The ``coaxing'' to higher pressure can be accomplished in more brutal fashion with a reciprocating compressor and much greater compression ratios can be achieved.

Large Systems

Despite the above difficulties, the advantages of reduced size, weight, and maintenance of rotary compressors compel their adoption in large systems to an increasing degree. It is fundamentally a question of development effort. Given the appropriate level of commercial application there is little doubt that satisfactory rotary compressors could be developed for all intermediate and large-scale applications. Substantial improvements have been made and work continues with a variety of compression equipment of the centrifugal, axial flow, and screw compressor variety.

Small Systems

Miniature, small, and intermediate cryogenic cooling systems almost invariably utilize reciprocating systems. Valiant efforts have been made with small turbosystems, but these remain of interest only in aircraft, spacecraft, and instrument applications, where size, weight, and dynamic balance are of primary significance.

Heat Exchangers

Heat exchangers are vital elements of all cryocoolers. Two principal types of heat exchanger are used: (i) recuperative and (ii) regenerative. A *recuperative* heat exchanger is a device where separate flow passages are provided for the hot and cold fluids. The fluids are separated by a solid wall and heat is transferred by conduction across the wall. The fluids may flow continuously or periodically. A *regenerative* heat exchanger has a single set of flow passages through which the hot and cold fluids flow alternately and periodically. The regenerative matrix, often a porous finely divided mass (of metal wire or spheres), may be thought of as a thermodynamic sponge alternately accepting or rejecting heat as the hot or cold fluid flows through it.

Both types of exchange may be used in counterflow or in parallel-flow operation. In counterflow operation the hot and cold fluids flow in opposite directions. In parallel flow operation the fluids flow in the same direction. Counterflow operation is much more effective than parallel flow operation and should always be used except in special circumstances.

The type of heat exchanger used, either regenerative or recuperative, is a key feature in identifying the particular family of machine. Within this family various machines may be further separated and classified by the thermodynamic cycle on which they operate or by some other feature.

Cryocoolers with regenerative heat exchangers include Solvay and Gifford–McMahon, Stirling, Ericsson, and Vuilleumier engines. Cryocoolers with recuperative heat exchangers include Linde, Hampson, Claude, Collins, and Joule–Thomson engines.

Flow Regulation

Another key feature is the manner in which the flow of working fluid is controlled, namely, by valves or by volume variations. All recuperative engines have valves. Therefore, classification by means of flow regulation is specific to regenerative engines. Here an arbitrary distinction is adopted:

i. Stirling engines where the flow is controlled by volume changes;
ii. Ericsson engines where the flow is controlled by valves.

This distinction between engines with and without valves is important and has a profound impact on the design and operation of regenerative machinery.

Application of This Text

This natural division of application, reciprocating engines to small coolers and rotary engines to large coolers, made the task of defining the limits of this presentation relatively simple. It is devoted to small and intermediate capacity systems, and so we have eliminated all discussion of rotary machines except for passing reference to the principal source materials. Attention will be concentrated wholly on reciprocating systems. A companion volume on rotary systems is contemplated.

The above definitions and classifications are not widely established, but are included as an attempt to bring some order to a confused field.

HISTORICAL DEVELOPMENT

Refrigeration as an aid to improving the quality of life has been known from early times. There are many Biblical references to "cool waters." The Romans had well-developed systems for hauling ice from distant Alpine regions and storing it underground in caverns insulated with straw. In England, Charles II was delighted by the ice cream confections of his Italian chef. All this depended on the natural refrigeration induced by environmental change.

It was not until the 19th century that serious development of mechanical refrigeration was attempted. Invention of the reciprocating steam engine made available the necessary power. It also triggered the development of improved metals and materials, better construction and manufacturing processes with consequent growth of populations and their expectations.

In 1815, Robert Stirling, a minister in the Church of Scotland, conceived the regenerative heat exchanger and the first engine now known as the Stirling engine. This was used for the production of power from furnace heat. However, in 1834, John Herschel, in a letter to "The Athenaeum" clearly anticipates application of the Stirling engine for making ice. It was not until 30 years later that Alexander Kirk (1874)* succeeded, about 1861, in reducing the concept to practice.

Collins and Cannaday (1958) have presented a good historical review of early refrigerative engines commencing about 1828 with the Cornish mine engineer, Richard Trevethick, better known for his work on steam power systems and his adventures in the silver mines of Peru. Collins makes the point that refrigeration expansion engines were not developed as

* References listed at the end of chapter.

well-defined independently built pieces of machinery. Instead they always appeared as components of integrated systems including compressors and the associated heat exchange equipment. Nearly all the early machines used air as the working fluid with the notable exception of the engine invented by the Australian, Davy Postle, in 1873. This machine used hydrogen and is further remarkable as the earliest known "free piston" engine and the forerunner of the class of cryocooler presently called Gifford–McMahon machines.

The first machines were used for making ice and for process refrigeration. The great impetus for large-scale refrigeration followed the opening of the Suez Canal and the development of the meat trade between Australia, New Zealand, and Great Britain. Lloyds Register for 1899/1900 records 285 ships with refrigerating facilities, 133 with air, 76 with ammonia, and 76 with carbonic acid as the working fluid. Twelve years later the corresponding figures were 66 with air, 241 with ammonia and 463 carbonic acid. This illustrates the growth of the meat and fruit trades and the decline of air cycle machines in favor of condensing cycles.

The development of cryocoolers followed much later. Collins (1958) writes:

> The equipment required for liquefying a gas, namely a compressor, a heat exchanger and an engine for expansion was fully appreciated as early as 1860. The lack of a compelling interest in the achievement of low temperature seems to have caused the many competent scientists and engineers of that day to neglect the subject.

Air was first liquefied in 1877 by Cailletet and, simultaneously, by Pictet. Working independently, both expanded high-pressure precooled air and oxygen in a nozzle to obtain the liquefied gas droplets.

The original engine expressly intended for the liquefaction of air was described by E. Solvay in a patent of 1887. The engine used hydrogen as the working fluid and air or other gases were expected to condense on the outer surface of the expansion cylinder. Solvay's engine failed to attain a sufficient reduction in temperature to achieve liquefaction of air. However, Collins attributes Solvay with "a real understanding of the problems of gas liquefaction in general" and refers to him "as the first to specify broadly the apparatus needed."

Karl von Linde, a professor at the Munich Polytechnical Institute in Germany, was, in 1895, the first to achieve the liquefaction of air in quantity. Linde's apparatus comprised a compressor, counterflow heat exchanger, and Joule–Thomson expansion valve. His work laid the foundation for the German Linde Company, now renowned in the design and construction of low-temperature equipment. Similar work was carried out in England contemporaneously by Hampson and led eventually to the British Oxygen Company, equally well known for innovative work in the cryogenic field.

The first use of counterflow heat exchangers was in 1849, by the American, John Gorrie of Apalachicola, Florida, in an air refrigerator for making ice. Later, in England, in 1854, Charles Siemens, in a provisional patent application, and elsewhere in a report on tests made with a British version of the Gorrie engine (Siemens, 1882), discussed the use of counterflow heat exchangers and clearly anticipated their use for the production of very low temperatures. In view of this early appreciation, Collins found it remarkable that more than 30 years were to elapse before Linde achieved the liquefaction of air in quantity.

In 1896, Kamerlingh Onnes of the Leiden Laboratory outlined schemes for the liquefaction of hydrogen by means of a heat exchange system employing liquid oxygen in the first stage and a hydrogen expansion engine for the final stage of cooling. The cycle he proposed was quite sound and, much later, was used for the liquefaction of hydrogen. However, Onnes failed in his early attempts and gave up. Later, of course, Onnes secured his niche in history with the first liquefaction of helium and the initial observation of the phenomenon of superconductivity.

In France, the resourceful Claude completed his first expansion engine in 1899. This was no better than the Solvay engine of twelve years earlier, but, Claude was made of sterner stuff and his perseverance and careful analysis eventually led to successful operation. For the first time a reasonably efficient expansion engine and heat exchanger were combined with compressed air free of water and carbon monoxide. When used in conjunction with an auxiliary liquefaction and separation element, the eagerly awaited continuous stream of liquid air was finally achieved in May, 1902. Claude's expansion engine development was accompanied by many improvements related to liquid air production and separation. His work led to the French industrial company L'Air Liquide, nowadays concerned with the construction and operation of plants for the production of industrial gases.

A noteworthy design feature for expansion engines introduced in 1912 by Heylandt, was the use of a "crowned" piston. A long cap of low-conductivity material was mounted on the piston to separate the cold expanding gas from the piston rings used to seal the high-pressure gas in the cylinder. This allows the seals to work at ambient temperatures and, more importantly, isolates the thermal energy resulting from frictional degradation of the seals thus preventing dissipation of the low-temperature refrigeration arising from expansion of gas in the cylinder. It is now a common feature of gas liquefaction machinery.

The first expansion engine for hydrogen was constructed by Claude in 1920 to separate hydrogen from coke oven gas. The hydrogen engine was similar in design to the earlier air engines. In his book Collins (1958) includes photographs of Claude expansion engines and an account of similar engines used in the United States by du Pont de Nemours.

Helium was liquefied in an apparatus incorporating a helium expansion engine described, in 1934, by Peter Kapitza, then a professor at the University of Cambridge. Kapitza, long resident in Russia, was recently (1978) awarded a Nobel prize for his extensive cryogenics activity. In the Kapitza engine, helium at 30 atmospheres pressure expanded to 2.2 atmospheres and achieved a temperature reduction from 20 to 10 K. The engine operated at about 100 strokes a minute and was operated by a hydraulic drive system. It was notable as the first use of a controlled leak of the pressurized working fluid to lubricate the piston during operation in the cylinder.

About the same time (1938) Collins completed a diaphragm expansion engine which operated sufficiently well on compressed air to justify incorporating the unit in a helium liquefaction system. The unit is described in detail by Collins in his book. Temperatures of 10 K were achieved in late 1939 but no liquid helium was produced because of leakage in the vacuum circuit. The war in Europe caused a redirection of effort to oxygen generation and work on the diaphragm engine was never resumed.

Turning now to a brief consideration of turbine systems, we return to the close of the 19th Century. This was the period when pioneer work on steam power turbines was in progress by Sir Charles Parsons and others. Possibilities for the use of turbines for gas liquefaction were not overlooked. An early discussion of the prospect was given in a letter written in 1898 by Lord Rayleigh to the Journal, *Nature*, wherein he suggested a turbine would obviate many of the practical difficulties of reciprocating engines. The same year, Edgar Thrupp, a British engineer, patented a liquefaction system using an expansion turbine. About the same time an American, Joseph Johnson, patented a liquefaction system incorporating a single-stage De Laval impulse turbine. U.S. patents for turbine expanders were granted in 1914 to the American inventors Charles and Orin Crummett. Later on, the American engineer Harvey Davis patented a compound expansion turbine for gas liquefaction in 1922.

However, it was not until the early 1930s that expansion turbines were applied commercially for gas liquefaction by the Linde Company. Initially, the turbine was an axial flow single-stage impulse machine but was subsequently replaced by an inward radial flow turbine invented by the Italian, Guido Zerkowitz. Other German refrigerative turbines of this era were described by Swearingen (1947) for both high-pressure and low-pressure applications. Few details of these machines are known but Collins (1958) describes them as "a commercial success and certainly matched the efficiency of any low temperature reciprocating expander available at the time."

An extensive theoretical treatment of refrigerative turbines was published in 1939 by Kapitza. He concluded that a low-pressure turbine liquefier was preferable to a high-pressure liquefier with reciprocating expander. Not only was the unit thermodynamically superior but it was also cheaper to build. Kapitza compared inward flow radial and axial flow impulse turbines and concluded the former was the preferred machine for gas liquefaction. Experimental results were presented for an inward flow radial expander capable of handling 570 kg of air per hour.

The above account of the early history of cryocoolers is a much abbreviated review of the historical survey given by Collins and Cannaday (1958). This excellent book will be found in any reasonably equipped library. Reference to it is strongly recommended.

RECENT DEVELOPMENT

Two important developments in cryocoolers occurred soon after the Second World War. Both developments have already had a major impact, but the principal ramifications are still developing and will eventually affect the course of human social development as we enter what people of vision already call "the coming age of superconductivity."

The first of these events was the development, in 1946, of a helium liquefier by Samuel Collins at the Massachusetts Institute of Technology. With two expansion cylinders and a compact contraflow heat exchanger it was possible to liquefy any gas without the need for supplementary refrigeration. Later, a large helium–hydrogen liquefier was fabricated in 1956 at MIT with special emphasis on a heavy-duty machine that would require minimum maintenance. Descriptions of the Collins engine are given in Collin's book (1958), and in his various other papers listed in the Bibliography. The Collins helium liquefier was developed for commercial application by the A. D. Little Co. of Cambridge, Massachusetts and was sold commercially by an associated company, Cryogenic Technology, Inc., of Waltham, Massachusetts.

Commercial development of the Collins Cryostat put the possibility for liquid helium research in the hands of any reasonably well equipped university or industrial laboratory. Moderate capital funds and will were the only prerequisites. This transformed a situation previously existing where helium research was confined to a zealous few prepared to put up with considerable difficulties and expense in their work. The transformation coincided more or less with an explosive growth in cryogenic engineering in the late 1950s. This arose as a consequence of long-range, rocket missile development, and the "Sputnik" era. Funds were virtually unlimited for military, education, and space-related activities in the United States at that time and all through the 1960s.

One consequence was the growth of interest about phenomena at helium temperatures, particularly in the field of superconductivity. Now, as a result, there exist or can be foreseen manifold military and civil applications for superconducting systems. These embrace the application of Josephson tunnel diodes to electronic computers, the use of SQUIDs (superconducting quantum interference devices) as sensitive magnetometers for all manner of dynamic measurements varying from magnetocardiograms of the heart to the detection of nuclear submarines. Superconducting magnet systems will be used in fusion and magnetohydrodynamic power systems, in magnetic levitation rapid transit systems, in high-speed ship propulsion, and in industrial electric motors of high power-to-weight ratios.

The other important development after the Second World War was the production of Stirling engines operating as refrigerators by the Philips Company in Holland. In the late 1930s, Philips started work on small combustion heated Stirling engines, to generate power for radios in areas without electricity. This development continued during the war. By 1946, an engine of one horsepower had been developed. The invention of transistors and improvements to storage batteries and tubes eliminated the need for the engine. However, work continued at Philips and has led to Stirling engines of 5-W output for artificial hearts to large 2000-kW engines burning solid industrial, municipal, and agricultural wastes. Engines of 150 kW have been installed in vehicles and boats with advantages in terms of efficiency, noise, multifuel capacity, and reduced environmental pollution (Walker, 1980).

Stirling engines can also be used as refrigerators. This was appreciated by John Herschel as early as 1834. Later, in 1874, Alexander Kirk described a Stirling cooling engine that had been used for ten years. However, little development of Stirling refrigerators had occurred until 1946, when one of the new Philips engines was driven with no heating. The unit operated so well as a refrigerator that the researchers were astonished to attain temperatures of the cylinder head so low that air liquefied on it. Thereafter, research on Stirling engine refrigerators was undertaken by a separate team led by Jan Köhler (1954). Under his direction, research proceeded apace and the machine was ready for commercial sale by 1956. Initially, the machine was used for air liquefaction. Shortly thereafter, a separation column was introduced to produce nitrogen. Other developments followed including the invention of multiple expansion machines to attain temperatures as low as 7 K incorporated in hydrogen and helium liquefiers.

Larger machines of intermediate capacity rating were developed for industrial liquefaction. Smaller machines of miniature and microminiature capacity were developed for military and electronic applications. Impetus for this work arose in the 1950s and 1960s from the development of military infrared night vision equipment and heat-seeking infrared missile guidance systems. Sensitive infrared devices were developed with appropriate optical characteristics at temperatures corresponding to liquid nitrogen (80 K) and liquid hydrogen (20 K). Their use in field applications required the development of miniature cooling engines capable of attaining these temperature levels with moderate electrical or thermal power input to obviate the need for liquid cryogens.

Much of the military interest and funding for research on infrared applications was centred in the United States. As a consequence, Philips research on small cooling engines was carried out by the North American Philips Laboratories at Briarcliff Manor, New York. Small cryocoolers were produced at the North American Philips factory at Ashton, Rhode Island. This concentration by North American Philips on small cryocoolers is thought to be the consequence of various "Buy American" acts of the U.S. Congress rather than a free choice by the Philips Company.

The manufacturing facilities at Ashton were not a commercial success and in the early 1970s were sold to a competitor, Cryogenic Technology Inc. (CTI), of Waltham, Massachusetts, an outgrowth of A. D. Little, Inc., Cambridge, Massachusetts, responsible for the commercial development of the Collins Cryostat. With the facilities at Rhode Island went the rights and privileges in North America of all Philips patents on Stirling cryocoolers. The sale included the agency for handling in the United States all the intermediate and large Stirling cooling engines made by the Philips company in Eindhoven. This consolidation established CTI as the leading cryogenic engineering company in North America and therefore the world.

The United States Government is never comfortable with a single source of supply for its requirements. It distributes knowledge of developments in technology to several suppliers and allocates purchases on considerations other than a straightforward commercial basis. One consequence has been the development of activity in small Stirling cryocoolers since the early 1960s.

Infrared night vision equipment was in the early stages of development at the end of the war. Missiles, notably the German V2, were at a similar stage. The obvious attractions of night vision equipment and heat-seeking missile guidance systems resulted in substantial funding for infrared (IR) equipment development. Good progress was made with IR sensors operating at cryogenic temperatures and by the early 1960s the real difficulty became the development of small, light-weight, low-power, low-interference cryocoolers capable of attaining and maintaining an IR sensor at 80 K and, later, 20 K. Sensors with favorable optical characteristics at these temperatures were developed in various laboratories using liquid-nitrogen and liquid-hydrogen cooling.

From the early 1960s substantial research funds for cryocooler development have been provided by the U.S. Government, principally the Air Force in a program from the Flight Dynamics Laboratories of the Wright-Patterson Air Force Base, Dayton, Ohio. Other substantial cooler development has been supported by the U.S. Army Night Vision and Electro-Optics Laboratory at Fort Belvoir, Virginia. The U.S. Navy also has interests in cryocooler development and various programs have been supported by the Naval Engineering Laboratory, Washington, D.C., the Office of Naval Research, Washington, and the Naval Weapons Research Laboratory at China Lake, California.

The National Aeronautics and Space Administration (NASA) is interested in IR systems and spacecraft cryocoolers. They have supported relatively small programs for development through the NASA Goddard Space Flight Center, Greenbelt, Maryland.

The principal United States Government participant in this work is the National Bureau of Standards Cryogenic Engineering Laboratory, Boulder, Colorado, recently reconstituted as the Thermophysical Properties Division of the National Engineering Laboratory.

Under the unique American system of easy access and wide dissemination of research information most of the achievements have been reported either in the open literature or in reports available through the National Technical Information Service (NTIS), Springfield, Virginia, or the Defense Documentation Center (DDC), Cameron Island, Virginia. A list of the available reports is contained in the Bibliography. Reports and patents are so numerous as to justify separate listing from the open literature.

A variety of exceptionally able and competent research and development teams for small cryocoolers have been assembled in the United States by the contractor companies. Most of the sponsoring agencies also maintain a small "in-house" research effort working in the frontier areas, to provide the technical capability to supervise the work of contractors. Among these, Philips Laboratories in Briarcliff Manor, New York, supported by the resources of the Philips Laboratories at Eindhoven, has to be one of the most experienced. They have a continuous program extending back to the earliest days of the pioneer Stirling engine work by Köhler at Eindhoven.

Similarly, Cryogenic Technology Inc. has a broad experience in both regenerative and recuperative cooling engines extending back through A. D. Little Inc., to the pioneer work of Samuel Collins at the Massachusetts Institute of Technology.

Another pioneer company in the Stirling cryocooler field in the 1960s was the Malaker Corporation of High Bridge, New Jersey. Under the technical leadership of John Daunt, now a professor at Stevens Institute, Hoboken, New Jersey, the company made many excellent, innovative cryocoolers, but failed commercially about the late 1960s/early 1970s.

Work on Stirling and Vuilleumier cycle cooling engines for missile infrared detector and guidance systems has been in progress since the mid 1960s at the Hughes Santa Barbara research center, and from about 1967 at the Hughes Aircraft Co., in Culver City, California. Production is concentrated at the Hughes plant at Torrance, California. The company has made a large number (literally thousands) of very advanced Stirling and Vuilleumier cryocoolers for missiles, far more than any other company. They have a policy of never publishing their work in the open literature. The NASA report by Russo (1976) is readily available from NTIS and is a good starting point for those interested in the Hughes activity.

Another California company active in the cryocooler field is Submarine Systems, Inc., now called Kinergetics Inc. of Tarzana. The company, led by Ken Cowans, formerly at Hughes, has done excellent work in the field of Vuilleumier and Stirling cryocoolers and in the development of super-compact recuperative heat exchangers.

More recent entrants are Texas Instruments Inc. of Dallas, (Anon, Texas Instruments, 1974), and the Martin-Marietta Corp. of Orlando, Florida. Both are active in the development and production of the small Stirling cryocooler like the one shown in Fig. 1.2. This is the standardized interservice cryocooler intended for forward-looking infrared (FLIR) systems. The same unit is also in production at CTI and Magnavox, the Dutch Philips Company subsidiary in North America, and in Europe at the German Company AEG-Telefunken A.G.

Woodard *et al.* (1978) of Martin-Marietta have reported a program of manufacturing methods for split-cycle Stirling cooling engines for the U.S. Army. A split-Stirling device was also featured in the Texas Instrument report cited above.

William Gifford, recently deceased, a coinventor at A. D. Little of the Gifford–McMahon cryocooler, contributed much to the cryocooler field. Gifford, a distinguished professor of engineering at the University of Syracuse, New York, founded Cryomech Inc. of Jamestown, New York. Cryomech manufactures Gifford–McMahon cryocoolers in a range of sizes for civil and military applications. Gifford–McMahon cryocoolers are rela-

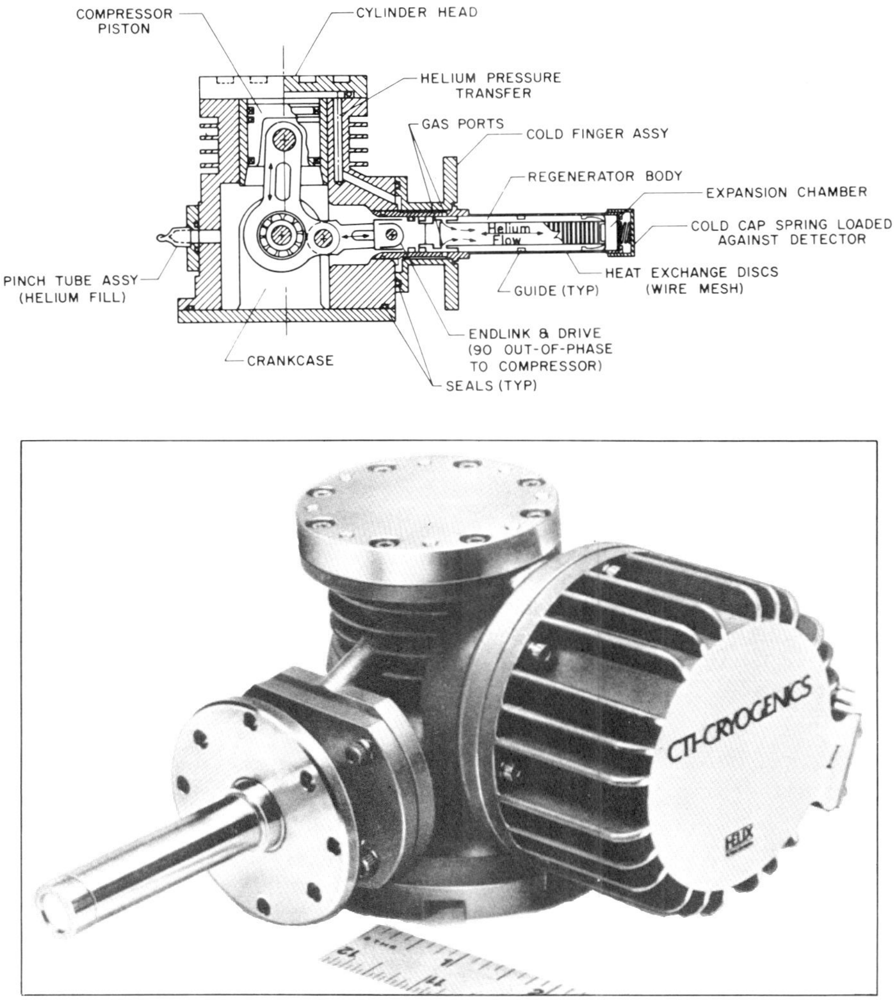

Fig. 1.2. Small integral Stirling cooling engine for forward looking infrared applications.

tively large, heavy machines with separate compressor and expander units characterized by exceptional durability and long life (Higa *et al.*, 1977). Gifford was a prolific writer and the Bibliography contains many entries indicative of his exceptional activity.

Gifford–McMahon machines are manufactured by CTI, Waltham, Massachusetts, and by Air Products and Chemicals Inc., Allentown, Pennsylvania, (APCI). Ralph Longsworth of APCI, a former graduate student and favored protégé of Professor Gifford, calls the APCI machine a "modified Solvay engine." It is indistinguishable from the Gifford–McMahon engine. Those wishing to clarify the distinction should refer to Chapter 2—Elementary Thermodynamics of Cryocoolers.

The exceptional sensitivity of infrared detectors and superconducting devices requires that the cold region of the cooling engine be isolated from mechanical shock and vibration, thermal variation, and electromagnetic interference. For systems mounted in gimbals to facilitate rotary or swivel action it is important to keep the mounted mass as small as possible. Much ingenuity has been exercised in efforts to satisfy these demands. Early success was achieved by condensing liquid nitrogen on a cold cylinder head and transferring the droplets along a tube to the cold region by the evaporative boiling process known as "Leidenfrost boiling." However, this technique appears to have been supplanted with the development of cooling engines comprised of separate or "split" expander and compressor systems.

Gifford–McMahon cryocoolers were always of this type. They have valves, two or four in number, with two connecting lines, one supply and one return between the expander and compressor units. Miniature split-Stirling cryocoolers are a comparatively recent development, although the arrangement whereby a displacer operates in a separate cylinder which is independently driven dates from the Stirling brothers' second engine in 1827. This was a standard beam engine intended for steam drive but modified to operate as a hot-air engine. Many later hot-air engines had separate cylinders and drive arrangements including the Robinson and Heinrici engines (see Walker, 1980). The Kirk engines of 1860 were also of this type, as was the free-displacer engine invented by Postle (1873) and by Ringbom (1905). However, it does not appear to have been widely used for cooling engines until quite recent times. Credit for the reinvention of free displacer cooling engines belongs to Walter Higa of the Jet Propulsion Laboratory, Pasadena (Higa, 1965). Similar concepts appear to have occurred to Stuart Horn at the Night Vision Laboratory and to Fred Chellis at CTI about the same time.

The great advantage of the split-Stirling cryocooler is that it has no valves and only one connecting tube. There are disadvantages, principally in moving the displacer in properly synchronized relation to the compressor piston. Despite this, the unit is sufficiently attractive to justify the present concentrated development effort.

Another company active in Vuilleumier cryocooler development was AiResearch Manufacturing Company of Phoenix, Arizona working on contract for the NASA Goddard Space Flight Center, Greenbelt, Maryland. Browning *et al.* (AiResearch, 1974) gave a comprehensive summary report on this program.

Military and spacecraft applications have severe weight and volume requirements. To attain these, the price must be paid in terms of regular maintenance and relatively limited life. There are many civil applications for cryogenic coolers of low capacity where the reverse is true. The important parameters are reliable operation with no maintenance and long lifetimes at low cost. There are virtually no restraints on size, weight, or efficiency. Applications include SQUID magnetometers for, say, routine magnetocardiograms in the offices of a physician and application of Josephson tunnel diodes in computers.

A cryocooler that has the potential to fulfil this function was described by Zimmerman and Radebaugh (1977). The unit is shown in Fig. 1.3. It was constructed of a readily available, low-cost fiberglass filament epoxy tube with a solid nylon displacer. The machine was capable of achieving a minimum temperature of 8.5 K. A subsequent special version of the engine achieved a minimum value of 3.5 K. This memorable achievement coincided with the visit by the author to the U.S. National Bureau of Standards Boulder laboratory of Zimmerman. At that low temperature, the helium working fluid (at a pressure substantially less than 1 atm)

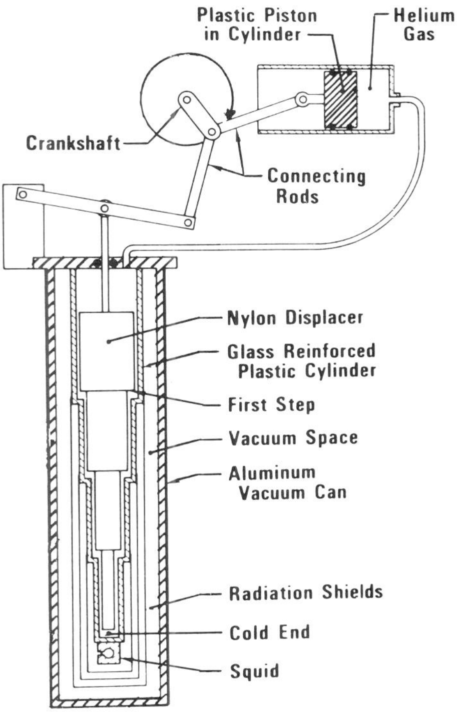

Fig. 1.3. Diagrammatic cross section of low-speed, low-cost, long-life cryogenic cooling engine (after Zimmerman and Radebaugh, 1977).

liquefied in the cylinder. It was the first time a two-phase single-component working fluid regenerative cryocooler had been successfully operated.

Such low temperatures are not necessary for the operation of superconducting devices. Transition to the superconducting state of niobium metal occurs at 9 K. Research on new superconducting materials is pushing the transition temperature towards 20 K.

The development of superconducting electronic devices has stimulated the interest of many companies with no previous history in the cryogenic field. There is every possibility that new concepts for low-cost maintenance-free cooling machines for civil applications will emerge from these companies untrammeled by the constraints imposed by military or space considerations.

Another important emerging civil field for cryogenic refrigeration is that concerned with superconducting power. The U.S. Navy has, for many years, been investigating and assessing the prospects for superconducting power for ship propulsion. There is increasing interest in electric power transmission by superconducting cable. This will entail a virtual pipeline of liquid or vapor phase helium with a liquid-nitrogen or natural gas shroud and having terminal and intermediate on-line cooling facilities.

Other superconducting power prospects with concomitant cryogenic cooling requirements include the intense magnetic fluxes required for fusion power and the magnetic levitation of trains and rapid transit systems.

The present explosive growth in the liquid natural gas (LNG) industry is likely to become a significant contributor to future needs for cryocoolers of intermediate capacity at relatively high temperatures. As LNG storage and pipeline transmission facilities develop, the need for intermediate cooling and reliquefaction systems will be established.

So far, attention has been concentrated principally on refrigerating cryocoolers. These are capable of attaining 80 K in a single stage of expansion. Multiple expansion machines of two, three, or four stages can attain temperatures of 20, 10, and 6–7 K, respectively. Most applications in the future will lie in this range. However, there will always be a demand for lower temperatures and so a brief outline of very low temperature cooling systems is included.

The best known system is the Joule–Thomson (JT) cryocooler. When a gas expands from a high pressure to a low pressure through a restricted orifice, the expansion occurs at constant enthalpy. If the initial temperature is below a critical value called the inversion temperature, the expansion will result in a decrease in temperature.

Air, and its components oxygen and nitrogen, can be readily liquefied by expansion in a JT system without precooling, although, of course, this helps. Hydrogen and helium can also be liquefied, but less readily, and need to be precooled (perhaps by an expansion engine) to a temperature below the inversion temperature before expanding. Many gases are still liquefied in commercial quantities by systems incorporating JT coolers, complex contraflow heat exchangers and reciprocating or turboexpansion engines.

Considerable effort has been invested in the development of miniature and microminiature capacity coolers for military infrared systems for aircraft, and for hand-carried portable and vehicle-mounted Army use. The systems have virtually instantaneous cool-down characteristics and the cold region is small, lightweight, and without vibration, and mechanical or thermal, and electromagnetic interference. Despite these advantages, JT coolers have some drawbacks that have inhibited their widespread use and which appear to be intractable. The first is that very low gas flow rates must be maintained since very small refrigerating effects are required (less than 1 W at 80 K). This involves use of an expansion orifice of minute dimension which can become blocked by small solids, either dust, debris, or frozen water and carbon dioxide. The expansion gas must, therefore, be ultraclean and free of contaminants. This eliminates use of atmospheric air and involves the use of nitrogen in compressed gas bottles. Compressed gas bottles are heavy and suitable for only short-term operation. The logistical and replacement problems are simply not practical for field use. Furthermore, in battle situations the high-pressure bottles can be dangerous if ruptured by a bullet or shrapnel. The British company, Hymatic Engineering Ltd. is well known in the field of portable JT coolers.

For temperatures below the liquefaction temperature of helium (4.2 K) there are three principal methods of refrigeration:

i. Dilution refrigerators.
ii. Magnetocaloric refrigeration.
iii. Electrocaloric refrigeration.

A review of these methods of very low temperature refrigeration can be found in Chapter 12, and in excellent books by Lounasmaa (1974) and Betts (1974), and in the survey papers by Steyart (1978).

This historical review is far from complete. We have entirely neglected the history and development of large-scale cryogenic refrigeration for gas liquefaction, tonnage oxygen production, and the like. This properly belongs to a future companion volume directed specifically to large-scale systems and turbines. We have also neglected many contributions to the development of cryogenic technology made in Europe, by the industrial organizations Linde Aktiengesellschaft Wiesbaden in Germany, L'Air Liquide Ltée in Paris, the British Oxygen Company in London, and cryogenic laboratories at the Universities of Oxford, Delft, and elsewhere.

There is a long tradition of cryogenic engineering in Russia evident by the recent (1978) award of a Nobel Prize to P. Kapitza. We have sought to fill this omission by inclusion of Chapter 13 contributed by Professor Ivan Evgeny Mikulin of the Moscow Baumann Technical High School. There is also an appreciable level of cryogenic activity in Japan. This is reviewed in Chapter 14, contributed by Yoshiro Ishizaki of the University of Tokyo.

In our review of the cryogenic literature, Chapter 16, sources of information are identified for those seeking to rectify other deficiencies.

In conclusion, mention must be made of other research not specifically concerned with cryocoolers, but so pertinent to it as to be of substantial interest. Over the past decade research on artificial hearts, using radioisotope heating and regenerative engines to produce work, has been funded by the U.S. National Institutes of Health. The problems encountered by companies engaged in this work are similar to those encountered with small regenerative cooling engines. The artificial heart program is described in a recent book on Stirling engines by Walker (1980).

Free-piston Stirling engines are of substantial research interest both as prime movers and as refrigerating machines. Chapter 11 in the above book on Stirling engines was contributed by William Beale, the inventor of the free-piston Stirling engine. A companion volume devoted entirely to free-piston Stirling engines is in course of production (Walker and Senft, 1983). Philips at Eindhoven already have in commercial production free-piston Stirling cryocoolers. ERG Inc., of San Francisco, and Sunpower Inc., Athens, Ohio, were recently awarded development contracts by the NASA Goddard Space Flight Center, Greenbelt, Maryland.

STATUS SURVEYS

Various surveys of cryocooler systems have been attempted. One of the most extensive and comprehensive compilations was made by Strobridge (1974) of the National Bureau of Standards Cryogenic Engineering Laboratory. He solicited data from all possible sources and compiled these in the form of charts reproduced herein. A total of 144 refrigerators and liquefiers are included, 21 from Europe, the remainder from the U.S.

He divided the systems into three groups on the basis of refrigeration temperature:

Group A—1.8 to 9 K
Group B—10 to 30 K
Group C—30 to 90 K

He further distinguished the various systems as follows:

i. Old, those included in a previous survey (1969);
ii. New, systems not included in the previous survey;
iii. Under development, as the name implies, systems not available commercially but at an advanced stage of development.

To maintain his guarantee of confidentiality to those providing the data, Strobridge offers no information permitting the identification of any of the systems shown, nor was the attempt made to compare the various types of system or distinguish between Stirling, Vuilleumier, Gifford–McMahon, etc.

The Strobridge report contains the four charts reproduced in Figs. 1.4–1.7. These show the efficiency, mass, and volume of over 100 cryocoolers as a function of the refrigerating capacity ranging from less than a watt to 1 MW. Mean characteristics have been drawn on all four charts. Costs are presented as a function of input power rather than refrigerating capacity.

The efficiency is presented as a percentage of the Carnot value defined as η = actual COP/Carnot COP (COP=coefficient of performance). The volume and mass are believed to be of a complete operational system. The cost is the quoted price (in 1974 U.S. dollars) for low-volume purchases, i.e., one to five units. Strobridge recommends that costs be estimated by the equation $C = 6000P^{0.7}$, where C is the cost in dollars and P is the installed power. An allowance of 10% per annum for inflation would raise

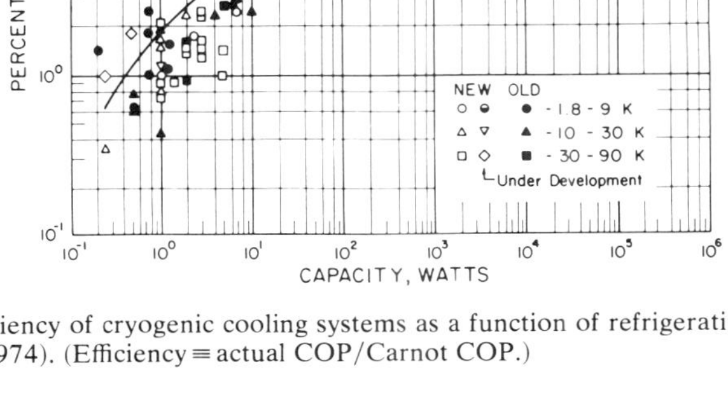

Fig. 1.4. Efficiency of cryogenic cooling systems as a function of refrigerating capacity (after Strobridge, 1974). (Efficiency ≡ actual COP/Carnot COP.)

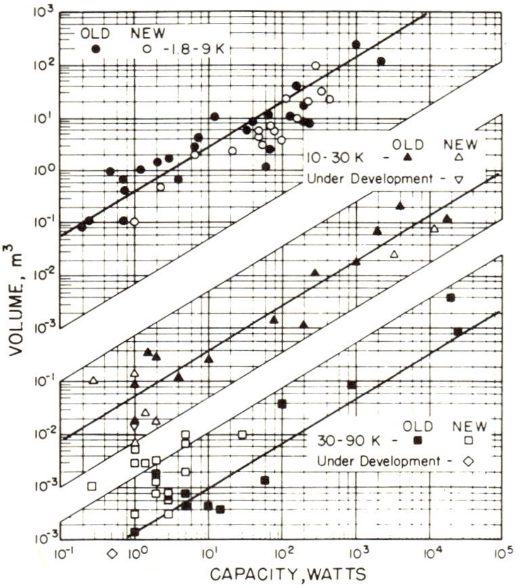

Fig. 1.5. Volume of cryogenic cooling system as a function of refrigerating capacity (after Strobridge, 1974).

the value of the constant from 6000 in 1974 to 10,000 in 1980, 17,000 in 1985, and 25,000 in 1990.

Figures 1.5 and 1.6 can be used to determine the volume and mass of cryocoolers as a function of refrigeration temperature as shown in Fig. 1.8. In this figure, characteristics are sketched in for refrigerating capacities of 1, 10, 100, and 1000 W.

Charts such as these have obvious attraction for planning purposes. It must be recognized their value is rather limited. At first glance there appears to be a surprisingly close similarity for all the systems. However, the charts have logarithmic scales and small chart separations can be differences of several times. For example, consider Fig. 1.4. At 1 W capacity the range of variation in efficiency is from a low value of 0.4% to a high value of 10%—a range of variation of 25 times. Similarly, at 1000 W capacity the range of variation is between high and low values of 50% and 7%, a range of 7 times.

A most important factor not addressed in the Strobridge charts is the useful life or the maintenance requirements of the different systems. A cryocooler for aircraft or space use would likely be small, lightweight, highly pressurized, fast acting, and consequently have a short life or operating period between maintenance. The same equation relating life, strain, size, weight, and cost applies inexorably to cooling engines as to all other machines or biological systems. Only the exact terms of the equation are unknown and must be defined by experience.

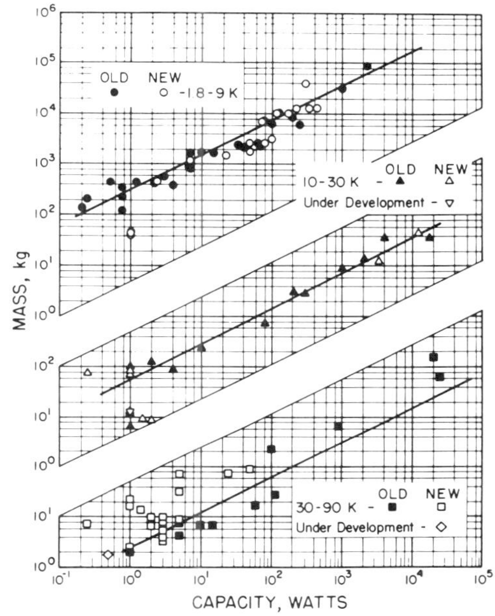

Fig. 1.6. Mass of cryogenic cooling systems as a function of refrigerating capacity (after Strobridge, 1974).

An extensive study of cryocoolers for spacecraft to provide long-term storage of cryogenic fluids was carried out by Jensen *et al.* (1971) for the NASA Manned Spacecraft Center. The four volumes of the Jensen study contain a wealth of information. The study is now somewhat dated but few fundamental changes have occurred so that the material remains a valid reference.

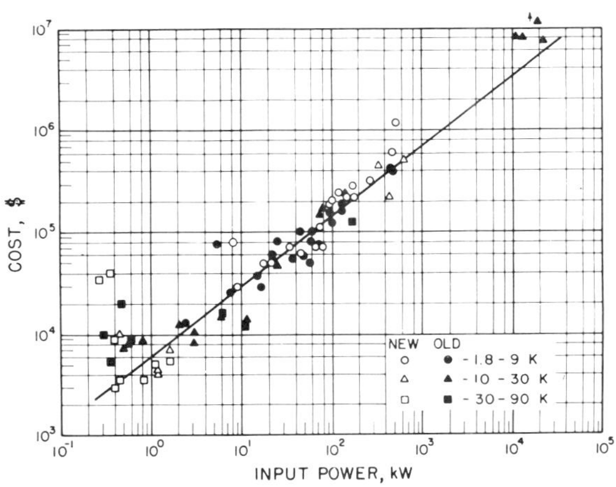

Fig. 1.7. Cost of cryogenic cooling system as a function of input power (after Strobridge, 1974).

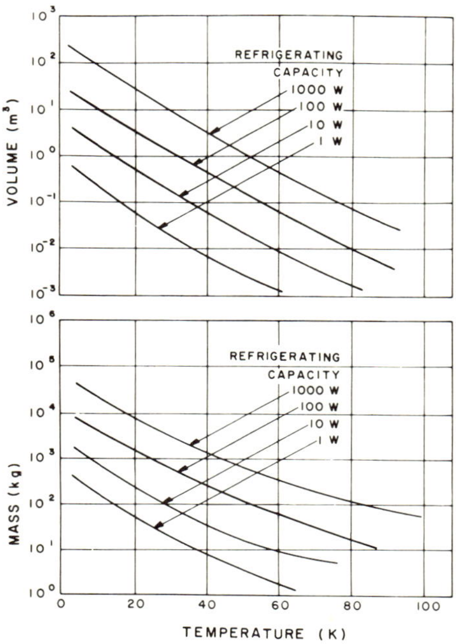

Fig. 1.8. Volume and mass of cryogenic cooling systems as a function of refrigeration temperature (prepared from Figs. 1.5 and 1.6) (after Strobridge, 1974).

Their findings are summarized on Figs. 1.9 and 1.10, reproduced from the Jensen report. These show the specific weight (kilograms per watt) and the coefficient of performance of a variety of cryogenic cooling systems as a function of refrigerating capacity at 77 K.

The cryocoolers considered include the following:

i. A group of Ericsson cycle systems, Gifford–McMahon, Taconis, and Solvay;
ii. Joule–Thomson systems;
iii. Brayton cycle systems;
iv. Vuilleumier cycle engines;
v. Stirling cycle engines.

In their report, Jensen *et al.* presented similar curves for cryocoolers providing refrigeration at 20 and 4.2 K. They paid particular attention to the problems of heat rejection by radiator, and assumed this to be included in the total system.

A feature of Figs. 1.9 and 1.10 is the relatively low specific weight and the high coefficient of performance of the Stirling cryocoolers. Another interesting comparison is the order-of-magnitude difference between

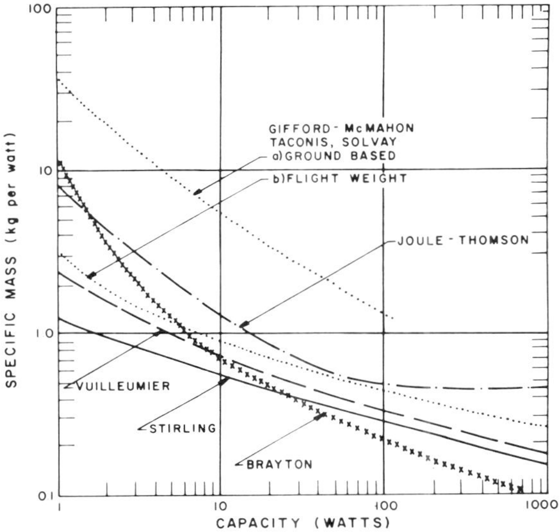

Fig. 1.9. Specific weight of cryogenic cooling systems as a function of capacity at 80 K (after Jensen *et al.*, 1971).

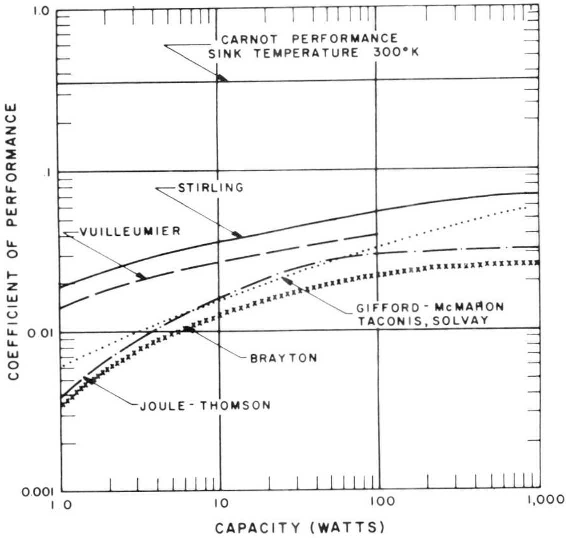

Fig. 1.10. Coefficient of performance of cryogenic cooling systems as a function of capacity at 80 K (after Jensen *et al.*, 1971).

ground-based and flight weight Gifford–McMahon, Taconis, and Solvay group of machines.

Earlier, Daunt and Goree (1969) carried out a survey of miniature cryogenic refrigerators for the U.S. Navy Office of Naval Research. This survey contains a large compendium of data for cooling engines then extant. The report is still of substantial interest but the information contained therein is now more than a decade old and could not be regarded as a contemporary review.

Another very comprehensive survey of cryogenic cooling techniques for space-borne systems was carried out by Donabedian (1972). This included an evaluation of mechanical closed-cycle cryocoolers, open-cycle expendable systems, passive radiation concepts utilizing the low temperature of deep space, and, finally, thermoelectric devices. This report is highly recommended. A survey of cryogenic coolers for space systems, carried out by Vance (1974), is not as comprehensive as the Jensen or Donabedian reports cited above. Cox (1974) reviewed potential cooling methods for an underwater naval detection system using a cryogenic SQUID. He concluded that the choice of cryocooler would lie between Claude or Stirling cycle engines in conjunction with a JT valve.

Morgan (1971) of Hughes Aircraft carried out a study for the U.S. Air Force of airborne air liquefiers. Basically, the study involved a comparison of Stirling and Vuilleumier cycle liquefiers capable of producing 2.3 kg per hour of liquid air for use in flight. Various parametric studies were completed and preliminary designs prepared for two systems. The Stirling system was found to be more efficient, lighter in weight, smaller in size, and cheaper to make than the Vuilleumier machine. The only advantage offered for the Vuilleumier machine was a presumed longer life due to reduced mechanical loading.

A very comprehensive survey of cryogenic refrigerators for superconducting transmission cables was made by Kadi and Longsworth (1976). This survey is of particular interest as it was directed to intermediate and large capacity systems involving both reciprocating and rotary systems. Much technical information of existing or proposed systems in Europe and North America is given.

REFERENCES

AiResearch Manufacturing Co. (1974). *Fractional Watt Vuilleumier Cryogenic Refrigerator Program Engineering Notebook*, Vol. 1, *Thermal analysis*. Contract NAS 5-21715, Goddard Space Flight Center, Greenbelt, Maryland, May.

Anon—Texas Instrument Equipment Group. (1974). "Final Engineering Report on the Design and Development of Two Miniature Cryogenic Refrigerators." Contract No. DAAK02-73-0495, Night Vision Lab., Ft. Belvoir, Virginia, NTIS Report No. AD 784436.

Betts, D. S. (1974). *Refrigeration and Thermometry Below 1 K*. Sussex Univ. Press, Falmer, Sussex (in the U.S., Crane Russak and Co. Inc., 347 Madison Avenue, New York).

Collins, S. C., and Cannaday, R. L. (1958). *Expansion Machines for Low Temperature Processes*. Oxford University Press, Oxford.

Cox, J. E. (1974). "Potential Cooling Methods for an ELF SQUID." Naval Res. Lab., NRL Memorandum Report 2899.

Daunt, J. G., and Goree, W. S. (1969). "Miniature Cryogenic Coolers." Office of Naval Research, NTIS AD 860866.

Donabedian, M. (1972). "Survey of Cryogenic Cooling Techniques." U.S. Air Force, SAMSO Report No. SAMSO-TR-73-74, AD No. 755 780, NTIS, Springfield, Virginia.

Higa, W. N. (1965). "A Practical Philips Cycle for Low Temperature Refrigeration." *Cryogenic Technol.* **8**, 203–209, (July–August).

Higa, W., and Wiebe, E. (1977). "One Million Hours at 4.5 Kelvin. Appl. of Closed Cycle Cryocooler to Small Superconducting Devices", Proc. of Conf. NBS Boulder, Oct., pp. 99–108 (issued as NBS Spec. Pub. 508, April 1978).

Jensen, H. L., Nast, T. C., Glassford, A. P. M., Vernon, R. M., and Ekern, W. F. (1971). "Investigation of External Refrigeration Systems for Long-Term Cryogenic Storage." Summary Report, Report No. NASA CR 115192, NASA Manned Spacecraft Center, NTIS Accession No. N71-37107 (Lockheed Missiles and Space Co., Sunnyvale, California) (February 1971), See also, same authors, same title: (a) System Review Report, NASA CR 115191, NTIS N71-37105 (May 1970); (b) Final Report, NASA CR 114920, NTIS N71-20279 (February 1971).

Kadi, F. J., and Longsworth, R. C. (1976). "Assessment and Study of Existing Concepts and Methods of Cryogenic Refrigeration of Superconducting Transmission Cables." E.R.D.A. Report No. C00-2552-6, NTIS, Springfield, Virginia.

Kirk, A. (1874). "On the Mechanical Production of Cold," *Proc. Inst. Civil Eng.* **37**, 244–315 (London).

Köhler, J. W. L., and Jonkers, C. O. (1954). "Fundamentals of the Gas Refrigeration Machine." *Philips Tech. Rev.* **16**(3), 69–78; See also "Construction of the Gas Refrigerating Machine," *Philips Tech. Rev.* **16**(5), 105–115.

Lounasmaa, O. V. (1974). *Experimental Principles and Methods below 1 K*. Academic Press, New York.

Morgan, N. E. (1971). "Analysis and Preliminary Design of Airborne Air Liquefiers." AFFDL-TR-71-171, WPAFB, Dayton, Ohio, NTIS AD 892500 (Hughes Aircraft Co., Electro-Optical Division, Culver City, California).

Russo, S. C. (1976). "Study of a Vuilleumier Cycle Cryogenic Refrigerator for Detector Cooling on the Limb Scanning Infrared Radiometer." Report No. NASA-CR-145078, NTIS N77-11211/8ST.

Siemens, C. W. (1882). *Proc. Inst. Civil Eng.* **68**, 179–186.

Solvay, E. (1887). Deutches Reichspatent No. 39280.

Steyart, W. A. (1978). "Magnetic Refrigerators for Use at Room Temperatures and Below." Report No. LA-UR 78-1764, Los Alamos Scientific Laboratory, New Mexico.

Strobridge, T. R. (1974). "Cryogenic Refrigerators—An Updated Survey." National Bureau of Standards Tech. Note 655 (Supt. Documents, U.S. Govt. Printing Off.) (12 pp.).

Swearingen, J. S. (1947). "Expansion Turbines for Low Temperature Processing." *Trans. Amer. Inst. Chem. Eng.* **43**, 83–90.

Vance, R. W. (1974). "Cryogenic Coolers for Space Systems." Report No. SAMSO-TR-74-180, Space and Missile Systems Organization, A.F. Systems Command, Los Angeles, A.F. Station, NTIS AD-785-083 (Aerospace Corp., El Segundo, California).

Walker, G. (1980). *Stirling Engines*. Oxford University Press, Oxford.

Walker, G., and Senft., J. (1983). *Free-Piston Stirling Engines* (in preparation).

Woodard, R. S., Welch, P. H., and Jansson, R. M. (1978). "Manufacturing Methods and Technology for the Establishment of Production Techniques for a Split-Cycle Stirling Cryogenic Cooler," Report No. 15181, Contract DAA B07-77-C-0631, U.S. Army Elec. R. and D. Comm., Fort Monmouth, New Jersey (Martin-Marietta Corp., Orlando, Florida).

Zimmerman, J. E., and Radebaugh, R. (1977). "Operation of a SQUID in a Very Low-Power Cryocooler. App. of Closed Cycle Cryocoolers to Small Superconducting Devices," Proc. of Conf. NBS, Boulder, October 1977, pp. 59–66 (Issued as NBS Spec. Pub. 508, April 1978).

Chapter 2

Elementary Thermodynamics of Cryocoolers

INTRODUCTION

The objective of this chapter is to review the various operating cycles of cryocoolers. For this to be meaningful it is necessary to be acquainted with the elements of thermodynamic analysis. For the benefit of those without such acquaintance we include a brief thermodynamics review. To the purists we offer apologies for liberties taken with the subject in the interest of brevity. A more rigorous treatment may be found in university engineering texts (see Rogers and Mayhew, 1967).

Part I: *Thermodynamics Review*

Matter can exist in three *phases*: gas, liquid, and solid. It can be changed from one phase to another by the transfer of energy to or from the *system* (the particular element of matter under consideration). It is a matter of common experience that to melt ice (water in the solid phase) we supply heat, whereupon the ice is converted to liquid. Further heating causes the liquid to boil and evaporate to vapor or gas. The difference between vapor and gas is that in one case (vapor) the temperature is below and in the other case (gas) is above the *critical temperature*.

STATE PROPERTIES

The *state* of a substance can be defined by the *state properties*. These are various properties of the substance that characterize it at any given condition. There are six state properties of primary interest in thermodynamics and others of less interest for specialized applications. The combination of any *two* properties is sufficient to completely specify the state of a pure (single component) substance. The six properties of general interest are: pressure, volume, temperature, internal energy, enthalpy, and entropy.

Consider now the simple closed system shown in Fig. 2.1 consisting of a piston and cylinder closed at one end. Assume the piston has no mass and is free to slide without friction in the cylinder. We assume further there are no leaks of the working fluid past the piston seal. Finally we assume that some known mass of working fluid is contained in the cylinder, say 1 kg. A weight of mass M placed on the piston will cause it to descend until it comes to rest as shown in Fig. 2.1, a distance L from the bottom of the cylinder.

Volume

For the system shown in Fig. 2.1 the state property *volume*, V, is the total volume of the working fluid contained in the cylinder with diameter, D, below the piston, i.e., $V = (\pi/4)D^2L$. We have unit mass of working fluid in the cylinder so the volume measured is the *specific volume*.

Pressure

Pressure is the force exerted on the piston to keep it "floating" in the cylinder a distance L from the bottom expressed as the force exerted per unit area of the piston. In the system shown in Fig. 2.1 the total downward force is the gravitational force acting on the mass M on the piston.

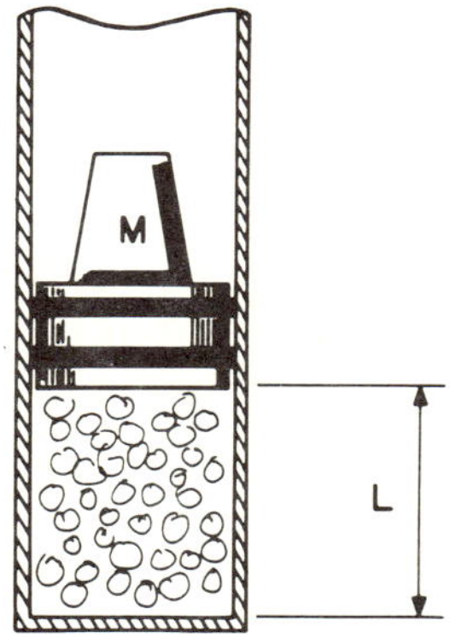

Fig. 2.1. Simple piston/cylinder apparatus.

From Newton's law, force = mass times acceleration, we know the downward force, F, is

$$F = Mg \tag{2.1}$$

The downward force is resisted (and balanced exactly) by the upward force exerted on the piston by the working fluid acting uniformly over the whole area of the piston, A:

$$pA = F \tag{2.2}$$

the pressure can therefore be calculated simply by dividing the known downward force by the area of the piston so that

$$p = F/A = Mg/A \tag{2.3}$$

Temperature

The *temperature*, T, indicates the energy level of the working fluid. Using the concepts of the kinetic theory of gases one can assume the working fluid to be composed of molecules consisting of "billiard balls" of matter either singly or in groups of two or three moving about in random fashion as well as vibrating and rotating. The temperature is the degree of activity of the billiard balls. Application of heat causes the billiard balls to become more agitated and move faster with increased random collisions. This is perceived as an increase in the temperature.

Temperature is measured on four scales: Centigrade, Fahrenheit, Kelvin, and Rankine. Two useful temperatures are the ice-point and boiling point of water at atmospheric pressures. These correspond to 0°C and 100°C for the Centigrade scale or 32°F and 212°F for the Fahrenheit scale.

The lowest temperature that can exist is −273.15°C. This is taken as the starting point or zero datum of the Kelvin scale. The Centigrade unit is the same as the Kelvin unit of temperature change. The absolute minimum temperature (−273.15°C) corresponds to −459.67°F. This is taken as the zero datum of the Rankine scale. The Fahrenheit unit is the same as the Rankine unit of temperature change.

Internal Energy

Internal energy, U, is similar to temperature as a measure of the energy level of the working fluid. According to Joule's law, the internal energy of a perfect gas varies only with the temperature. In the special case of a perfect gas the change in internal energy is the change in temperature multiplied by the specific heat at constant volume, C_v, i.e.,

$$dU = C_v \, dT \tag{2.4}$$

Enthalpy

The *enthalpy*, H, is property of state defined by three other state properties in combination:

$$H = U + pV \tag{2.5}$$

It is a useful measure to facilitate calculations for continuous flow systems where combinations of the internal energy and the "flow work", pV, occur frequently.

For a perfect gas the change in enthalpy is simply

$$dH = C_p\,dT \tag{2.6}$$

where C_p is the specific heat at constant pressure.*

Entropy

Entropy, S, is a property of state useful in calculating the quantities of energy with respect to temperature exchanged in changing from one state to another. It is defined by the equation

$$dS = \int (dQ)/T \tag{2.7}$$

The change in entropy is the integral sum of the quantity of heat transferred, Q, divided by the temperature at which it is transferred.

Books have been written about entropy and students become confused by it. Everything that happens causes an increase in the entropy. It becomes difficult to understand how this incessant aggregation does not result in the whole world getting top-heavy with entropy. To the newcomer not of philosophical bent, entropy is best regarded simply as a handy way to calculate heat transfers.

Equations of State and Thermodynamic Tables

The above six properties of state (temperature, pressure, volume, internal energy, enthalpy, and entropy) are interrelated, and any two, say the pressure and temperature, completely define the state of the substance. In practice gases never are perfect, but many approximate this condition. A perfect gas is one that obeys the simple equation of state:

$$pV = mRT \tag{2.8}$$

where m is the number of moles of gas present and R is a constant that

* Since $pV = RT$ then $dH = C_v\,dT + p\,dV = C_v\,dT + R\,dT$. Since $(C_v + R) = C_p$ then $dH = C_p\,dT$.

is characteristic of the gas and equal to $(\bar{R}/M)$, where $\bar{R}$ is the universal gas constant and M the molecular weight of the fluid.

Many gases follow an equation of state:

$$pV = ZmRT \tag{2.9}$$

Z is the coefficient of compressibility, a factor which varies widely as the condition of the gas changes. It is determined experimentally. More complicated equations of state exist, or the properties may be determined experimentally and summarized in tables of thermodynamic data.

Properties of state for various fluids have been measured over many levels of temperature and pressure. The results are best presented in tabular form called thermodynamic tables. The tables are usually in two parts. The first part contains properties of the *saturated* fluid, the second part contains properties of the *superheated* fluid.

THE TEMPERATURE–ENTROPY (*T–S*) PLANE

The distinction between saturation and superheated fluid is shown in Fig. 2.2, where the temperature, T, is plotted as function of the entropy, S. It is called the T–S plane. Assume the simple piston/cylinder system, shown in Fig. 2.1, is at such a low temperature that the working fluid is all liquid. We can represent this by point A on Fig. 2.2. If the cylinder is heated, the temperature will rise and the working fluid will expand and cause the piston to move up the cylinder. The pressure will remain the same because the mass on the piston is unchanged. This will continue until, at point B, the liquid working fluid is saturated (*saturated liquid*). Further addition of heat will cause evaporation of the working fluid. This evaporated vapor will be at the same tempearture and pressure as the liquid but will have radically different values of specific volume, entropy, enthalpy, and

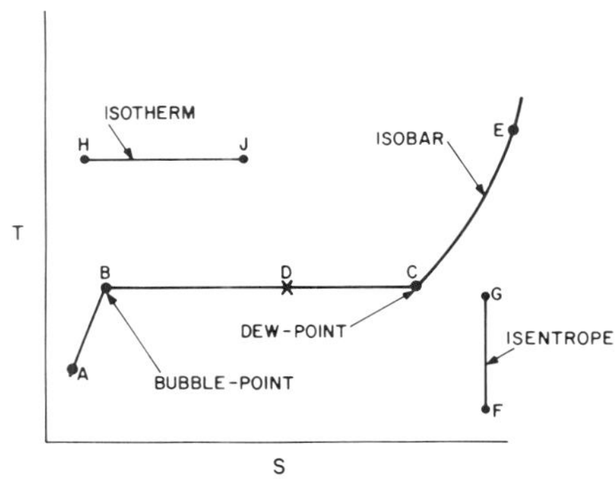

Fig. 2.2. Heating and evaporation process displayed on the temperature–entropy plane.

internal energy than the liquid. Let that state be represented by point C. As more heat is added an increasing fraction of the working fluid will evaporate. The temperature and pressure remain the same but the piston will move a long way up the cylinder. The volume of the evaporated state is much greater than the liquid state. The heat added to cause evaporation of the working fluid is called the latent heat (or enthalpy) of vaporization.

In the process of vaporizing we can think of the fluid moving from point B to point C. At some intermediate point, say D, we can talk about the fluid having a "vapor fraction" x. This is the fraction of fluid at the saturated vapor state C. It may be calculated simply as

$$\text{vapor fraction } x = \frac{\text{length } BD}{\text{length } BC} = \frac{S_D - S_B}{S_C - S_B} \tag{2.10}$$

The fluid is never actually at any intermediate state for it can only exist at *either* state B or state C. It is impossible for any particular element of the fluid to be at state D. The intermediate state is simply a concept to help visualize the fraction of fluid that has evaporated. The vapor fraction x is simply the fraction of the total mass that has been evaporated from state B to state C. Eventually all the liquid evaporates to vapor at state C called *saturated vapor*. Further heat supplied to the cylinder causes the fluid to increase in temperature along the line CE. It is then said to be *superheated*.

The line $ABDCE$ is a line of constant pressure called an *isobar*. A line of constant entropy, say FG, is called an *isentrope* and a line of constant temperature, say HJ, is called an *isotherm*. If necessary, we could also show lines of constant volume (*isometric*) or constant enthalpy (*isenthalp*) and constant internal energy.

Point B is called the *bubble point*, where vapor first appears when ascending the isobar from A to E. Point C is called the *dew point* where liquid droplets first appear when descending the isobar from E to A.

If a larger mass were to be placed on the piston the pressure would increase. If the above experiment were repeated, the same process of evaporation would occur but at different values of temperature and entropy. This is illustrated in Fig. 2.3. The new high-pressure isobar KLMN is shown along with the original isobar $ABDCE$. Evaporation at the higher pressure occurs at a higher temperature and starts, at point L, at an increased value of the entropy. The increase in entropy during evaporation $(S_M - S_L)$ is less than the increase $(S_C - S_B)$ at the lower pressure.

Further isobars could be obtained for pressures greater and smaller than the two considered so far. As the pressure increases, vaporization occurs at higher temperatures. The loci of the bubble points BL and the dew points CM may be drawn to obtain the envelope $BLPMC$. These loci

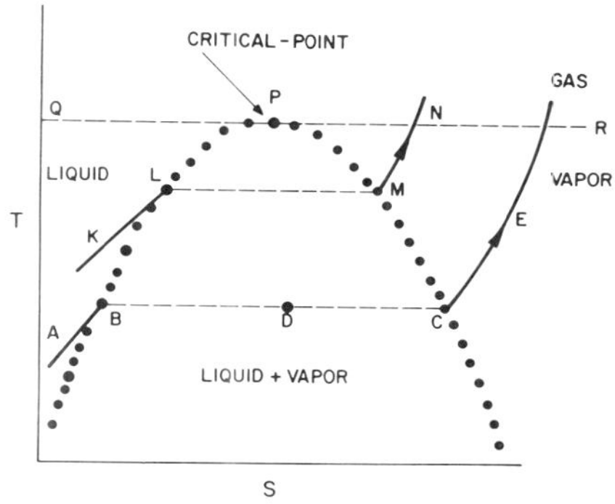

Fig. 2.3. Elements of the temperature–entropy diagram for a fluid in the liquid–vapor region.

are called the *saturation curves. BLP* is the saturated liquid line. *CMP* is the saturated vapor line. The area encompassed by the saturation curves is called the liquid–vapor envelope. Within the envelope, fluid exists partly as saturated liquid and partly as saturated gas. Fluid at any state to the right side of the envelope exists as a superheated vapor or gas. Fluid on the left side of the envelope exists only as liquid.

The apex of the liquid–vapor envelope, point *P*, is called the *critical point.* Properties of state at that condition are called the *critical properties*, critical pressure, critical temperature, etc. At the critical point no distinction can be made between a saturated liquid and a saturated vapor—they are the same fluid.

Above the critical temperature, no liquid can exist whatever the pressure exerted on it. The critical temperature is therefore taken as the breakpoint distinguishing gas from vapor. The critical isotherm *QPR* is shown on Fig. 2.3. Fluid above this temperature is said to be gaseous. Fluid in the evaporated state below this temperature is said to be vapor.

THE *T–S* DIAGRAM

Temperature–entropy diagrams of *T–S* planes have been produced for most known fluids. On such charts it is customary to reproduce lines of constant enthalpy and volume as well as isobars.

T–S diagrams are widely used in the analysis of engines and thermodynamic processes and will be referred to many times in the ensuing discussion. They are particularly useful for estimating quantities of energy or heat transferred when a fluid changes from one condition to another. For example, consider the change in state from condtion *A* to condition *B* represented on Fig. 2.4. There are an infinite number of process paths

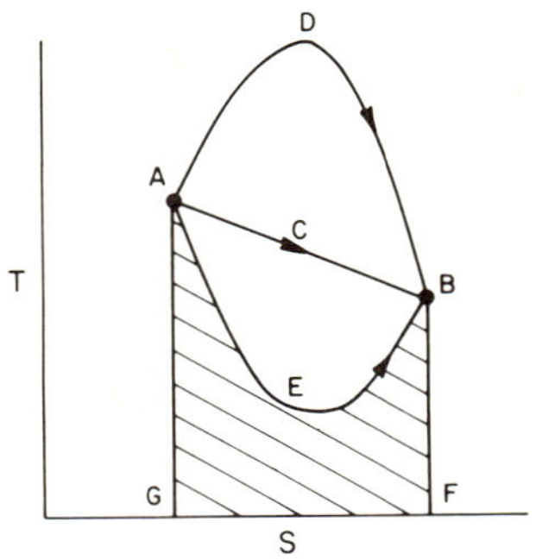

Fig. 2.4. Representation of thermodynamic processes on the *T–S* diagram.

which the system may change from state A to state B. Three such paths are shown. The most direct route between A and B is the straight line process ACB. An alternative involving some high temperatures is the path ADB. Another alternative involving low temperatures is the process AEB. Whatever the path followed the state properties at B will always be the same. The final pressure, temperature, volume, or entropy is independent of the process. However, the energy involved in accomplishing the change depends greatly on the process.

It is useful to recall that entropy was defined as $dS = \int dQ/T$. From this it can be readily seen that rearrangement of the equation results in $dQ = \int T\,dS$, the heat transfer is equal to the integrated area of the *T–S* diagram. Thus, the energy transfer of the process can be estimated by computing the *area* enclosed on the *T–S* diagram. In Fig. 2.4 the energy transfer associated with the process AEB is the shaded area enclosed by $AEBFG$. This is the least energetic process of the three shown. Process ACB would encompass a greater area and process ADB a still greater area with increased energy flows.

THE PRESSURE–VOLUME (*P–V*) DIAGRAM

Another graph of exceptional importance to thermodynamic analysis is the pressure–volume diagram or *P–V* plane frequently called the *work diagram*. Areas on the *P–V* plane represent work done just as areas on the *T–S* plane represent heat transfers.

Consider again the piston/cylinder system shown in Fig. 2.1. At the start let the temperature be low so the volume is small. This could be represented on the *P–V* and *T–S* planes shown in Fig. 2.5 by point A. Now let the cylinder be heated. This will cause the temperature to increase or some liquid to evaporate. The fluid will expand and the piston will move

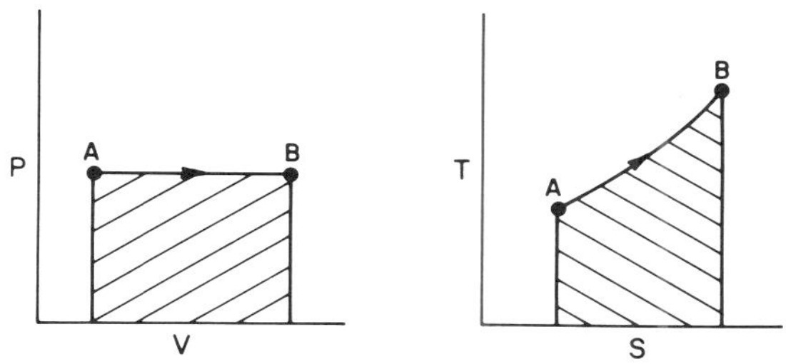

Fig. 2.5. Constant pressure expansion on the pressure–volume and temperature–entropy diagram.

up the cylinder to a new state B. The pressure remains constant because the mass M on the cylinder is unchanged. We can represent that constant pressure process on the P–V and T–S diagrams by the line A–B. We know that the amount of heat added is the shaded area on the T–S diagram under the line AB. In the same way it can be shown that the work done by the piston is the shaded area on the P–V diagram under the line AB. The area can be readily calculated as

$$\text{area of } P\text{–}V \text{ diagram} = P_{\text{const}}(V_B - V_A) \tag{2.11}$$

The piston moved a distance dL so the change in volume is

$$(V_B - V_A) = A(dL) \tag{2.12}$$

and

$$W = pA(dL) \tag{2.13}$$

The mass M was elevated by dL as the piston moved. The work done was therefore

$$W = M(dL) \tag{2.14}$$

Now the pressure of the working fluid is $p = M/A$. Then

$$W = pA\, dL = M\, dL \tag{2.15}$$

The work done is therefore equal to the area of the P–V diagram. It can be expressed mathematically as

$$\text{(a)} \qquad dW = \int p\, dV \qquad \text{(for a nonflow process)} \tag{2.16}$$

and

$$\text{(b)} \qquad dW = \int V\, dp \qquad \text{(for a flow process)} \tag{2.17}$$

THE FIRST AND SECOND LAWS OF THERMODYNAMICS

There are two important laws of thermodynamics controlling the operation of all thermal processes and machines whether they act as prime movers converting high-temperature heat to work or as refrigerators pumping heat from a low temperature to a higher temperature. The first law is simply a statement of the conservation of energy: Energy can be neither created nor destroyed, only changed from one form to another.

By way of illustration consider heating of the piston/cylinder system. This increases the temperature of the working fluid causing it to expand and lift up the piston and the mass upon it, thereby doing some work. The heat supplied was equal to the sum of the work done and the change in internal energy:

$$dQ = dW + dU \tag{2.18}$$

This is the mathematical statement of the first law for a nonflow system.

The first law is important to refrigerator operation. To produce refrigeration, work must be supplied to drive the system. According to the first law the amount of energy to be dissipated at the upper temperature is the sum of the work supplied and the refrigeration generated.

The first law is also important in refrigeration where rubbing friction takes place in the low-temperature region. The rubbing friction will require input of additional work to keep the machine working and that work will be converted (degraded) by frictional action into heat energy. Furthermore the heat is produced in the very region one is trying to refrigerate. The additional friction work degrades the refrigeration generated and increases the heat to be dissipated.

The second law of thermodynamics is more subtle. The simplest statement particularly apt for refrigerators is that "heat will not flow from a low-temperature area to a high-temperature area without the application of work." Enunciation of the second law in this form is attributed to Clausius. It sounds so commonplace as to appear trivial: everyone knows that heat will flow only from hot to cold and that to produce refrigeration, work must be done; but one must never be deceived by the apparent triviality of the second law.

Concepts for machines which contravene the two laws of thermodynamics are called perpetual motion machines—*perpetua mobilae*. Those which contravene the first law are called perpetual motion machines of the first class. Those which contravene the second law are of the second class. Angrist (1968) in a brilliant review of perpetual motion machines provides examples of exceedingly subtle contraventions of the second law.

Unfortunately, there is no way to demonstrate the inevitability of the first and second laws. Equally however, there is no record in all human experience of a single contravention of either law. Belief in the first and second laws is therefore like religious faith. If one has it, all can be explained. If one lacks faith, nothing can be explained. With faith, we continue to look at cooling cycles!

THE CARNOT CYCLE

The operation of thermal engines is best analyzed by comparison with ideal engines. These are assumed to work so that a succession of idealized thermodynamic processes occur within the system. The succession is endlessly repeated so the system is in a state of quasisteady operation. The succession of thermodynamic processes is called a cycle. All the processes are assumed to be reversible, i.e., all the fluid involved is assumed to be instantaneously at the same state so there are never any pressure, temperature, or velocity gradients, never any local heat transfers, and never any frictional effects. One cycle that has the highest performance between given temperature limits, is the Carnot cycle. Sadi Carnot, a young French Army artillery cadet first expounded the cycle in 1841 without the slightest idea of the niche in history he was thereby carving for himself.

The cycle consists of four thermodynamic processes, two isothermal and two isentropic (or adiabatic—without heat transfer). Operation of the Carnot cycle is presented here for a refrigerating machine and for both condensing and noncondensing working fluids. The cycle is illustrated in Fig. 2.6. The simple piston/cylinder system used for our earlier discussion is now "improved" by the addition of a piston rod coupled to a mechanism that can absorb or provide work as required. The end of the cylinder, the cylinder head, can be infinitely thermal conducting or infinitely thermal insulating as required. The cylinder walls are thermally insulated and the piston moves in the cylinder without friction or leaks.

To start the cycle assume the piston is at the outer dead point (ODP) so that the volume contained is a maximum. The pressure and temperature of the working fluid are at their minimum values, represented on Fig. 2.6a by the point 1. In Fig. 2.6 the P–V and T–S planes for a *gaseous working fluid* are drawn on the left-hand side. Corresponding P–V and T–S planes for the *condensing working fluid* are drawn on the right-hand side.

Let the piston move towards the cylinder head, assumed to be perfectly insulating. The compression process (1–2) is therefore isentropic (adiabatic) since no heat is transferred to or from the cylinder. Now let the cylinder head become perfectly conducting so that the further compression

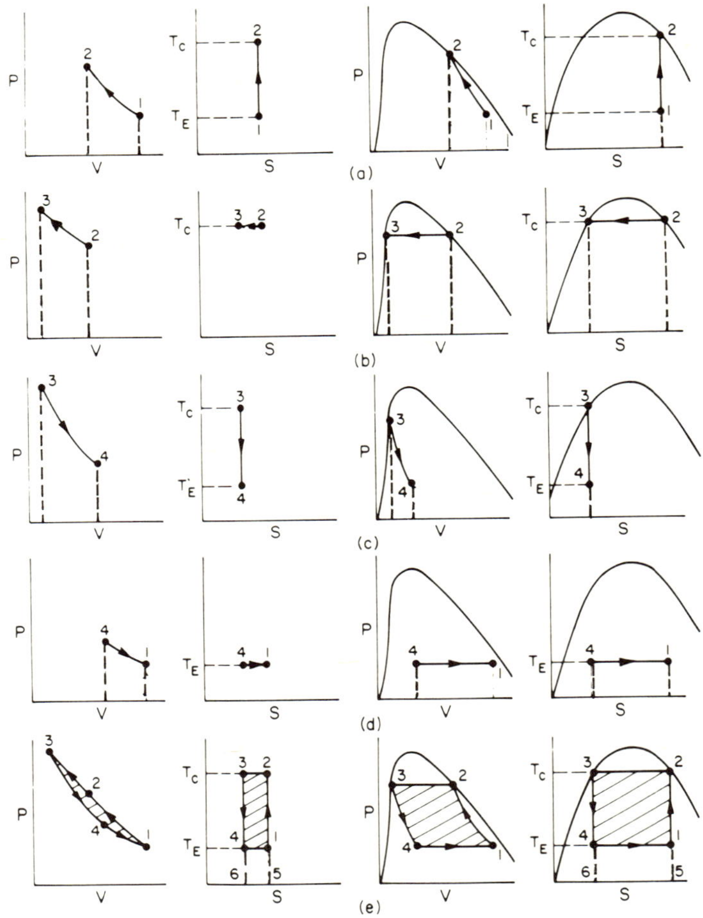

Fig. 2.6. The Carnot cycle. The series of diagrams on the left side are the *P–V* and *T–S* diagrams for operation with a gaseous working fluid. The diagrams on the right side are the *P–V* and *T–S* diagrams for operation with a condensing working fluid. (a) Process 1–2, isentropic compression; (b) process 2–3, isothermal compression; (c) process 3–4, isentropic expansion; (d) process 2–3, isothermal expansion; (e) combined diagram for the cycle.

(2–3), occurs isothermally at temperature T_C. Heat Q_C is rejected from the working fluid to the surroundings. At point (3) the head changes to the perfectly insulating state and as the piston moves on the outstroke the expansion process (3–4) occurs isentropically. The temperature of the working fluid decreases to T_E. Finally the cylinder head becomes perfectly conducting and the expansion continues. Process (4–1) occurs isothermally and heat Q_E is abstracted from the surroundings of the cylinder to maintain the temperature at T_E. This is the refrigerating effect of the cycle and the useful product.

The four separate processes are combined in Fig. 2.6e and the closed cycle 1–2–3–4 is obtained. The shaded area of the P–V diagram is the work supplied to operate the cycle. The shaded area on the T–S diagram is also the work necessary to drive the cycle. It is the difference in the heat rejected, Q_C, in process (2–3) and the heat supplied, Q_E, in process (4–1), i.e.,

$$W = Q_C - Q_E \tag{2.19}$$

The refrigerating effect Q_E is the area 1–4–5–6 on the T–S plane. The heat to be rejected from the engine is the sum of the work done and the heat lifted or refrigeration effect $Q_C = W + Q_E$.

The work diagrams for gaseous and condensing working fluids are much different in shape and area. The diagram for the gaseous engine has a small area even though the range of pressures $(p_{\max}/p_{\min})$ is greater than for the condensing engine. The small area is due to the fact that, with a gaseous working fluid, there is little difference in the slope of an isothermal process and an adiabatic process. Although the pressure range and volume compression ratios are high the actual amounts of heat transfer and work done are small. The cycle is very efficient but the amount of refrigeration, area 1–4–5–6, is limited because the expansion process 4–1 is constrained by the need to keep the maximum pressures and volumes within tolerable bounds.

The condensing system has no such difficulty. There the isothermal processes occur at constant pressure during condensation and evaporation. As a consequence, the work diagram has a large area and large quantities of heat are transferred in phase change processes (4–1) and (3–2).

The condensing Carnot cycle is the basis for all refrigeration plants for food preservation, air conditioning, domestic refrigeration, etc. In practical machines the adiabatic expansion process (3–4) is replaced by an isenthalpic expansion, except in very large plants, because it is cheaper and easier to accomplish. The effect is a reduction in performance with both a loss in the refrigeration and an increase in the work required.

Carnot engines cannot be contemplated for cryocoolers. With gaseous working fluids the large temperature range requires impractically high pressures and volumes for the amount of refrigeration produced. Similarly, no known fluid has the characteristics for the working medium in a condensing Carnot cryocooler.

It is evident from Fig. 2.6e that the cycle work increases as the difference in temperature widens. T_C is the temperature of the cooling air or water available for heat rejection from the cycle. In practice T_C would be greater than the atmospheric ambient temperature to accomplish the heat transfer. Temperature T_E is the refrigeration temperature and is fixed by the application. In practice T_E would be less than the refrigeration temperature actually required so as to induce heat to tranfer to the cylinder. It is thermodynamically (and therefore economically) wasteful to operate the engine at lower refrigeration or higher rejection temperatures than the application requires or the ambient conditions allow.

Reflection on Fig. 2.6e will lead to the very important conclusion that when operating *between given temperature limits* the cycle with the maximum refrigerating effect per unit of work done is that with *isothermal heat addition and rejection* to the cycle. Given upper and lower temperatures of operation the best refrigeration performance *always* comes from the cycle with isothermal heat transfers to and from the cycle.

COEFFICIENT OF PERFORMANCE

The performance of a refrigerator is expressed in terms of the coefficient of performance, COP. This is the ratio of the heat extracted to the work done:

$$\mathrm{COP} = Q_E/W = Q_E/(Q_C - Q_E) \tag{2.20}$$

If we have a perfect gas the heat transfers can be represented as

$$Q_E = C_P T_E \quad \text{and} \quad Q_C = C_P T_C \tag{2.21}$$

Therefore the coefficient of performance is

$$\mathrm{COP} = T_E/(T_C - T_E) \tag{2.22}$$

This is called the Carnot coefficient of performance and is the maximum theoretical coefficient of performance attainable by a refrigerator operating between given temperature limits. The second law of thermodynamics restrains the maximum theoretical COP to the Carnot value.

The Carnot value is the reference index for all other refrigerators. A refrigerator "efficiency" is sometimes quoted as the ratio of the actual COP to the Carnot COP, i.e.,

$$\eta_{\mathrm{ref}} = (\mathrm{COP}_{\mathrm{actual}})/(\mathrm{COP}_{\mathrm{carnot}}) \tag{2.23}$$

Practical refrigerators never approach the Carnot value. Figure 2.7 is a comparison of various refrigerating systems published by Köhler (1965), and attributed to Professor Nesselman. The highest efficiencies are attained by vapor compression machines at the relatively high temperatures for

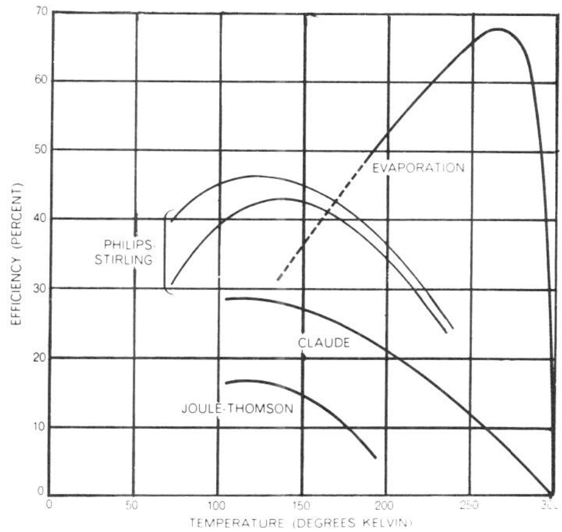

Fig. 2.7. Comparison of the efficiency of fractical refrigerating systems as a function of refrigeration temperature. Efficiency is the ratio actual COP/Carnot COP (after Köhler, 1965).

domestic refrigeration, air conditioning, and the like. These are basically Carnot systems with a condensing working fluid and are modified to replace the expansion engine process.

The work per unit refrigeration increases dramatically as the refrigeration temperature decreases. Values calculated for a Carnot cycle system with T_C of 300 K are quoted in Figure 1.1 (see Chapter 1). The work per unit of refrigeration at 100 K is 2 watts per watt. At 10 K the ratio is 29 watts per watt and at 1 K it is 299 watts per watt.

Exactly the same cycle may be used as a *heat pump* to abstract heat at ambient temperatures and to reject it from the system (say for heating a building) at higher temperatures. It is customary to quote the coefficient of performance of a heat pump as the ratio of heat *rejected* to work done:

$$\text{COP} = Q_C/(Q_C - Q_E) \tag{2.24}$$

This coefficient is greater, by unity, than the coefficient of performance defined as a refrigerator. This is true even though the *same* machine may be working over the *same* temperature limits, i.e.,

$$(\text{COP})_{\text{heat pump}} = Q_C/(Q_C - Q_E) \tag{2.25}$$

$$(\text{COP})_{\text{refrigerator}} = Q_E/(Q_C - Q_E) \tag{2.26}$$

so that

$$(\text{COP})_{\text{heat pump}} = 1 + (\text{COP})_{\text{refrigerator}} \tag{2.27}$$

It is fashionable to talk about refrigerators as heat pumps even though they are used as refrigerators. In such unfortunate circumstances it is necessary to be pedantic about the nomenclature to avoid the evident problems that will arise without a precise definition of terminology.

Part II: Regenerative Cycles

THE REITLINGER CYCLE

The coefficient of performance for the Carnot engine was found above to be $COP = T_E/(T_C - T_E)$. It is the maximum value that can be achieved by a refrigerating engine operating between temperature limits T_E and T_C. However, it is less well known that other thermodynamic cycles can attain the same performance. In fact, there are an infinite number of thermodynamic cycles that can achieve the Carnot value. All these cycles incorporate *regenerative heat transfer processes* in place of the adiabatic compression and expansion of the Carnot cycle. The high-performance regenerative cycles all embrace isothermal processes of heat addition and rejection from the cycle. These are the necessary key features of high-performance thermodynamic cycles.

The most general form of idealized thermodynamic cycle, the Reitlinger cycle, shown in Fig. 2.8, consists of two isothermal processes where the heat is supplied to and drawn from the cycle and two polytropic regenerative processes. It is essential in the ideal reversible engine that external heat exchange occurs only at the upper and lower temperature limits of the cycle. Heat reception and heat rejection during the polytropic phase must be provided by thermal storage in a regenerator. This can be conceived as a thermodynamic sponge, absorbing and releasing heat as required.

The generalized regenerative cycle was first analyzed by Reitlinger (1876) a century ago. It can exist in an infinite variety of forms distinguished by the character of the polytropic regenerative phases. The Carnot cycle is simply a special case of the Reitlinger cycle in which the regenerative polytropic phases are replaced by adiabatic, isentropic phases. Two other special cases have been given names: the Stirling and Ericsson cycles where

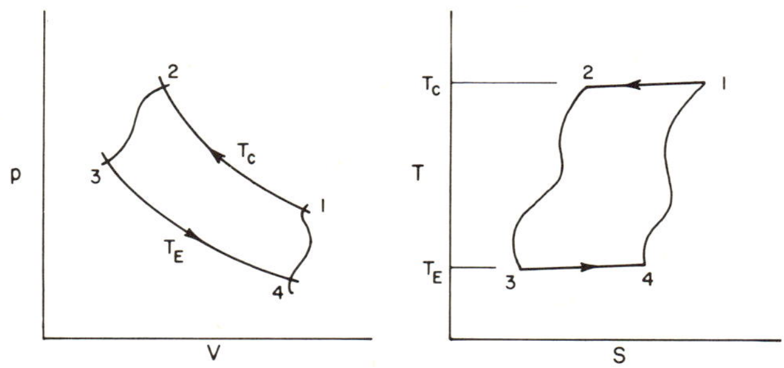

Fig. 2.8. The generalized Reitlinger isothermal cycle.

the regenerative processes occur at constant volume and constant pressure, respectively. An infinite number of variations may be postulated in which the regenerative processes are not constant volume, constant pressure, or isentropic. All these cycles are characterized by the same thermodynamic refrigeration performance:

$$\mathrm{COP} = \frac{T_E}{T_C - T_E} \tag{2.28}$$

THE STIRLING CYCLE

The Stirling cycle illustrated in Fig. 2.9 is similar to the Carnot cycle. Operation is as follows. Consider a cylinder containing two opposed pistons, with a regenerator between them. The regenerator may be thought of as a thermodynamic sponge, alternatively releasing and absorbing heat. In practice this is accomplished with a matrix of finely divided metal in the

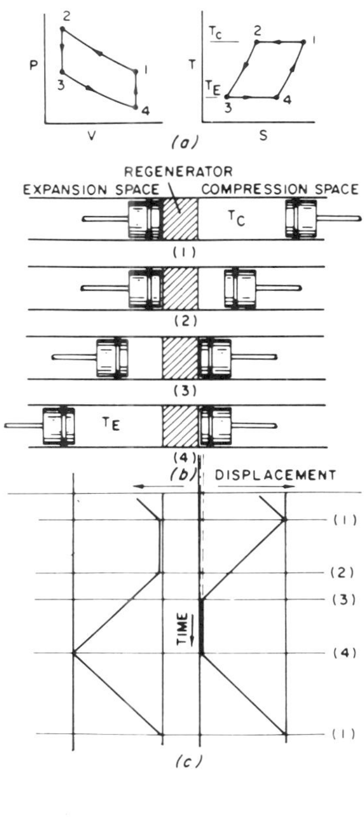

Fig. 2.9. The Stirling cycle with conceptual engine. (a) P–V and T–S diagrams. (b) Piston arrangement at the terminal points of the cycle. (c) Time-displacement diagram.

form of wires or strips. One of the two volumes between the regenerator and the pistons is called the *expansion space*, and is maintained at low temperature T_E. The other volume is called the *compression space*, and is maintained at ambient temperature T_C. There is a temperature gradient $(T_C - T_E)$ across the transverse faces of the regenerator and it is assumed there is no thermal conduction in the longitudinal direction. The pistons are assumed to move without friction or leakage of the working fluid enclosed between them.

Assume that the compression-space piston is at the outer dead point, and the expansion-space piston is at the inner dead point, close to the face of the regenerator. All the working fluid is then in the compression space at ambient temperature. The volume is a maximum, and the pressure and temperature are represented by (1) on the *P–V* and *T–S* diagrams, shown in Fig. 2.9. During compression (process (1–2), the compression piston moves towards the inner dead point, and the expansion-space piston remains stationary. The working fluid is compressed in the compression space, and the pressure increases. The temperature is maintained constant because heat Q_C is abstracted from the compression-space cylinder.

In the transfer process (2–3), both pistons move simultaneously, the compression piston towards, and the expansion piston away from, the regenerator. The volume between them remains constant. The working fluid is transferred through the porous metallic matrix of the regenerator to the expansion space. In passing through the regenerator, the working fluid is cooled from T_C to T_E, by heat transfer to the matrix, and emerges from the regenerator into the expansion space at temperature T_E. The decrease in temperature passing through the matrix at constant volume causes a decrease in pressure.

In the expansion process (3–4), the expansion piston moves to the outer dead point; the compression piston remains stationary at the inner dead point, adjacent to the regenerator. As the expansion proceeds, the pressure decreases as the volume increases. The temperature remains constant because heat Q_E is abstracted from the cylinder surroundings. This is the useful refrigeration of the cycle.

Finally both pistons move simultaneously to transfer the working fluid (at constant volume) back through the regenerative matrix from the expansion space to the compression space. In passing through the matrix heat is transferred from the matrix so the working fluid increases in temperature and emerges at T_C into the compression space.

The cycle is therefore composed of four heat-transfer processes:

Process (1–2)—Isothermal compression; heat transfer from the working fluid to the external sink at ambient temperature, T_C.

Process (2–3)—Constant volume; heat transfer from the working fluid to the regenerative matrix.

Process (3–4)—Isothermal expansion; heat transfer to the working fluid from an external source at the refrigerating temperature T_E.

Process (4–1)—Constant volume; heat transfer to the working fluid from the regenerative matrix.

The heat transferred in process (2–3) is the same as in process (4–1). The only heat transfer process between the engine and its surroundings are (a) heat supply at T_E and (b) heat rejected at T_C. Heat supply and rejection at constant temperature satisfies the requirement of the second law of thermodynamics for maximum thermodynamic performance. The coefficient of performance of the Stirling cycle is the same as for the Carnot cycle:

$$\mathrm{COP} = T_E/(T_C - T_E) \tag{2.29}$$

The advantage of the Stirling cycle over the Carnot cycle lies in the replacement of two *isentropic* processes by two *constant-volume regenerative* processes. This greatly increases the area of the *P–V* diagram and the quantities of heat transfer to and from the system so that it is not necessary to resort to very high pressures and swept volumes.

The *P–V* and *T–S* diagrams of Carnot and Stirling cycles between given limits of pressure, volume, and temperature are compared in Fig. 2.10. The shaded areas (4–1–5) and (2–3–6) represent the additional work made available by substituting constant-volume processes for isentropic processes. The isothermal processes (2–5) and (6–4) of the Carnot cycle

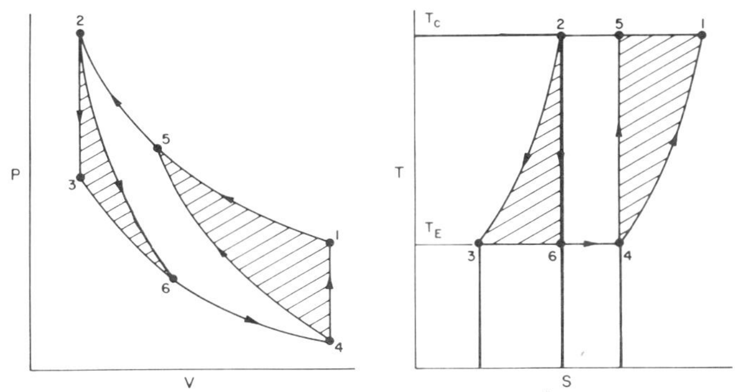

Fig. 2.10. Stirling and Carnot cycles superimposed with common values for the minimum and maximum temperatures, pressures, and volumes. Cross-hatched areas on the *P–V* plane represent the increased work output of the Stirling cycle. Cross-hatched areas on the *T–S* plane represent increased heat transfer of the Stirling cycle.

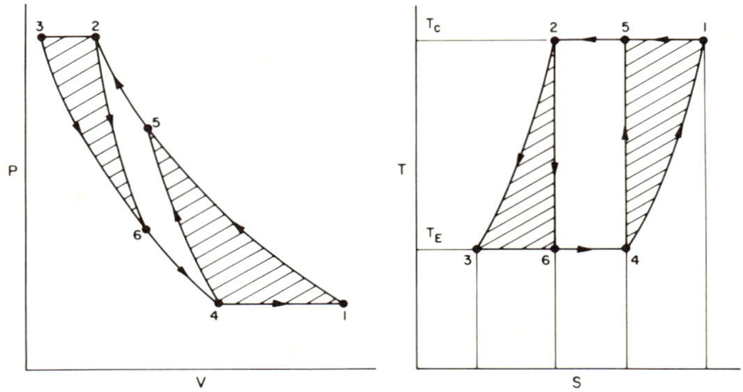

Fig. 2.11. Ericsson and Carnot cycles superimposed with common values for the maximum and minimum values of temperature, pressure, and volume. Cross-hatched areas represent the increased work output and heat transfer of the Ericsson cycle.

are extended to processes (1–2) and (3–4), respectively. The quantities of heat transfer for the Stirling cycle are increased by the same proportion as the work. The coefficient of performance (heat extracted per unit of work) is the same for both.

THE ERICSSON CYCLE

For the Ericsson cycle, the processes of constant-volume regenerative heat transfer are replaced by constant-pressure regenerative processes. This leads to the *P–V* and *T–S* diagrams shown in Fig. 2.11. The coefficient of performance of the cycle is the same as that of the Carnot cycle but for given limits of pressure, volume, and temperature the work and heat transfer are much greater.

THE STIRLING CYCLE AS A PRIME MOVER, HEAT PUMP, AND PRESSURE GENERATOR

When the Stirling engine is used as a prime mover, heat at high temperature T_{max} is supplied to the cycle. In the expansion process (3–4) (Fig. 2.9) part of the heat is converted to work, and part is rejected, as heat, at ambient temperature T_C. The machinery is precisely the same as a refrigerator. Comparison of the cycles for operation as a prime mover and refrigerating machine is given in Fig. 2.12. In operating as a prime mover the pressure and temperature are greater in expansion than in

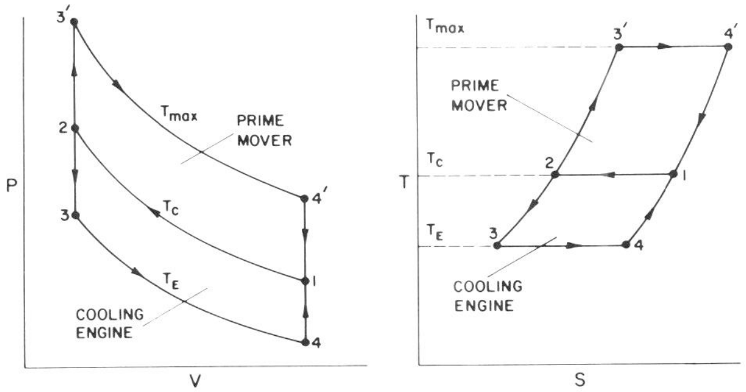

Fig. 2.12. Stirling cycle as cooling engine and as prime mover. In both applications the compression process occurs at ambient temperature T_C. The expansion process occurs at a low temperature T_E in the cooling engine and at a high temperature T_{max} in the prime mover. In both cases heat is absorbed during expansion and rejected during compression.

compression. In operating as a refrigerator the reverse is true: the pressure and temperature are both less in expansion than in compression.

As a heat pump the Stirling cycle operates exactly as the refrigerator. The temperature of the expansion space T_E is less than the compression space T_C. The difference between operation as a heat pump and refrigerator is that both T_E and T_C are increased. In the prime mover and refrigerator application, T_C is the ambient atmospheric temperature at which cooling water is available. For a heat pump, T_C is the temperature at which heat is rejected from the system, and is the useful output for heating a concert hall or office building. Therefore T_C is *above* the ambient atmospheric temperature. Heat is supplied to the cycle (at T_E), from atmospheric air or river water, at approximately the ambient atmospheric temperature.

A comparison of the Stirling cycle as a heat pump and a refrigerator is drawn in Fig. 2.13. In both cases, work from an external source is required to drive the cycle, and is equivalent to area (1–2–3–4). In the case of the heat pump, the useful product is the heat rejected at temperature T_C, and the performance of a heat pump is therefore assessed as

$$(\mathrm{COP})_{\mathrm{HP}} = \frac{\text{heat rejected}}{\text{work done}} = \frac{T_C}{T_C - T_E} \tag{2.30}$$

This is the inverse of the thermal efficiency. The coefficient of the performance of a refrigerator

$$(\mathrm{COP})_{\mathrm{ref}} = T_E/(T_C - T_E) \tag{2.31}$$

is not the inverse of thermal efficiency.

Stirling engine systems have been used to pump a fluid and increase its pressure. When the fluid to be pumped was a liquid or gas, separated

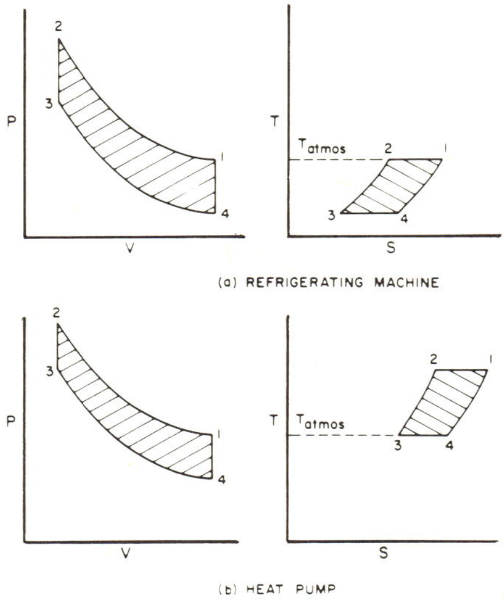

Fig. 2.13. Stirling cycle as a heat pump and as a cooling engine. When operating as a cooling engine or as a heat pump the expansion process occurs at a temperature less than that of the compression process and an input of work to operate the cycle is required. The cooling engine absorbs heat during expansion at temperatures less than the ambient and rejects heat at near ambient temperatures. The heat pump absorbs heat during expansion at ambient temperatures and rejects heat at temperatures greater than ambient.

by a diaphragm or piston from the working fluid in the Stirling cycle device, the system can be classified as a Stirling cycle working as a prime mover. In other instances, where the working fluid itself was the fluid to be compressed and pumped, there are, invariably, valves or other flow-controlling devices. According to the defintion given in Chapter 1 these systems cannot be classified as Stirling engines but rather as Ericsson engines. Nevertheless, when discussed in the literature they are frequently called Stirling engines. In most cases, fluid is added to the cycle when the pressure is low and withdrawn at a higher pressure. Heat is supplied at high temperature and rejected at ambient temperature. Work during expansion is greater than work during compression by an amount equivalent to the "pump" work of the compressed fluid.

A VERSATILE DEMONSTRATION ENGINE

In the late 1950s when the Philips Stirling engine liquefiers were introduced it was common for them to demonstrate the versatility of the Stirling engine for operation as a prime mover, cooling engine, or a heat pump. A simplified diagram of the engine used for the demonstration is shown in Fig. 2.14. The engine was of a type known as the piston-displacer machine rather than the opposed piston version used above to describe the cycle.

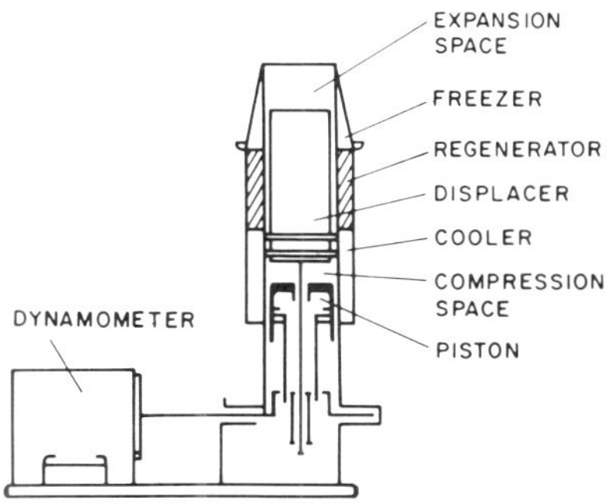

Fig. 2.14. Diagram of the versatile prime mover/refrigerator/heat pump Stirling engine demonstrated by Philips.

In this engine the space above the displacer, below the cylinder head, is the expansion space. Between the underside of the displacer and the top of the piston is the compression space. The two spaces were in communication through a water-cooled heat exchanger, regenerative heat exchanger, and a finned freezer heat exchanger. The machine was mounted on a bed-plate and coupled to an electric dynamometer capable of operating as a motor or as a generator in either direction of rotation. Helium or hydrogen gas was used as the working fluid. A thermocouple was attached to the cylinder head and a large temperature scale was projected onto the wall of the lecture room with a moving light spot to indicate the cylinder head temperature.

The engine would be started initially as a refrigerator with the dynamometer working as a motor driving the machine in the forward direction. The cylinder head around the expansion space would cool rapidly and on the temperature scale one could observe the decrease in temperature. At about 80 K air condensed on the cylinder head and ran into a collector at the base of the head. After a short time power to the motor would be switched off and in less than a revolution the machine would stop and self-start *in the reverse direction.*

It was now operating as a prime mover generating power. With the reversal of rotation the expansion and compression spaces switched locations so the expansion space was between the displacer and piston. Heat was drawn *from* the cooling water to drive the engine and *rejected* from the cylinder head to the pool of liquid air serving as the engine coolant. The reservoir of liquid air was soon dissipated and soon the temperature of the cylinder head started increasing. The engine continued to run but with progressively less vigor as the temperature increased until finally it slowed to a very low speed.

At that point power would once more be supplied to the dynamometer causing it to run again as a motor but preserving the reverse direction of rotation. With this power input the temperature of the cylinder head continued to increase until eventually it attained and then exceeded the ambient temperature. The machine was now operating as a heat pump drawing heat from the "cooling" water and rejecting it at higher temperature to the surroundings of the cylinder head now acting as the compression space. The temperature of the head would increase rapidly, whereupon the lighting would be lowered and the cylinder head could be observed glowing cherry red. At that point Jan Köhler would break his discourse, reach into his pocket, withdraw a cigar and casually lean down to light it on the hot cylinder head. It was beautifully executed and done so matter of factly. Following this, power to the motor would be switched off and again, immediately, the engine would stop and starting running in the opposite direction.

With this second reversal the direction of rotation was now as it had been initially. The expansion and compression spaces were restored to the configuration shown in Fig. 2.14. The engine now ran again as a prime mover, withdrawing heat from the hot cylinder head, rejecting it to the cooling water. After some time the cylinder head cooled and, eventually, there was insufficient temperature difference between the expansion and compression spaces and the engine stopped.

It was a remarkable demonstration, beyond the capability of any other machine. The versatility for the same unit to operate as a prime mover, refrigerator, and heat pump has practical implication for the future. A Stirling refrigerator driven by a Stirling prime mover has already been demonstrated by William Beale at Sunpower Inc. Use of a common working fluid has outstanding attractions to overcome seal problems. The combination may be a solar, isotope, or combustion heated cryocooler. More likely in the near future it will be developed as a gas-fired air conditioning refrigerator. The combination or duplex Stirling engine should not be confused with the Vuilleumier cycle machines we shall consider later (see Chapter 4).

Another concept to utilize the versatility of the Stirling engine is to use it as a prime mover extracting heat from a thermal battery (lithium fluoride) as an automotive propulsion engine. For downhill operation, or when braking, the engine could be converted to function as a heat pump feeding thermal energy into the thermal battery. In practice it is not necessary to physically reverse the engine. The same effect can be gained in some forms of engine by simply switching fluid connections (see Walker 1979). The combined regenerative braking/propulsion concept is attractive in mining applications where trucks are used on loaded downhaul operation and for underground mines. It will become attractive for road vehicles as the fossil fuel supplies dwindle and the world moves into the "electric economy." The Stirling engine with thermal storage will likely become the preferred automotive propulsion unit. Commuters will plug in their car for an electric (thermal) recharge on arrival at work or on arrival at home. This consumer behavior prototype already exists in Canada, where, in winter, it is customary when parking to "plug in" an engine block heater.

A. J. Oxley of the U.K. Rutherford Laboratory has explored use of liquid nitrogen as the coolant for Stirling engines. Engines could be energized from an ambient temperature source to operate as a prime mover for underground mining applications with no emission of noxious fumes. Other possibilities for large-scale power generation exist using the temperature difference between sea water and liquid natural gas now coming into the United States, Japan, and Europe in very large quantiites.

THE RALLIS CYCLE

Generalized reference cycles for regenerative power engines were developed by Rallis *et al.* (1977). The basic cycle was developed in two forms characterized by

i. Isothermal processes of compression and expansion (called the Rallis isothermal regenerative cycle);
ii. Isentropic processes of compression and expansion (called the Rallis adiabatic regenerative cycle).

In both forms the regenerative processes occur partly at constant pressure and partly at constant volume in any given combination.

The Stirling and Ericsson cycles are both special cases of the Rallis isothermal regenerative cycle and many other special cases can be defined which have no identifying names. The cycles were developed by Rallis for prime movers, but the same approach appears well suited for application to refrigeration applications.

Analytical expressions for the general form of the Rallis cycle are developed below for both the isothermal and adiabatic versions. They are then reduced to various special cases and, finally, a comparison is drawn with some computed results for the ideal Stirling cycle and the pseudo-Stirling cycle with adiabatic compression and expansion processes.

RALLIS ISOTHERMAL REGENERATIVE CYCLE

A conceptual arrangement of an engine capable of operating on the Rallis isothermal regenerative cycle is shown in Fig. 2.15. The *P–V* and

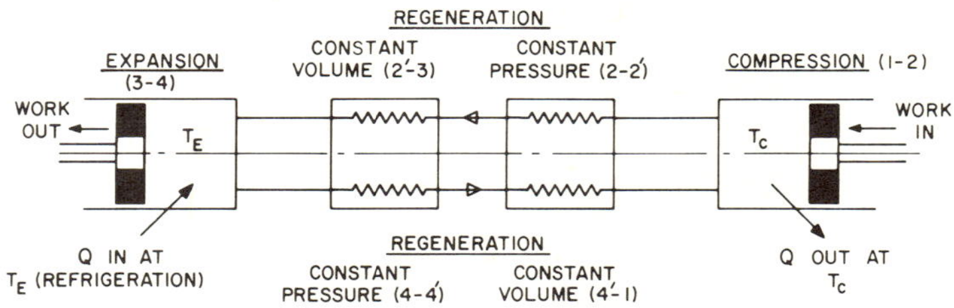

Fig. 2.15. Conceptual arrangement of an engine for operating on the Rallis isothermal regenerative cycle.

T–S planes for the cycle are shown in Fig. 2.16. The cycle is defined by two isothermals, temperatures T_C and T_E, bounded by constant volume and constant pressure regenerative processes at the two extremes as shown above. T_C is the ambient temperature at which cooling is available, T_E is the refrigeration temperature.

Processes (1–2)—Isothermal compression, heat is rejected from the system to the surroundings.
Process (2–2′)—Constant pressure regenerative cooling.
Process (2′–3)—Constant volume regenerative cooling.
Process (3–4) —Isothermal expansion, heat is extracted from the surroundings, (this is the useful refrigerating effect of the cycle).
Process (4–4′)—Regenerative heating at constant pressure.
Process (4′–1)—Regenerative heating at constant volume.

For convenience in analysis we define the following nondimensional parameters:

$\tau = T_C/T_E$
$r_c = V_1/V_2$, volume compression ratio
$r_e = V_4/V_3$, expansion compression ratio
$a = T_{2'}/T_2$, regenerative cooling constant pressure ratio
$b = T_3/T_{2'}$, regenerative cooling constant volume ratio
$d = T_{4'}/T_4$, regenerative heating constant pressure ratio
$e = T_1/T_{4'}$, regenerative heating constant volume ratio
$\varepsilon = (T_5 - T_2)/(T_3 - T_2)$, regenerator effectiveness
COP = coefficient of performance defined as the ratio Q/W
W = specific external work required per cycle
Q = specific external heat extracted at the low temperature (refrigeration)
$\bar{p}_m = W/\text{swept volume}$, mean effective pressure
$P = \bar{p}_m/p_{\max}$, mean effective pressure ratio of the cycle
$p_{\max} = p_2 = p_{2'}$, maximum cycle pressure

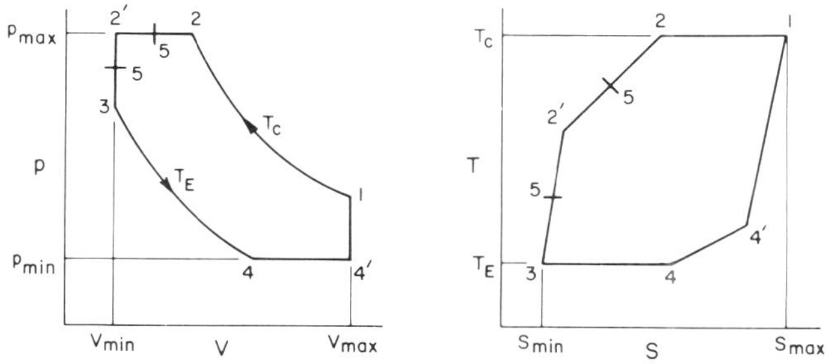

Fig. 2.16. Rallis isothermal regenerative cycle.

Now

$$ab = (T_{2'}/T_2)(T_3/T_{2'}) = (T_E/T_C) = 1/\tau \tag{2.32}$$

$$de = (T_{4'}/T_4)(T_1/T_{4'}) = (T_E/T_C) = 1/\tau \tag{2.33}$$

$$(r_c/a) = (dr_e)$$

therefore

$$r_c = adr_e$$

and

$$T_5 = T_2\left[1 - \varepsilon\left(\frac{\tau - 1}{\tau}\right)\right] \tag{2.34}$$

The external work required per cycle is [from Eq. (2.19)]

$$\begin{aligned} W &= p_2(V_2 - V_{2'}) + RT_2 \ln r_c - p_4(V_{4'} - V_4) - RT_4 \ln r_e \\ &= p_2V_2\left(1 - \frac{V_{2'}}{V_2}\right) + RT_2 \ln r_c - p_4V_4\left(\frac{V_{4'}}{V_4} - 1\right) - RT_4 \ln r_e \\ &= RT_2(1-a) + RT_2 \ln r_c - RT_4(d-1) - RT_4 \ln r_e \\ &= R[(1 - a + \ln r_c)T_2 - (d - 1 + \ln r_e)T_4] \\ &= RT_C[(1 - a + \ln r_c) - (1/\tau)(d - 1 + \ln r_e)] \\ &= \frac{RT_C}{\tau}\{\tau[1 - a + \ln(adr_e)] - (d - 1 + \ln r_e)\} \\ &= RT_E\{\tau[1 - a + \ln(adr_e)] - (d - 1 + \ln r_e)\} \end{aligned} \tag{2.35}$$

The heat transfer *to* the working fluid during the expansion (3–4) is the refrigerating effect of the cycle. With imperfect regeneration the refrigerating effect is partially or wholly consumed in cooling the working fluid from T_5 to T_3.

There are two cases to consider:

$$\text{(a)}\quad T_3 \leq T_5 \leq T_{2'}$$

and

$$\text{(b)}\quad T_{2'} \leq T_5 \leq T_2$$

For case (a) the external refrigerating effect is

$$Q = RT_3 \ln r_e - C_v(T_5 - T_3)$$

and for case (b) the external refrigerating effect is

$$Q = RT_3 \ln r_e - C_v(T_{2'} - T_3) - C_p(T_5 - T_{2'})$$

Now for case (a) $T_3 \leq T_5 \leq T_{2'}$, so that

$$\frac{T_2}{\tau} \leq T_2\left[1 - \varepsilon\left(\frac{\tau - 1}{\tau}\right)\right] \leq aT_2$$

$$(1/\tau) \leq \left[1 - \varepsilon\left(\frac{\tau - 1}{\tau}\right)\right] \leq a$$

and for case (b) $T_{2'} \leq T_5 \leq T_2$, so that

$$aT_2 \leq T_2\left[1 - \varepsilon\left(\frac{\tau - 1}{\tau}\right)\right] \leq T_2$$

$$a \leq 1 - \varepsilon\left(\frac{\tau - 1}{\tau}\right)$$

Thus for case (a), the refrigeration is (where $\gamma = C_p/C_v$):

$$Q = RT_E \ln r_e - C_v T_c\left[1 - \varepsilon\left(\frac{\tau - 1}{\tau}\right) - \frac{1}{\tau}\right]$$

$$= RT_C\left[\frac{\ln r_e}{\tau} - \frac{1}{(\gamma - 1)}\,\frac{\tau - \varepsilon(\tau - 1) - 1}{\tau}\right]$$

$$= \frac{RT_C}{\tau(\gamma - 1)}\{(\gamma - 1)\ln r_e - [\tau - \varepsilon(\tau - 1) - 1]\}$$

$$= \frac{RT_E}{(\gamma - 1)}\{(\gamma - 1)\ln r_e - [\tau - \varepsilon(\tau - 1) - 1]\} \qquad (2.36)$$

The coefficient of performance (Q/W) is

$$\mathrm{COP} = \frac{\dfrac{RT_E}{\gamma-1}\{(\gamma-1)\ln r_e - [\tau - \varepsilon(\tau-1)-1]\}}{RT_E\{\tau[1-a+n(adr_e)]-(d-1+\ln r_e)\}}$$

$$= \frac{1}{(\gamma-1)}\,\frac{(\gamma-1)\ln r_e - [\tau-\varepsilon(\tau-1)-1]}{\{\tau[1-a+\ln(adr_e)]-(d-1+\ln r_e)\}} \qquad (2.37)$$

The mean effective pressure

$$\bar{p}_m = W/\text{stroke volume}$$

$$= \frac{RT_E\{\tau[1-a+\ln(adr_e)-(d-1+\ln r_e)]\}}{V_{4'}-V_3}$$

$$= \frac{RT_E\{\tau[1-a+\ln(adr_e)-(d-1+\ln r_e)]\}}{(dr_e-1)V_3} \qquad (2.38)$$

and in dimensionless form with respect to the maximum pressure

$$P = \frac{\bar{p}_m}{p_{\max}} = \frac{RT_E[\tau(1-a+\ln(adr_e)-(d-1+\ln r_e)]}{(dr_e-1)V_3\times p_{2'}}$$

but $p_{\max} = p_{2'} = bp_3$ and since $p_3V_3 = RT_E$ then

$$P = \frac{\tau[1-a+\ln(adr_e)-(d-1+\ln r_e)]}{b(dr_e-1)} \qquad (2.39)$$

Similarly for case (b) where $T_{2'} \leq T_5 \leq T_2$ the external refrigerating effect is

$$Q = RT_3\ln r_e - C_v(T_{2'}-T_3) - C_p(T_v - T_{2'})$$

$$= RT_E\ln r_e - C_vT_E\left[\left(\frac{1}{b}\right)-1\right] - C_pT_c\left[1-\varepsilon\left(\frac{\tau-1}{\tau}\right)-a\right]$$

$$= RT_E\left\{\ln r_e - \left(\frac{1}{\gamma-1}\right)\left(\frac{1-b}{b}\right) - \tau\left(\frac{\gamma}{\gamma-1}\right)\left[1-\varepsilon\left(\frac{\tau-1}{\tau}\right)-a\right]\right\}$$

$$= \frac{RT_E}{(\gamma-1)}\left\{(\gamma-1)\ln r_e + \left(\frac{b-1}{b}\right) - \gamma[\tau-\varepsilon(\tau-1)-a\tau]\right\} \qquad (2.40)$$

The coefficient of performance is

$$\mathrm{COP}=\frac{\dfrac{RT_E}{\gamma-1}\left\{(\gamma-1)\ln r_e+\left(\dfrac{b-1}{b}\right)-\gamma[\tau-\varepsilon(\tau-1)-\alpha\tau]\right\}}{RT_E\{\tau[1-a+\ln(adr_e)]-(d-1+\ln r_e)\}}$$

$$=\frac{\left\{(\gamma-1)\ln r_e+\left(\dfrac{b-1}{b}\right)-\gamma[\tau-\varepsilon(\tau-1)-a\tau]\right\}}{(\gamma-1)\{\tau[1-a+\ln(adr_e)]-(d-1+\ln r_e)\}} \tag{2.41}$$

The mean effective pressure and the specific power are not dependent on the regenerator effectiveness and so are the same as case (a). Now consider some special cases.

Ideal Stirling Cycle

In this cycle the regenerative processes (2–3) and (4–1) occur wholly at constant volume as shown in Fig. 2.17. Therefore $a=d=1$, since $T_{2'}=T_2$ and $T_{4'}=T_4$, $b=e=\tau$, $r_e=r_c=r$ and $T_3\le T_5\le T_2$ only. Then the coefficient of performance is given by [from Eq. (2.37)]

$$\mathrm{COP}=\frac{1}{(\gamma-1)}\left\{\frac{(\gamma-1)\ln r-[\tau-\varepsilon(\tau-1)-1]}{\tau[1-1+\ln(r)]-(1-1+\ln r)}\right\}$$

$$=\frac{1}{\gamma-1}\left\{\frac{(\gamma-1)\ln r-[\tau-\varepsilon(\tau-1)-1]}{(\tau-1)\ln r}\right\}$$

$$=\frac{1}{(\gamma-1)}\left\{\frac{(\gamma-1)\ln r-(\tau-1)(1-\varepsilon)}{(\tau-1)\ln r}\right\}$$

$$\frac{1}{(\gamma-1)}\left[\frac{(\gamma-1)\ln r-(\tau-1)(1-\varepsilon)}{(\tau-1)\ln r}\right] \tag{2.42}$$

and if $\varepsilon=1$

$$\mathrm{COP}=\frac{1}{\tau-1}=\frac{T_E}{T_C-T_E}\ \text{the Carnot value} \tag{2.43}$$

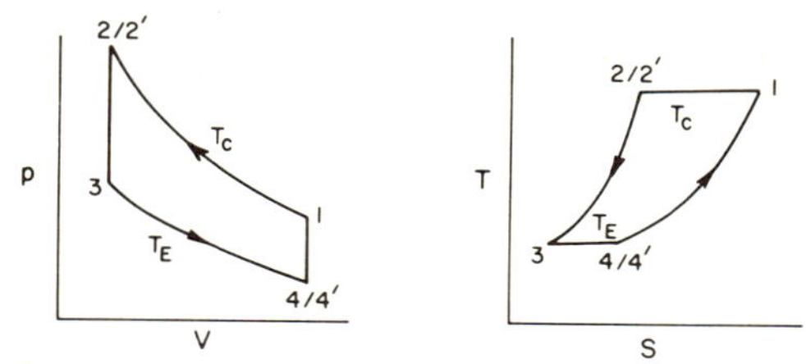

Fig. 2.17. Ideal Stirling cycle.

The mean effective pressure of the Stirling cycle is [from Eq. (2.38)]

$$\bar{p}_m = RT_E \frac{\tau[1-1+\ln(1\times 1\times r)]-(1-1+\ln r)]}{(r-1)V_3}$$

$$= \frac{RT_E}{V_3}\frac{(\tau-1)\ln r}{(r-1)} \tag{2.44}$$

and the pressure ratio $P = \bar{p}_m/p_{\max}$

$$P = \left(\frac{\tau-1}{\tau}\right)\left(\frac{\ln r}{r-1}\right) \tag{2.45}$$

Ideal Ericsson Cycle

In this cycle the regenerative processes (2–3) and (4–1) occur wholly at constant pressure as shown in Fig. 2.18. Therefore $b = e = 1$ since $T_{2'} = T_3$ and $T_{4'} = T_1$, $1/a = d = \tau$, $r_e = r_c = r$ and $T_3 \leq T_5 \leq T_2$ only.

Then the coefficient of performance is:

$$\mathrm{COP} = \frac{(\gamma-1)\ln r + \left(\frac{1-1}{1}\right) - \gamma\left[\tau - \varepsilon(\tau-1)\frac{-\tau}{\tau}\right]}{(\gamma-1)\{[1-1+\ln(1\times 1\times r)]-(\tau-1+\ln r)\}}$$

$$= \frac{(\gamma-1)\ln r + \gamma(\tau-1)(1-\varepsilon)}{(\gamma-1)(\tau-1)\ln r}$$

and when $\varepsilon = 1$

$$\mathrm{COP} = \frac{1}{(\tau-1)} = \frac{T_E}{T_C - T_E}, \qquad \text{the Carnot value} \tag{2.46}$$

The mean effective pressure of the Ericsson cycle is

$$\bar{p}_m = \frac{RT_E\left\{\tau\left[1-\frac{1}{\tau}+\ln\left(\frac{\tau}{\tau}r\right)\right]-(\tau-1+\ln r)\right\}}{V_{4'}-V_3}$$

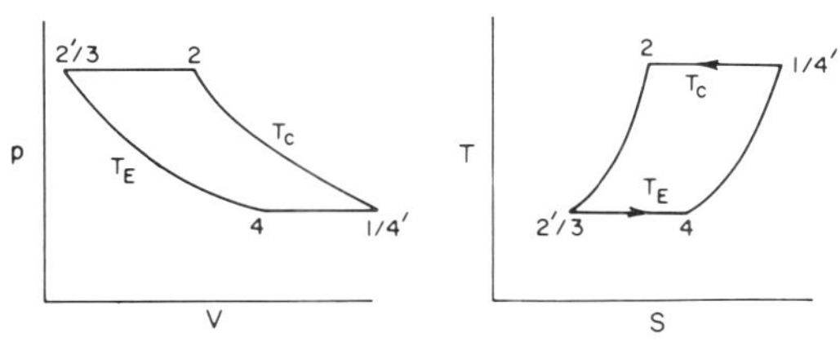

Fig. 2.18. Ideal Ericsson cycle.

and since $V_{4'} = dV_4 = \tau V_4$

$$V_{4'} - V_3 = V_3(\tau r - 1)$$

$$\bar{p}_m = \frac{RT_E}{V_3}\left|\frac{\tau(\tau - 1 + \tau \ln r)/[\tau - (\tau - 1 + \ln r)]}{(\tau r - 1)}\right|$$

$$= \frac{RT_E}{V_3}\frac{(\tau - 1)}{(\tau r - 1)}\ln r \tag{2.47}$$

and the pressure ratio $P = \bar{p}_m/p_{\max}$ is

$$P = \left(\frac{\tau - 1}{\tau}\right)\frac{\ln r}{(\tau r - 1)} \tag{2.48}$$

Ideal Cycle with Constant-Volume Regenerative Cooling and Constant-Pressure Regenerative Heating

In this cycle the regenerative process of cooling, (2–3), occurs at constant volume, and of heating, (4–1′), occurs at constant pressure, as shown in Fig. 2.19. (This cycle has no given name so far as is known.) Therefore $a = e = 1$ since $T_{2'} = T_2$, and $T_{4'} = T_1$, $1/b = d = \tau$, and $T_3 \leq T_5 \leq T_2$ only. The coefficient of performance is

$$\text{COP} = \frac{1}{(\gamma - 1)}\left|\frac{(\gamma - 1)\ln r_e - [\tau - \varepsilon(\tau - 1) - 1]}{\tau[1 - 1 + \ln(1 \times d \times r_e)] - (d - 1 + \ln r_e)}\right|$$

$$= \frac{1}{(\gamma - 1)}\left|\frac{(\gamma - 1)\ln r_e - [\tau - \varepsilon(\tau - 1) - 1]}{(\tau \ln dr_e) - \ln(r_e) - d + 1)}\right| \tag{2.49}$$

The mean effective pressure $\bar{p}_m$ of this cycle is

$$\bar{p}_m = \frac{RT_E\{\tau[1 - 1 + \ln(1 \times \tau \times r_e)] - (\tau - 1 + \ln r_e)\}}{V_{4'} - V_3}$$

and since $V_{4'} - V_3 = (V_4 - V_3) = V_3[\tau V_4/(V_3 - 1)]$

$$\bar{p}_m = \frac{RT_E}{V_3}\frac{\tau \ln(\tau r_e) - \ln(r_e) - \tau + 1}{(\tau r_e - 1)} \tag{2.50}$$

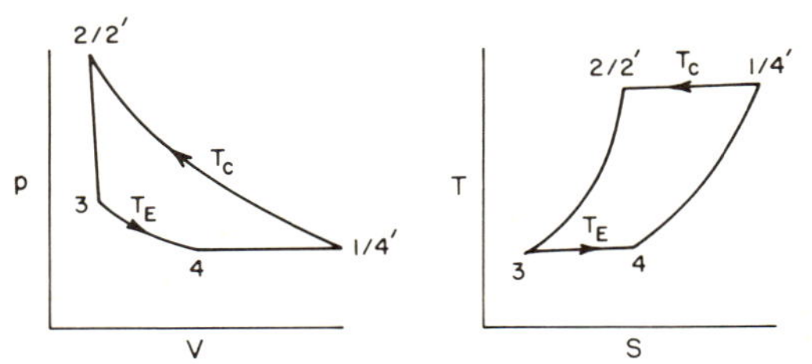

Fig. 2.19. Cycle with constant-volume regenerative cooling and constant-pressure regenerative heating.

and the pressure ratio $P = \bar{p}_m/p_{\max}$ is

$$P = \frac{[\tau \ln(\tau r_e) - \ln(r_e) - \tau + 1]}{\tau(\tau r_e - 1)} \tag{2.51}$$

Ideal Cycle with Constant-Pressure Regenerative Cooling and Constant-Volume Regenerative Heating

In this cycle the regenerative process of cooling, (2–3), occurs at constant pressure, and of heating, (4′ − 1), occurs at constant volume as shown in Fig. 2.20. (This cycle also has no given name as far as is known.) In this cycle $b = d = 1$ since $T_2' = T_3$ and $T_{4'} = T_4$; $(1/a) = e = \tau$ and $T_{2'} \leq T_5 \leq T_2$. The coefficient of performance is [from Eq. (2.41)]

$$\mathrm{COP} = \left| \frac{(\gamma - 1)\ln r_e\left[\frac{(1-1)}{1}\right] - \gamma\left[\tau - \varepsilon(\tau - 1) - \frac{\tau}{\tau}\right]}{(\gamma - 1)\left\{\tau\left[1 - \frac{1}{\tau} + \ln(r_e/\tau)\right] - (1 - 1 + \ln r_e)\right\}} \right|$$

$$= \frac{[(\gamma - 1)\ln r_e - \gamma(\tau - 1)(1 - \varepsilon)]}{(\gamma - 1)[\tau - 1 - \tau\ln(r_e/\tau) - \ln r_e]} \tag{2.52}$$

The mean effective pressure is [from Eq. (2.38)]

$$\bar{p}_m = \frac{RT_E}{V_3}\frac{\{\tau[(1-1)/\tau + \ln(r_e/\tau)] - (1 - 1 + \ln r_e)\}}{(r_e - 1)V_c}$$

$$= \frac{RT_E}{V_3}\frac{[\tau - 1 + \ln(r_e/\tau) - \ln r_e]}{(r_e - 1)} \tag{2.53}$$

and the pressure ratio $P = \bar{p}_m/p_{\max}$ is

$$P = \frac{\tau - 1 + \ln(r_e/\tau) - \ln r_e}{r_e - 1} \tag{2.54}$$

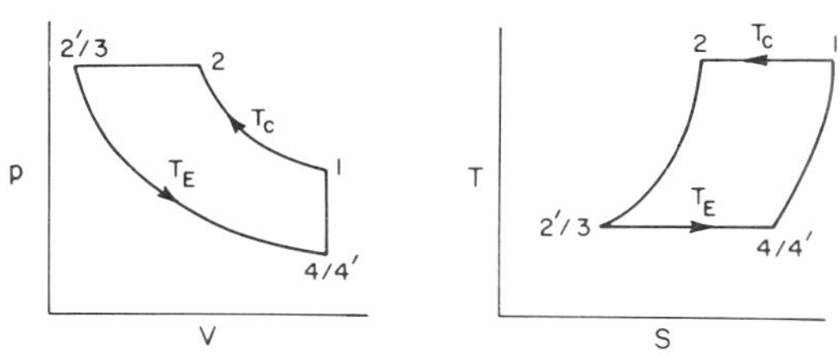

Fig. 2.20. Cycle with constant-pressure regenerative cooling and constant-volume regenerative heating.

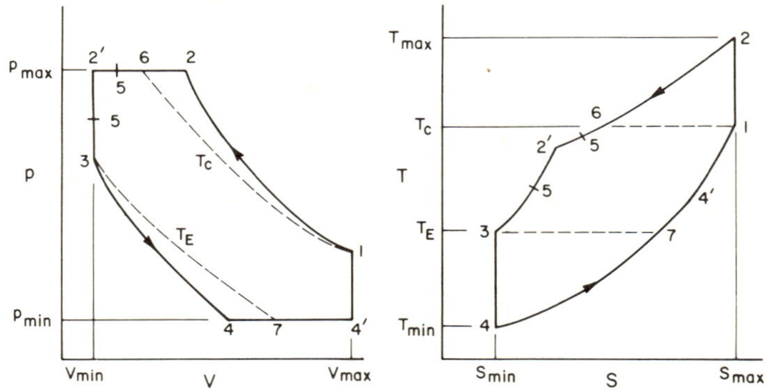

Fig. 2.21. Rallis adiabatic regenerative cycle.

RALLIS ADIABATIC REGENERATIVE CYCLE

The pressure volume and temperature entropy planes for the Rallis adiabatic cycle are shown in Fig. 2.21. The cycle is defined by two adiabatic processes having initial values of temperature defined by given temperatures T_C and T_E and bounded by constant-volume and constant-pressure regenerative processes and external heat transfers to and from the system. T_C is close to the ambient temperature at which cooling is available. T_E is the refrigeration temperature.

Process (1–2)—Adiabatic compression, no heat transfer to or from the working fluid.
Process (2–6)—Constant-pressure cooling. Temperature is reduced from T_2 to T_6 by heat rejection from the system to the surroundings.
Process (6–2′)—Constant pressure regenerative cooling.
Process (2′–3)—Constant-volume regenerative cooling.
Process (3–4)—Adiabatic expansion, no heat transfer to or from the working fluid.
Process (4–7)—Constant-pressure heating. Temperature increased from T_4 to T_7 by heat abstracted from the surroundings. This is the useful refrigerating effect of the cycle.
Process (7–4′)—Constant-pressure regenerative heating.
Process (4′–1)—Constant-volume regenerative heating.

For convenience in the analysis we define the following nondimensional parameters different from those in the Rallis isothermal regenerative cycle:

$a = T_{2'}/T_6$, regenerative cooling constant-pressure ratio

$d = T_{4'}/T_7$, regenerative heating constant-pressure ratio

$f = T_6/T_2 = 1/r_e^{(\gamma-1)}$, constant-pressure cooling ratio

$g = T_7/T_4 = r_e^{(\gamma-1)}$, constant-pressure heating ratio (the refrigerating effect)

$\varepsilon = \dfrac{T_5 - T_6}{T_3 - T_6}$, regenerator effectiveness so that

$$T_5 = T_6 + \varepsilon(T_3 - T_6) = T_6 - \varepsilon\left(\frac{\tau - 1}{\tau}\right)T_6$$

COP is the coefficient of performance (Q/W), where Q is the specific external heat extracted in the low-temperature process (4–7) less the heat necessary to reduce the working fluid temperature from T_5 to T_3 (due to imperfect regeneration). The specific external work input required per cycle, W, the mean effective pressure, $\bar{p}_m$, and the mean effective pressure ratio of the cycle, P, are defined as before.

Now

$$ab = (T_{2'}/T_6)(T_3/T_{2'}) = (T_E/T_C) = 1/\tau$$

$$de = (T_{4'}/T_7)(T_1/T_{4'}) = (T_C/T_E) = \tau$$

$$T_2 = r_c^{(\gamma-1)}$$

$$T_6 = fT_2 = fr_c^{(\gamma-1)}T_1 = T_C = T_1$$

therefore

$$fr_c^{(\gamma-1)} = 1$$

or

$$f = 1/r_c^{(\gamma-1)}$$

$$T_{2'} = aT_6 = aT_1$$

$$T_3 = bT_{2'} = abT_1 = T_1/\tau$$

$$T_4 = \frac{T_3}{r_e^{(\gamma-1)}} = \frac{T_1}{\tau r_e^{(\gamma-1)}}$$

$$T_7 = gT_4 = \frac{g}{r_e^{(\gamma-1)}}T_3 = T_E$$

therefore

$$g/r_e^{(\gamma-1)} = 1 \text{ or } g = r_e^{(\gamma-1)}$$

$$T_{4'} = dT_7 = dT_3 = (d/\tau)T_1$$

but since $T_{4'} = T_1/e$ then $1/e = d/\tau$ or $ed = \tau$.

The external work required per cycle is

$$W = p_2(V_2 - V_{2'}) + \frac{(p_2V_2 - p_1V_1)}{(\gamma-1)} - p_4(V_{4'} - V_4) - \frac{(p_3V_3 - p_4V_4)}{(\gamma-1)}$$

$$= R\left[(T_2 - T_{2'}) + \frac{(T_2 - T_1)}{(\gamma-1)} - (T_{4'} - T_4) - \frac{(T_3 - T_4)}{(\gamma-1)}\right]$$

$$= \frac{R}{\gamma-1}[(\gamma-1)(T_2 - T_{2'}) + (T_2 - T_1) - (\gamma-1)(T_{4'} - T_4) - (T_3 - T_4)]$$

$$= \frac{R}{\gamma-1}[(\gamma T_2 - \gamma T_{2'} + T_{2'} - T_1) - (\gamma T_{4'} - \gamma T_4 - T_{4'} + T_3)]$$

$$= \frac{RT_1}{\gamma-1}\left[(\gamma r_c^{(\gamma-1)} - \gamma a + a - 1) - \left(\gamma\frac{d}{\tau} - \gamma\frac{1}{\tau r_e^{(\gamma-1)}} - \frac{d}{\tau} + \frac{1}{\tau}\right)\right]$$

$$= \left(\frac{RT_1}{\gamma-1}\right)\left[\gamma(r_c^{(\gamma-1)} - a) + (a-1) - \frac{\gamma}{\tau}\left(d - \frac{1}{r_e^{(\gamma-1)}}\right) + \frac{1}{\tau}(d-1)\right]$$

$$= \frac{RT_1}{\gamma-1}\left[\gamma(r_c^{(\gamma-1)} - a) + (a-1) - \frac{\gamma}{\tau r_e^{(\gamma-1)}}(dr_e^{(\gamma-1)} - 1) + \frac{1}{\tau}(d-1)\right]$$

$$= \frac{RT_1}{\tau(\gamma-1)}\left[\tau\gamma(r_c^{(\gamma-1)} - a) + \tau(a-1) - \frac{\gamma}{r_e^{(\gamma-1)}}(dr_e^{(\gamma-1)} - 1) + (d-1)\right] \tag{2.55}$$

$$= \frac{RT_E}{(\gamma-1)}\left[\tau\gamma(r_c^{(\gamma-1)} - a) + \tau(a-1) - \frac{\gamma}{r_e^{(\gamma-1)}}(dr_e^{(\gamma-1)} - 1) + (d-1)\right] \tag{2.56}$$

The mean effective pressure is

$$\bar{p}_m = \frac{RT_E}{(\gamma-1)(V_{4'} - V_3)} \times \left[\tau\gamma(r_c^{(\gamma-1)} - a) + \tau(a-1) - \frac{\gamma}{r_e^{(\gamma-1)}}(dr_e^{(\gamma-1)} - 1) + (d-1)\right]$$

but

$$(V_{4'} - V_3) = V_3\left(\frac{V_{4'}}{V_3} - 1\right) = V_c(d-1)$$

so that

$$\bar{p}_m = \frac{RT_E}{(\gamma-1)(d-1)V_3}$$

$$\times\left[\tau\gamma(r_c^{(\gamma-1)} - a) + \tau(a-1) - \frac{\gamma}{f_e^{(\gamma-1)}}(dr_e^{(\gamma-1)} - 1) + (d-1)\right] \quad (2.57)$$

In dimensionless form with respect to the maximum pressure, $p_{\max} = p_2 = bp_3$ and since $RT_E/V_3p_3 = 1$ then

$$P = \frac{1}{b(\gamma-1)(d-1)}\left[\tau\gamma(r_c^{(\gamma-1)} - a) + \tau(a-1) - \frac{\gamma}{r_c^{(\gamma-1)}}(dr_e^{(\gamma-1)} - 1) + (d-1)\right] \quad (2.58)$$

Thermal Regeneration

With perfect regeneration $Q_{(7-4')} + Q_{4'} = Q_{(6-2')} + Q_{(2'-3)}$ so that T_5, the temperature of the gas at the end of regenerative cooling, is T_c or T_E. With less than perfect regeneration, $\varepsilon < 1$ (due to heat leaks or friction effects) or with less than perfect external system cooling (so that $T_6 > T_C$ at the start of regenerative cooling) then T_5, temperature at the end of regeneration, will be greater than T_3. The energy to reduce the temperature of the gas from T_5 to T_3 will therefore have to be drawn from the constant-pressure refrigeration process Q_{4-7}.

The heat extracted or available refrigeration will therefore be reduced to

$$Q = Q_{(4-7)} - Q_{(5-3)}$$

There are two cases to consider:

(a) $T_3 \leq T_5 \leq T_{2'}$

(b) $T_{2'} \leq T_5 \leq T_6$

For case (a)

$$\frac{T_1}{\tau} \leq T_6 - \varepsilon\left(\frac{\tau-1}{\tau}\right)T_6 \leq aT_6$$

$$\frac{1}{\tau} \leq -\varepsilon\left(\frac{\tau-1}{\tau}\right) \geq a$$

For case (b)

$$aT_6 \leq T_6 - \varepsilon\left(\frac{\tau-1}{\tau}\right)T_6 \leq T_6$$

$$a \leq 1 - \varepsilon\left(\frac{\tau-1}{\tau}\right) \leq 1$$

For case (a)

$$\begin{aligned}
Q &= Q_{(4-7)} - Q_{(5-3)} \\
&= C_p(T_7 - T_4) - C_v(T_5 - T_3) \\
&= C_p(T_7 - T_4) - C_v\left\{T_6\left[1 - \varepsilon\left(\frac{\tau-1}{\tau}\right) - T_3\right]\right\} \\
&= C_pT_E\left(1 - \frac{1}{r_e^{(\gamma-1)}}\right) - C_vT_E\left\{\tau\left[1 - \varepsilon\left(\frac{\tau-1}{\tau}\right) - 1\right]\right\} \\
&= T_E\left\{\frac{C_p}{r_e^{(\gamma-1)}}(r_e^{(\gamma-1)} - 1) - C_v[\tau - \varepsilon(r-1) - 1]\right\} \\
&= T_E\left[\frac{C_p}{r_e^{(\gamma-1)}}(r_e^{(\gamma-1)} - 1) - C_v(\tau-1)(1-\varepsilon)\right]
\end{aligned}$$

The coefficient of performance (Q/W) is, with W from Eq. (2.56)

$$\text{COP} = \frac{T_E\left[\dfrac{C_p}{r_e^{(\gamma-1)}}(r_e^{(\gamma-1)} - 1) - C_v(\tau-1)(1-\epsilon)\right]}{\dfrac{RT_E}{(\gamma-1)}\left[\tau\gamma(r_c^{(\gamma-1)} - a) + \tau(a-1) - \dfrac{\gamma}{r_e^{(\gamma-1)}}(dr_e^{(\tau-1)} - 1) + (d-1)\right]}$$

and since $C_p = [\gamma/(\gamma-1)]R$ and $C_v = [R/(\gamma-1)]$, then

$$\text{COP} = \frac{\dfrac{\gamma}{r_e^{(\gamma-1)}}(r_e^{(\gamma-1)} - 1) + (\tau-1)(1-\varepsilon)}{\tau\gamma(r_c^{(\gamma-1)} - a) + (a-1) - \dfrac{\gamma}{r_e^{(\gamma-1)}}(dr_e^{(\gamma-1)} - 1) + (d-1)} \tag{2.59}$$

For case (b)

$$\begin{aligned}
Q &= Q_{4-7} - Q_{2'-3} - Q_{5-2'} \\
&= C_p(T_7 - T_4) - C_v(T_{2'} - T_3) - C_p(T_5 - T_{2'}) \\
&= C_p(T_7 - T_4 - T_5 + T_{2'}) - C_v(T_{2'} - T_3) \\
&= T_E\left\{Cp\left[1 - \frac{1}{r_e^{(\gamma-1)}} - \left(1 - \varepsilon\frac{\tau-1}{\tau}\right) + \frac{a}{\tau}\right] - C_v\left(\frac{a}{\tau} - 1\right)\right\}
\end{aligned}$$

and since $C_p = [\gamma/(\gamma-1)]R$ and $C_v = R/(\gamma-1)$

$$Q = \frac{RT_E}{(\gamma-1)}\left\{\gamma\left[1 - \frac{1}{r_e^{(\gamma-1)}} - \tau + \varepsilon(\tau-1) + \frac{a}{\tau}\right] - \left(\frac{a-\tau}{\tau}\right)\right\}$$

Therefore the coefficient of performance is

$$\text{COP} = \frac{\dfrac{RT_E}{(\gamma-1)}\left\{\gamma\left[1 - \dfrac{1}{r_e^{(\gamma-1)}} - \tau + \varepsilon(\tau-1) + \dfrac{a}{\tau}\right] - \left(\dfrac{a-\tau}{\tau}\right)\right\}}{\dfrac{RT_E}{(\gamma-1)}\left[\tau\gamma(r_c^{(\gamma-1)} - a) + \tau(a-1) - \dfrac{\gamma}{r_e^{(\gamma-1)}}(dr_e^{(\gamma-1)} - 1) + (d-1)\right]}$$

$$= \frac{\gamma\left[1 - \dfrac{1}{r_e^{(\gamma-1)}} - \tau + \varepsilon(\tau-1) + \dfrac{a}{\tau}\right] - \left(\dfrac{a-\tau}{\tau}\right)}{\tau\gamma(r_c^{\gamma-1} - a) + \tau(a-1) - \dfrac{\gamma}{r_e^{(\gamma-1)}}(dr_e^{(\gamma-1)} - 1) + (d-1)} \tag{2.60}$$

Pseudo-Ericsson Cycle

Now consider the special case where the regenerative processes occur only at constant pressure so that $r_c = r_e = r$, $b = e = 1$, $(1/a) = d = \tau$ and $T_{2'} \leq T_5 \leq T_6$ only. The *P–V* and *T–S* planes are shown in Fig. 2.22. This cycle may be called (after Rallis *et al.*, 1977) the reversed pseudo-Ericsson cycle. From Eq. (2.60)

$$\text{COP} = \frac{\gamma\left[1 - \dfrac{1}{r^{(\gamma-1)}} - \tau + \varepsilon(\tau-1) + \dfrac{\tau}{\tau}\right] - \left(\dfrac{\tau}{\tau} - 1\right)}{\tau\gamma\left(r^{(\gamma-1)} - \dfrac{1}{\tau}\right) + \tau\left(\dfrac{1}{\tau} - 1\right) - \dfrac{\gamma}{r^{(\gamma-1)}}(\tau r^{(\gamma-1)} - 1) + (\tau-1)}$$

$$= \frac{\gamma\left[2 - \dfrac{1}{r^{(\gamma-1)}} - \tau + \varepsilon(\tau-1)\right]}{\gamma(\tau r^{(\gamma-1)} - 1) - \dfrac{\gamma}{r^{(\gamma-1)}}(\tau r^{(\gamma-1)} - 1)}$$

$$= \frac{2 - \left(\dfrac{1}{r^{(\gamma-1)}}\right) - \tau + \varepsilon(\tau-1)}{(1-\gamma)} \tag{2.61}$$

and when $\varepsilon = 1$, perfect regeneration,

$$\text{COP} = \left(\frac{r^{(\gamma-1)} - 1}{r^{(\gamma-1)}(1-\gamma)}\right) \tag{2.62}$$

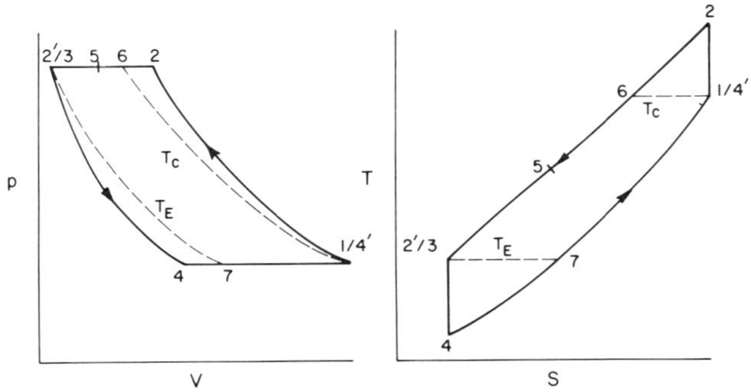

Fig. 2.22. Pseudo-Ericsson cycle.

The mean effective pressure is

$$\bar{p}_m = \frac{RT_E}{(\gamma -)(\tau - 1)V_{2'}}\left\{\tau\gamma\left(r^{(\gamma-1)} - \frac{1}{\tau}\right) + \tau\left[\frac{1}{\tau} - \frac{\gamma}{r^{(\gamma-1)}}(\tau r^{(\gamma-1)} - 1) + (\tau - 1)\right]\right\}$$

$$= \frac{RT_E}{(\gamma - 1)(\tau - 1)V_{2'}}$$

$$\times\left[\gamma(\tau r^{(\gamma-1)} - 1) - (\tau - 1) - \frac{\gamma}{r^{(\gamma-1)}}(\tau r^{(\gamma-1)} - 1) + (\tau - 1)\right]$$

$$= \frac{RT_E\gamma}{(\gamma - 1)(\tau - 1)V_{2'}}\left[(\tau r^{(\gamma-1)} - 1)\left(1 - \frac{1}{r^{(\gamma-1)}}\right)\right]$$

$$= \frac{\gamma RT_E}{(\gamma - 1)(\tau - 1)V_{2'}}\left[\left(\frac{r^{(\gamma-1)} - 1}{r^{(\gamma-1)}}\right)(\tau r^{(\gamma-1)} - 1)\right] \tag{2.63}$$

and the mean effective pressure rate becomes

$$P = \frac{\gamma(r^{(\gamma-1)} - 1)(\tau r^{(\gamma-1)} - 1)}{\tau r^{(\gamma-1)}(\gamma - 1)(\tau - 1)} \tag{2.64}$$

Pseudo-Stirling Cycle

Another special case arises when the regenerative processes occur wholly at constant volume. This may be thought of as the reversed pseudo-Stirling cycle. In such a case, however, it is most unlikely that the heat transfers to and from the system $Q_{(2-6)}$ and $Q_{(4-7)}$ would in fact occur at *constant pressure*. Rather, it is more likely they would occur at constant volume and therefore the equations derived above would not be applicable.

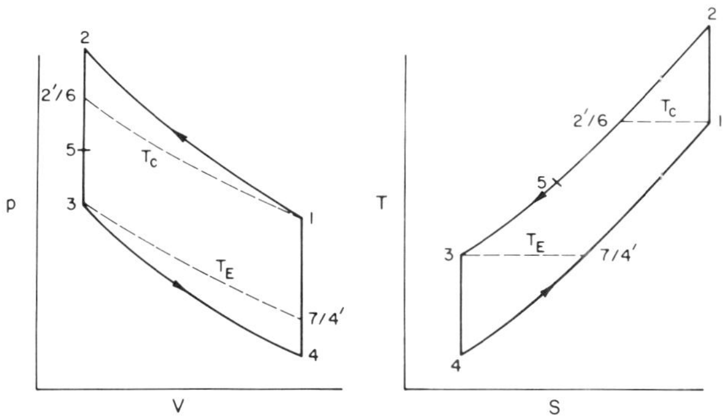

Fig. 2.23. Pseudo-Stirling cycle.

Let us consider the special case of the pseudo-Stirling engine with constant volume processes of heat transfers to and from the system. The T–S and P–V planes for this cycle are shown in Fig. 2.23. In this cycle let $r_c = r_e = r$ and the other ratios are as defined above with $a = d = 1$, $(1/b) = e = \tau$ and $T_3 \leq T_5 \leq T_6$ only. The external work input to the cycle, W, is

$$W = \frac{p_2V_2 - p_1V_1}{(\gamma-1)} - \frac{p_3V_3 - p_4V_4}{(\gamma-1)}$$

$$= \frac{R}{(\gamma-1)}[(T_2 - T_1) - (T_3 - T_4)]$$

$$= \frac{RT_3}{(\gamma-1)}\left[\tau(r^{(\gamma-1)} - 1) - \left(1 - \frac{1}{r^{(\gamma-1)}}\right)\right]$$

$$= \frac{RT_3}{(\gamma-1)}\left[(r^{(\gamma-1)} - 1)\left(\tau - \frac{1}{r^{(\gamma-1)}}\right)\right]$$

$$= \frac{RT_3}{(\gamma-1)(r^{(\gamma-1)} - 1)}[(r^{(\gamma-1)} - 1)(\tau r^{(\gamma-1)} - 1)] \tag{2.65}$$

The mean effective pressure is

$$\bar{p}_m = \frac{RT_3}{(\gamma-1)(r^{(\gamma-1)})(r-1)V_3}[(r^{(\gamma-1)} - 1)(\tau r^{(\gamma-1)} - 1)] \tag{2.66}$$

and since $p_{\max} = \tau r^{(\gamma-1)} p_3$ and $RT_3/p_3V_3 = 1$ the dimensionless mean effective pressure ratio becomes

$$P = \frac{\bar{p}_m}{p_{\max}} = \frac{(r^{(\gamma-1)} - 1)(\tau r^{(\gamma-1)} - 1)}{\tau r^{2(\gamma-1)}(\gamma-1)(r-1)} \tag{2.67}$$

The refrigerating effect is

$$Q = Cv(T_7 - T_4) - Cv(T_5 - T_3)$$

It is assumed that the gas is cooled from T_2 to T_6 by heat transfer from the system to the surroundings prior to thermal regenerative cooling from T_6 to T_5. With perfect regeneration $T_5 = T_3$. With imperfect regeneration $T_5 \geq T_3$. Therefore if $\varepsilon = (T_5 - T_6)/(T_3 - T_6)$ then

$$T_5 = T_6 + \varepsilon(T_3 - T_6) = \left[1 - \varepsilon\left(\frac{\tau - 1}{\tau}\right)\right] T_6$$

Then the heat Q extracted is

$$Q = C_v(T_7 - T_4) - C_v\left[1 - \varepsilon\left(\frac{\tau - 1}{\tau}\right) T_6 - T_3\right]$$

but $T_7 = T_E$, $T_4 = T_E/r^{(\gamma-1)}$, and $T_6 = T_E/\tau$, $C_v = R/(\gamma - 1)$ and $C_p = \gamma R/(\gamma - 1)$. Therefore

$$Q = \frac{RT_E}{(\gamma - 1)}\left\{\left(1 - \frac{1}{r^{(\gamma-1)}}\right) - \gamma\left[\tau\left(1 - \varepsilon\frac{\tau - 1}{\tau} - 1\right]\right\}$$

$$= \frac{RT_E}{(\gamma - 1)}\left\{\frac{(r^{(\gamma-1)} - 1)}{(r^{(\gamma-1)})} - \gamma[\tau - \varepsilon(\tau - 1) - 1]\right\}$$

$$= \frac{RT_E}{(\gamma - 1)}\left[\frac{r^{(\gamma-1)} - 1}{r^{(\gamma-1)}} - \gamma(\tau - 1)(1 - \varepsilon)\right] \tag{2.68}$$

The coefficient of performance COP is

$$\text{COP} = \frac{\dfrac{RT_E}{(\gamma - 1)}\left[\dfrac{r^{(\gamma-1)} - 1}{r^{(\gamma-1)}} - \gamma(\tau - 1)(1 - \varepsilon)\right]}{\dfrac{RT_E}{\gamma - 1(r^{(\gamma-1)})}[(r^{(\gamma-1)} - 1)(\tau r^{(\gamma-1)} - 1)]}$$

$$\text{COP} = \frac{(r^{(\gamma-1)} - 1) - \gamma r^{(\gamma-1)}(\tau - 1)(1 - \varepsilon)}{(r^{-1} - 1)(r^{-1} - 1)} \tag{2.69}$$

and if $\varepsilon = 1$, ideal regeneration, then

$$\text{COP} = \frac{r^{(\gamma-1)} - 1}{(r^{(\gamma-1)} - 1)(\tau r^{(\gamma-1)} - 1)} = \frac{1}{\tau r^{(\gamma-1)} - 1} \tag{2.70}$$

Many other special cases with different combinations of constant pressure and constant volume regeneration and adiabatic heat transfers to and from the system may be devised. Three are shown in Fig. 2.24. None of these cycles are known to be identified with a particular name of an inventor.

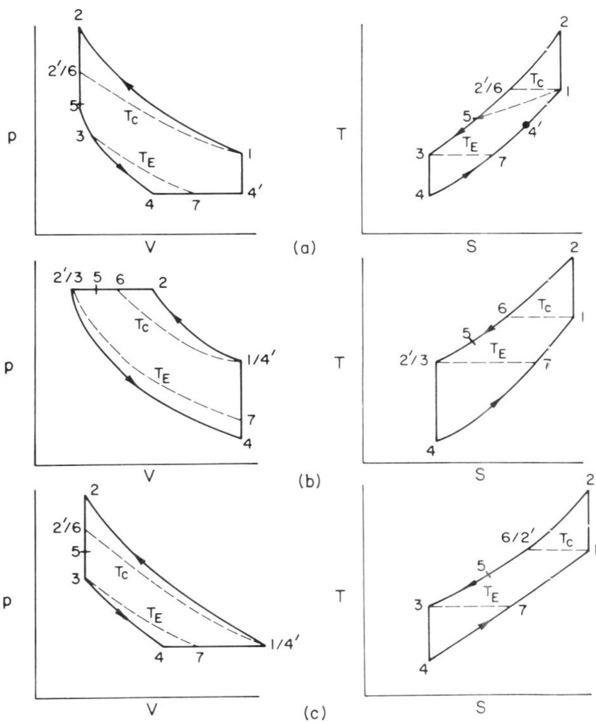

Fig. 2.24. Three adiabatic cycles with various combinations of constant-pressure and constant-volume regenerative processes.

PARAMETRIC EFFECTS

The above equations provide a basis for useful and interesting studies of the effects of various parameters of the cycle. By way of example, results computed for the ideal Stirling cycle are presented in Fig. 2.25. The compression temperature T_C was assumed to be 320 K and the refrigeration temperature, T_E, 80 K. The figure shows the coefficient of performance as a function of the volume compression ratio with different values of regenerator effectiveness. It is interesting to note the marked deterioration in coefficient of performance with even the slightest departure from perfect regeneration, particularly at low volume compression ratios. The volume compression ratio in actual Stirling cryocoolers rarely exceeds 2. At this value of volume compression ratio in Fig. 2.25 the COP with a regenerator effectiveness of 0.9 is only 0.11, a third of the Carnot value (0.33 with perfect regeneration).

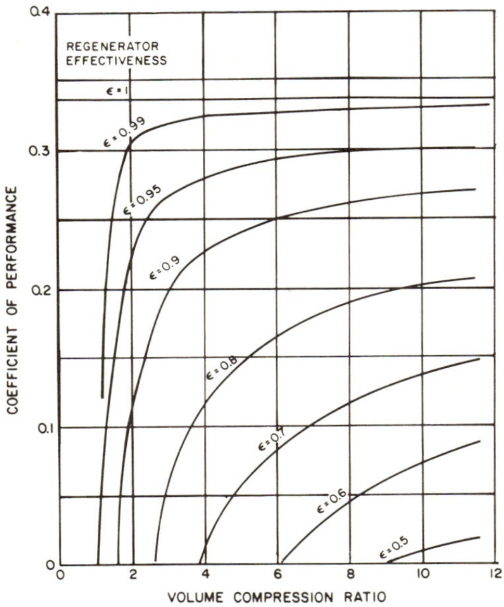

Fig. 2.25. Coefficient of performance of the ideal Stirling cycle as a function of the cycle volume compression ratio with different values of regenerator effectiveness.

Similar results calculated for the Rallis adiabatic cycle are presented in Fig. 2.26. This shows the coefficient of performance as a function of the volume compression ratio at different values of the regenerator effectiveness. Temperatures of the working fluid at the beginning of compression and expansion were assumed to be 320 K and 80 K as before.

An appreciation of the drastic consequence of replacing isothermal processes by adiabatic processes may be gained by comparing Figs. 2.25 and 2.26. With adiabatic expansion the available refrigeration is substantially consumed in completing the regenerative cooling process and the effect is progressively cumulative as the volume compression ratio is increased. It is sobering to realize that in practice, with cryocoolers operating at normal speeds (say 10 to 20 Hz), the compression and expansion are more likely to be adiabatic than isentropic processes.

In a cryocooler the maximum pressure is important for it determines the strength requirement of the cylinder wall and the forces acting on the drive mechanism. To a first approximation the maximum pressure is indicative of the weight of the engine. The mean effective pressure is indicative of the cycle power input per unit of piston displacement and hence, to a first approximation, of the refrigeration capacity per unit volume of machine. The ratio of mean effective pressure to maximum cycle pressure is therefore an important guide to the refrigeration output per unit weight, the specific capacity of the cryocooler.

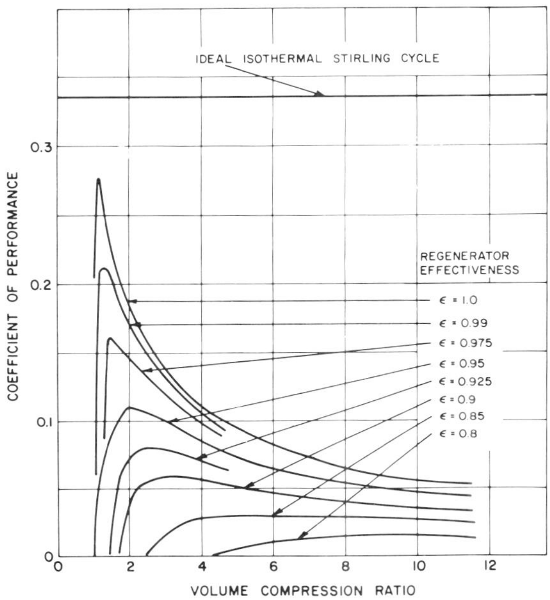

Fig. 2.26. Coefficient of performance of the pseudo-Stirling cycle as a function of the cycle volume compression ratio with different values of regenerator effectiveness.

In Fig. 2.27 the pressure ratio $(\bar{p}_m/p_{max})$ is compared for the adiabatic and the isothermal reference cycles as a function of the volume compression ratio. It can be seen from this figure there is a marked and progressive decline in the output per unit weight for both cycles with increase in the volume compression ratio. The isothermal cycle is superior to the adiabatic cycle.

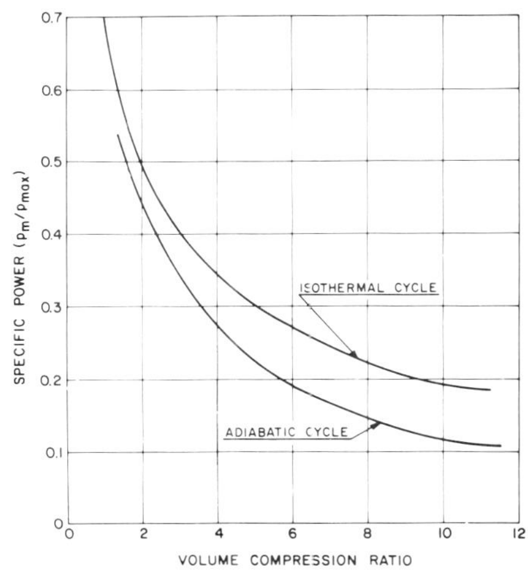

Fig. 2.27. Comparison of the pressure ratio: mean effective pressure/maximum cycle pressure $(\bar{p}_m/p_{max})$ for the ideal and the pseudo-Stirling cycles as a function of the volume compression ratio.

OTHER REGENERATIVE CYCLES

Figure 2.28 shows four different diagrammatic arrangements for cryocoolers using regenerative heat exchangers to separate the ambient and low-temperature regions of the machine. This diagram was reproduced directly from Radebaugh (1977). There are slight but important differences in terminology compared with our previous usage where quantities of work, heat transfer, and temperature at low temperature were referred to without a subscript. In Radebaugh's diagrams the subscript 0 is used for ambient temperatures. At high temperatures the subscript h is used.

The four systems shown are representative of the two principal classes of regenerative engine characterized by the method of fluid flow control, namely, by volume changes or by valves. Machines of the type where flows are controlled by volume changes may be classified conveniently as Stirling engines (note the word cycle is not used). Similarly, machines of the type

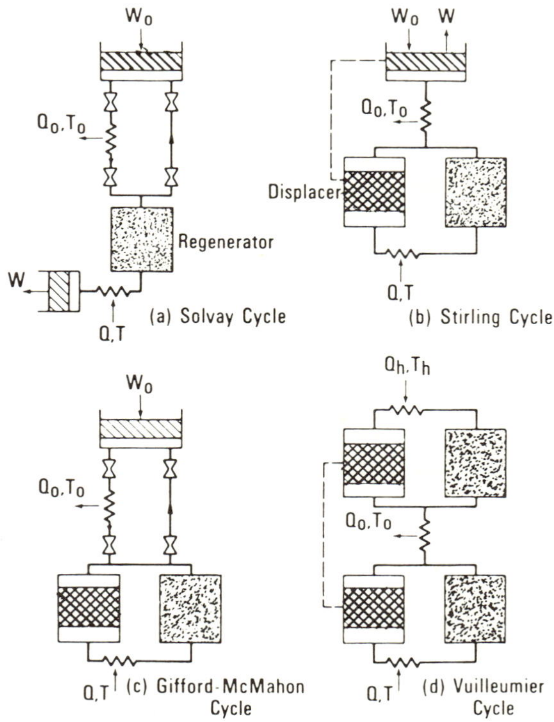

Fig. 2.28. Composite diagram of four different regenerative engines (after Radebaugh, 1977) where 0 represents ambient temperature and h high temperature.

where fluid flows are controlled by valves may be classified as Ericsson engines.

In Fig. 2.28 there are two Ericsson engines (with valves), namely, the Solvay machine, diagram (a) and the Gifford–McMahon machine, diagram (c). There are also two Stirling engines (without valves), namely, the Stirling machine, diagram (b), and the Vuilleumier machine, diagram (d).

Stirling Cycle

On the Radebaugh diagram the Stirling cycle is shown in Fig. 2.28b. It consists of the following:

i. A *piston* receiving work, W_0, from an external source during the piston upstroke and generating work, W, during the piston downstroke. When operating as a refrigerator the net cycle input work is $(W_0 - W)$.
ii. A *cooler* rejecting heat Q_0 at temperature T_0 *from* the working fluid to the surroundings.
iii. A *displacer* mechanically coupled to the piston but operating out of phase with it so that fluid is displaced between the ambient temperature compression space and the low-temperature expansion space.
iv. A *regenerative* heat exchanger acting as a thermodynamic sponge and receiving heat from the working fluid when the fluid is passing from the ambient to low-temperature region. It releases heat to the working fluid when the fluid is passing back from the low to ambient temperature region.
v. A *freezer* receiving heat, Q, or abstracting it from the surroundings at low temperature, T. This is the useful refrigeration product of the system.

Vuilleumier Cycle

Another regenerative cycle related to the Stirling cycle is the Vuilleumier (pronounced Viamay) cycle shown in Fig. 2.28d. This is is sometimes described as the Stirling cycle with a thermal compressor rather than a mechanical compressor.

Comparison of the diagrams for the Stirling and Vuilleumier cycles shows the lower halves for both systems (the low-temperature region) to be identical. The principal difference is that the Vuilleumier cycle has no mechanical compressor but has another displacer and regenerator arrangement with an additional heat exchanger called the heater.

The heater receives heat, Q_h, at high temperature, T_h. Motion of the displacer causes fluid to pass through the upper regenerator from the heated space above the displacer to the cold space below the displacer. This transfer of the fluid *at constant total volume* between a heated and cooled space causes the pressure to change. The pressure is greatest when the fluid is concentrated in the hot space and least when the fluid is concentrated in the cold space. The lower displacer assembly is directly connected to the upper displacer assembly through the cooler. Therefore pressure changes in the upper assembly are also experienced in the lower assembly.

The cyclic pressure change will produce a refrigerating effect as in the Stirling engine cycle provided that the fluid is concentrated in the cold expansion space at the time when the pressure is decreasing from the maximum to minimum value. The two displacers are therefore coupled as indicated by the broken line connecting them in Fig. 2.28d. In practice the displacers are not synchronized as the figure suggests but operate about 90° out of phase, with the low-temperature displacer *leading* the high-temperature displacer.

Vuilleumier cycle engines can be made to operate with no input of work from an external source. However, it is customary to provide a small work input to facilitate starting and maintain the proper frequency of operation. Vuilleumier cycle systems are well suited to produce refrigeration from a high-temperature thermal input derived from isotope degradation, concentrated solar energy, exhaust heat or combustion heating of fossil fuels and solid wastes.

Solvay Cycle

The Solvay cycle engine shown in Fig. 2.28a consists of a compressor, operating at ambient temperatures, compressing gas and receiving work input W_0 from an external source. The heat of compression Q_0 is rejected from the system in the cooler at T_0.

The compressed fluid is expanded in an expansion engine to produce work output, W, and a low temperature of the fluid. The low temperatures produced by expansion induce a flow of heat Q to the working fluid at temperature T, in the freezer. A regenerative heat exchanger is interposed between the ambient temperature compressor and the low-temperature expander. The regenerator facilitates the attainment of very low temperatures in the expansion space.

The Solvay engine is virtually identical to the opposed piston Stirling engine shown in Fig. 2.9 used for the detail description of the Stirling cycle. The only difference between that engine and the Solvay machine in Fig. 2.28a is the presence of valves to control flow in the Solvay machine. The use of valves to isolate the compressor and the expander is an important factor. It allows the frequency of operation of the compressor and the expander to be selected independently and for the possibility of higher pressure ratios (p_{max}/p_{min}) than can be achieved in Stirling engines.

The use of a slow-speed expander with consequent low noise and vibration levels is particularly important in the miniature cryocoolers used for electronic applications at cryogenic temperatures. Some of the detector elements used in these applications are incredibly sensitive and cannot function reliably in the presence of the slightest mechanical vibration or electromagnetic interference.

The isolated compressor can operate at high speed with all the heat, vibration, and noise necessary without affecting the sensitive detectors used in the cold region. It is customary to use standard commercial air or gas compressors suitably modified for the compressor units of both Solvay and the Gifford-McMahon cryocoolers. Standard compressors are available at comparatively low cost and their use allows development effort to be concentrated on the cryogenic region of the engine.

Gifford–McMahon Cycle

The Gifford–McMahon cycle system shown in Fig. 2.28c is similar to the Solvay unit with the difference that a displacer is used instead of the expansion engine of the Solvay system. The Gifford–McMahon engine is very like the piston/displacer Stirling engine shown in Fig. 2.28b but has valves isolating the compressor from the regenerator and displacer unit. Use of valves requires the provision of two connecting ducts rather than the single connecting duct of a Stirling engine. Separation of the compressor unit and the use of valves to control the flow gives the Gifford–McMahon engine the same advantage of independent compressor/expander operation and high-pressure ratios as the Solvay engine.

The Gifford–McMahon engine has an advantage over the Solvay engine. It uses a displacer rather than a piston. A displacer requires only a low-pressure seal. A piston requires the use of a high-pressure seal whereas this apparently trivial distinction is the biggest single reason why Gifford–McMahon engines are used very widely for cryogenic cooling engines whereas not a single Solvay engine is in production. Air Products and Chemicals, Inc. sometimes refer to their cryocoolers as modified Solvay engines. They are indistinguishable from Gifford–McMahon engines and do not include the low-temperature expansion engine characteristic of Solvay machines.

Parenthetically, it may be noted that all regenerative cryocoolers include a recuperative heat exchanger for cooling and sometimes to improve heat transfer at the expansion space.

Part III: *Recuperative Cycles*

CRYOCOOLER WITH RECUPERATIVE HEAT EXCHANGER

Many cryocoolers utilize recuperative heat exchangers rather than the regenerative exchangers considered above. In a recuperative exchanger separate flow passages are provided for both the hot and cold fluid streams. Usually the fluids flow continuously at steady state. The exchanger is used more effectively if the fluids flow in opposite directions (counterflow) rather than in the same direction (parallel flow).

The teminology of recuperative cryogenic cooling systems is unfortunately as confused as that for regenerative systems. Here we shall adopt terms in general usage although it should be understood these are not accepted universally.

A cryocooler utilizing counterflow recuperative heat exchangers is shown in Fig. 2.29. The corresponding *P–V* and *T–S* planes are shown in Fig. 2.30. The system consists of the following:

i. A compressor operating at ambient temperature. We assume compression to be the isentropic process (1–2) shown on the *P–V* and *T–S* planes.
ii. A water-cooled heat exchanger for cooling the gas to ambient temperature in the constant-pressure cooling process (2–3).
iii. A counterflow recuperative heat exchanger cooling the gas to a low temperature in the constant-pressure cooling process (3–4).
iv. An expansion engine operating at low temperature wherein the gas expands, process (4–5). The process is assumed isentropic and the gas at the exit from the engine is at temperature T_5, the minimum temperature attained in the cycle.

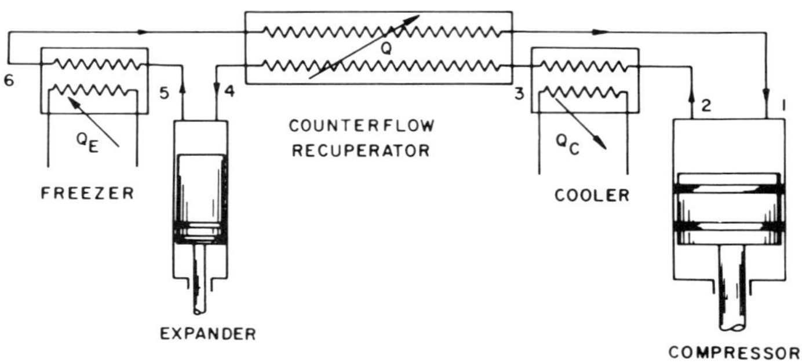

Fig. 2.29. General arrangement for a Joule–Brayton cycle refrigerating engine with recuperative heat exchanger.

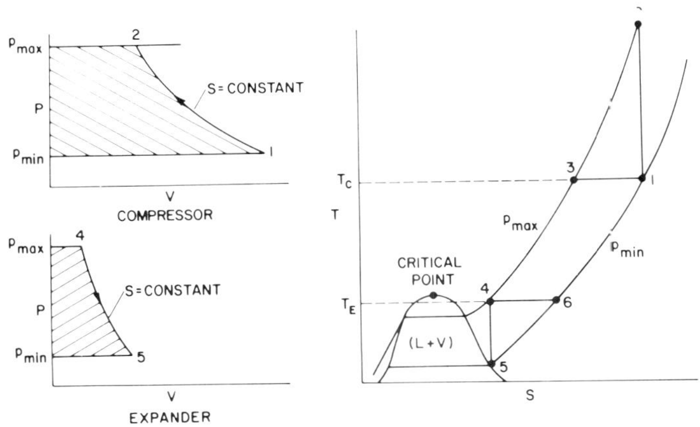

Fig. 2.30. Pressure–volume and temperature–entropy diagrams for a Joule–Brayton cycle refrigerating engine with recuperative heat exchanger.

v. A "freezer" heat exchanger wherein the cold gas leaving the engine is heated in the constant-pressure heating process (5–6). The heat is abstracted from the surroundings of the freezer and is the useful refrigeration of the cycle.

vi. The low-pressure fluid returns through the low-pressure flow passages of the counterflow recuperative heat exchanger and is heated in the constant pressure process (6–1).

It is possible for the expander and cooler to be coupled so the output work of the expander helps to drive the compressor. This reduces the work required by the compressor. However, in many systems it is not convenient for the expansion engine and the compression engine to be coupled. They may be physically separated. The machines may even be of different types. It is not unusual to find a high-speed expansion turbine supplied with fluid compressed in low-speed reciprocating compressors. Furthermore, the expansion work may be negligible compared with a compression work. It is simply dissipated in a friction brake or used to drive a small generating set or pump.

JOULE–THOMSON EXPANSION

Frequently no expansion engine is used. Rather the fluid is expanded from a high pressure to a low pressure in a *throttled* expansion. This occurs when fluid expands through a small orifice with high frictional degradation. It is called a "Joule–Thomson expansion" and occurs at constant *enthalpy*.

It is discussed in detail in Chapter 6. Here it is sufficient to indicate that a JT expansion shown in Fig. 2.31 results in a decrease in pressure from p_1 to p_2, a decrease in temperature from T_1 to T_2, and an increase in entropy from S_1 to S_2. A *decrease* in temperature occurs when the expansion takes place near the liquid/vapor envelope of the T–S diagram. At higher temperatures the expansion occurs without change in temperature or may even result in an *increase* in the temperature.

Joule–Thomson expanders are used in many liquefaction systems, as shown in Fig. 2.31b. The pressure decreases from p_1 to p_2 along the constant enthalpy line (a–b) and the fluid becomes a two-phase mixture of saturated liquid and saturated vapor. At the end of the expansion, point (b), the fluid exists in the two saturated states (c) and (d). The fraction of fluid existing as saturated vapor is the vapor quality:

$$x = \Delta S_{(c-b)}/\Delta S_{(c-d)} \tag{2.71}$$

where $\Delta S_{(c-b)}$ is the change in entropy per unit mass between (c) and (b) and $\Delta S_{(c-d)}$ is the change in entropy between the saturated vapor and liquid states (d) and (c), respectively.

The density of the saturated liquid is greater than the vapor so the two phases readily separate by natural gravitational action. The liquid is drawn off as the useful product for refrigerative cooling. The saturated vapor is returned to the compressor through a recuperative heat exchanger for recycling. With a liquefier it is necessary to add new fluid at low pressure and ambient temperature equal to that drawn off as saturated liquid.

It is possible to accomplish the liquefaction directly in an expansion engine. This is also shown in Fig. 2.31b by the isentropic process (a–f). The "yield" of liquid product is very much higher than in the constant enthalpy (JT) process (a–b). This is because in the isentropic process there is a decrease in the enthalpy.

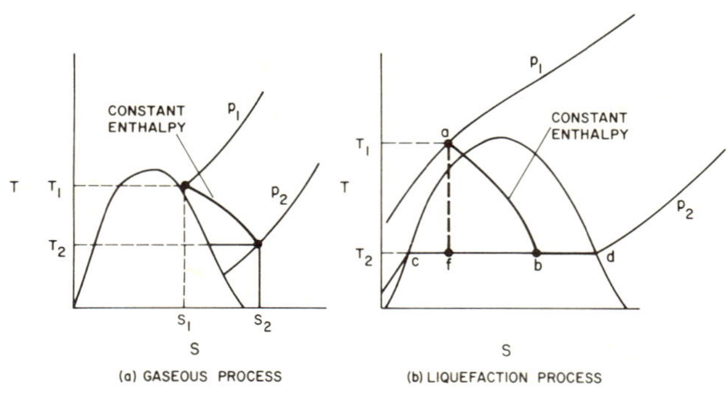

Fig. 2.31. Joule–Thomson expansion.

In practice it is undesirable to have a two-phase system present in the engine. In turbine systems the liquid droplets rapidly erode the turbine blades. In reciprocating engines the liquid tends to accumulate on the cylinder walls and piston crown. When a fresh charge is admitted the change in temperature from T_2 to T_1 (Fig. 2.31b) causes the liquid droplets remaining to evaporate and reduce the intake of fresh charge.

The preferred alternative to direct liquefaction in the engine is shown in Fig. 2.32. This utilizes an expansion engine for gaseous isentropic expansion to a low temperature and a JT valve to accomplish the liquefaction process.

This combined system operates as follows:

i. Low-pressure gas enters the compressor at state (1). The gas stream comes from the high-temperature recuperative exchanger (A) and is supplemented with new gas entering at state (0).
ii. The compressed gas at state (2) is cooled in a water cooler to state (3) and in the high-temperature recuperator to state (4).
iii. Part of the cooled gas expands in the expansion engine to state (5). Here we assume the process is isentropic.
iv. The remaining fraction of high-pressure gas is cooled in the low-temperature recuperator (B) to state (6).

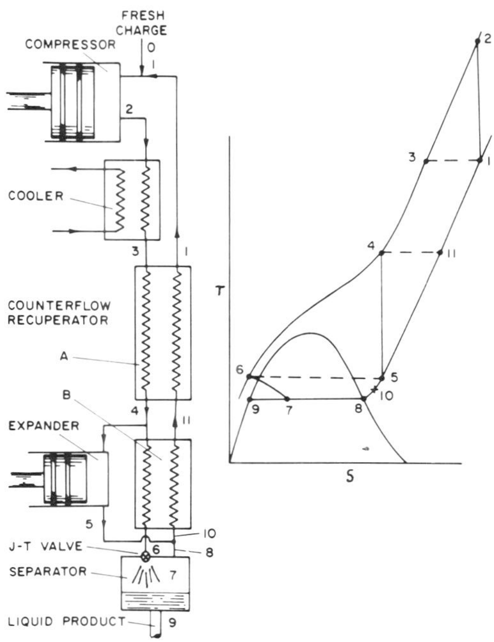

Fig. 2.32. Claude cycle refrigeration and gas liquefaction system.

v. The low-temperature gas at state (6) expands in a JT cooler to state (7), a two-phase fluid comprised of saturated liquid at state (9) and saturated vapor at state (8).
vi. The two-phase fluid separates in the separator and saturated liquid is drawn off (state 9) at low pressure as the useful product of the cycle.
vii. The saturated vapor at state (8) combines with the engine exhaust stream at state (5) to form a mixture at state (10).
viii. The low-pressure fluid at (10) enters the low-temperature recuperator (B) and is heated to state (11).
ix. Finally the low-pressure stream enters the high-temperature recuperator (A) and is heated to ambient conditions at state (11).

JOULE–BRAYTON CYCLE

The system shown in Figs. 2.31 and 2.32 operates on a thermodynamic reference cycle known variously as the Joule cycle (in England) and the Brayton cycle (in North America). Here we shall not adjudicate the merits of this historical controversy but will recognize common practice to accord the credit equally and use the term Joule–Brayton cycle.* The cycle was originally conceived as the reference cycle for power systems and is widely used for gas turbines. In that application the gas is compressed at ambient temperature from a low pressure to a high pressure [process (5–4) in Fig. 2.30], heated in a combustion chamber to the maximum temperature [process (4–2) in Fig. 2.30] and then expanded in the turbine [process (2–1) in Fig. 2.30]. This is the simple Joule–Brayton cycle. It is possible to utilize heat contained in the exhaust gas to heat the compressed gas and thus improve the engine thermal efficiency. When this is done the reference cycle is called the regenerative Joule–Brayton cycle. The term regenerative is widely used even though the actual heat exchanger may be either a recuperative or regenerator.

It is correct to use the term Joule–Brayton cycle for refrigerating applications as well as power systems. The cycle is made up of reversible processes without regard to the direction of heat flow. However, the cycle is never used in refrigerating applications without a recuperative or regen-

* The matter is beyond dispute for Joule's paper of 1851 ("On the Air Engine," *Phil. Trans.* 1852, 65–77, London) clearly precedes Brayton's work of 1873 ("Combustion Engine Processes," Lichty L. C. 1967, McGraw Hill, New York; also "The Brayton Ready Motor or Hydrocarbon Engine," *Sc. Am.* **34**, 1876). Actually both Joule and Brayton were well behind John Barber, whose English patent of 1791 clearly describes a turbo expander driving a reciprocating air compressor, operating on what is now known as the Joule or Brayton cycle. Barber's work is virtually unknown but is described briefly by Norman Davey in "The Gas Turbine," 1914, D. Van Nostrand Co. Inc., Princeton, New Jersey.

erative heat exchanger. The term Joule–Brayton cycle used here should therefore be understood as the abbreviated form of "Joule–Brayton cycle with recuperative constant pressure heat exchange."

When the cycle is used with a *regenerative* heat exchanger it is the cycle we have earlier called the pseudo-Ericsson cycle (see Part II, Chapter 2).

SIEMENS CYCLE

The Siemens cycle is a reference cycle similar to the Joule–Brayton cycle but with an isothermal compression (process 1–3 in Fig. 2.30). The cycle was named after Sir Charles Siemens who was granted a patent in 1857 for a refrigerator working approximately on the reference cycle. The isothermal work of compression is substantially less than the isentropic work. Rotary machines are always assumed to operate with isentropic processes but reciprocating compressors run slower than rotary systems and have better means of cooling. Sometimes it is more appropriate to assume their ideal compression process to be isothermal rather than isentropic. We could also assume the expansion process to be isothermal [process (4–6) in Fig. 2.30] rather than adiabatic [process (4–5) in Fig. 2.30]. This cycle with isothermal compression and expansion and constant pressure recuperative heat exchange has not been named. With *regenerative* heat exchange the cycle is of course the ideal Ericsson cycle.

LINDE–HAMPSON CYCLE

A cryocooler utilizing a Joule–Thomson expansion is shown in Fig. 2.33. This is known as the Hampson, the Linde–Hampson, the Linde, or

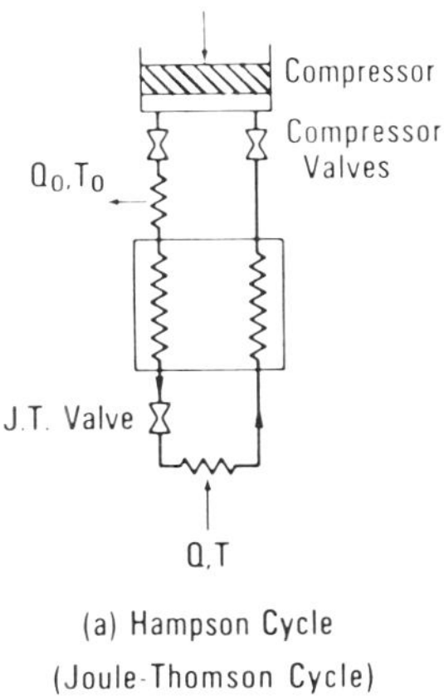

Fig. 2.33. Hampson liquefaction cycle (after Radebaugh, 1977).

simply the Joule–Thomson cycle. It consists essentially of a compressor, a water-cooled cooler, a recuperative heat exchanger, a Joule–Thomson expansion valve, and a freezer heat exchanger or a liquid/vapor separator depending whether or not the fluid condenses.

The preferred terminology is to call the system shown in Fig. 2.33 the Linde–Hampson cycle if there is simply a single stage of expansion through a single JT valve as shown.

LINDE DUAL-PRESSURE CYCLE

More efficient systems can be made with two or more stages of expansion in multiple JT valves. These are called Linde dual-pressure cycle systems. The essential difference between Linde–Hampson and Linde dual-pressure cycles is that the Hampson cycle has only one JT valve and a single stage of expansion. The Linde dual-pressure cycle uses several JT valves and at least two levels of pressure are used.

A system where a single compressor serves a multiplicity of JT valves all expanding over the same stage of expansion is a compound Linde–Hampson cycle rather than a Linde dual-pressure arrangement. To minimize confusion it is recommended that the term Joule–Thomson be restricted to the actual expansion valve rather than the complete ensemble of equipment in the Linde–Hampson cooler.

CLAUDE CYCLE

The addition to a Linde–Hampson cycle of an expansion engine and additional stage of recuperative exchanger converts the system to a Claude cycle. This was the system discussed above with reference to Fig. 2.32 and arranged for a condensing working fluid downstream of the JT valve. Figure 2.34 shows a Claude system operating as a refrigerator with no condensation downstream of the JT valve. A Claude cycle without the JT expansion reverts to the Joule–Brayton system.

Multiple stages of expansion engines are used in Claude cycle systems. The various expansions may occur in several individual engines or a long piston of successively smaller diameters may operate in a long cylinder with appropriate diameters to match the piston. This latter *tandem compound* arrangement was used by Collins in the helium liquefier engines operating in the three- or four-stage Claude cycle developed at the Massachusetts Institute of Technology after the Second World War.

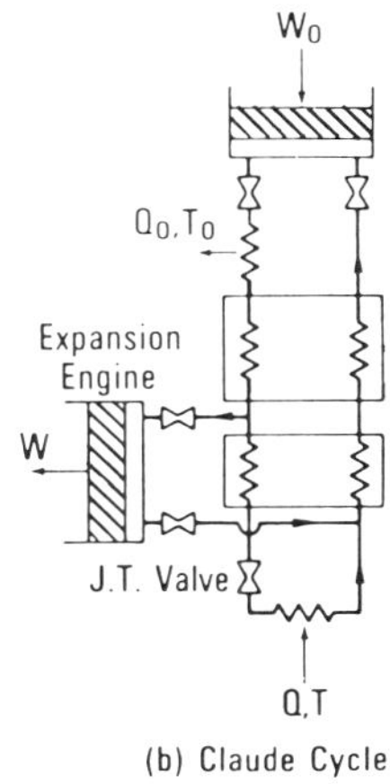

Fig. 2.34. Claude cycle refrigeration system with no liquefaction of the working fluid (after Radebaugh, 1977).

ANALYSIS OF RECUPERATIVE SYSTEMS

The elementary analysis of recuperative systems with continuous, steady flow of the working fluid is relatively simple compared with regenerative systems with continuously variable conditions.

Recuperative system analysis is best approached by use of the general energy equation for steady flow. Consider the thermodynamic "black box" shown in Fig. 2.35 with a variety of energy inputs to the box, including kinetic energy, internal energy, flow work, and heat transfer. There are also a variety of energy outputs from the box. When the system is in steady state operation, the energy input must be equal to the energy output. If this equality were not so, an accumulation or depletion of energy would occur and the system would not be in steady state.

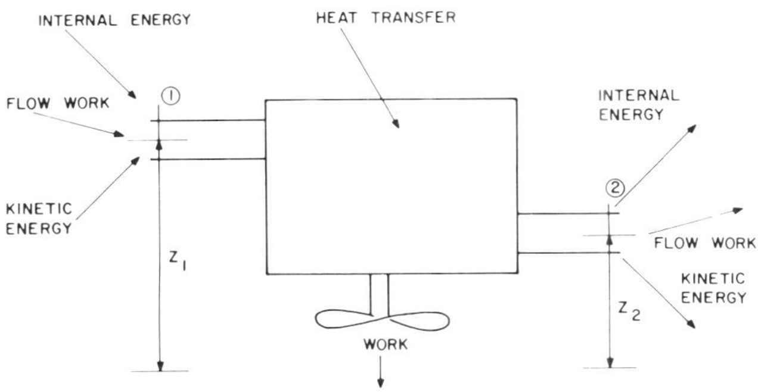

Fig. 2.35. Steady flow thermodynamic black box.

Summarizing the energy inputs and equating these to the energy outputs we can express the general steady flow energy equation in mathematical terms as

$$\dot{m}Z_1+\dot{m}C_1^2+\dot{m}U_1+p_1V_1+Q=\dot{m}Z_2+mC_2^2+\dot{m}U_2+p_2V_2+W \tag{2.72}$$

where $\dot{m}$ is the mass rate of flow of the fluid, Z is the height of fluid above some prescribed datum, C is the velocity of the fluid, U is the internal energy of the fluid, p is the pressure of the fluid, V is the volume of fluid flow, Q is the net heat transfer to the fluid, and W is the net work output. Frequently the terms $(\dot{m}U+pV)$ can be combined to the new term $\dot{m}H$ where H is the enthalpy of the fluid. In many cases the change in the datum heat $[\dot{m}(Z_1-Z_2)]$ is negligibly small.

Then Eq. (2.72) can be restated as

$$\dot{m}(H_1-H_2)=\dot{m}(C_2^2-C_1^2)+W-Q \tag{2.73}$$

The change in *enthalpy* from state 1 to state 2 is therefore the sum of the change in kinetic energy, the work done, and the heat transferred.

Elementary analysis of recuperative cryocoolers can be approached by considering enthalpy values at the different locations of the system. Consider the Joule–Brayton cycle system shown in Fig. 2.30. The change in enthalpy per unit mass flow (H_2-H_1) during compression is the work of compression that must be supplied to the cycle. This assumes the velocity of the gas at 2 is about the same as at 1. The heat transfer Q during compression is zero because the process is assumed to be isentropic (i.e., $\Delta S=\Delta Q/T=0$).

Similarly, the heat transfer $Q_{(3-2)}$ from the high-pressure fluid in the water-cooled heat exchanger is (H_2-H_3) and in the recuperative heat exchanger $Q_{(3-4)}=(H_3-H_4)$. The work produced in the expander W_{4-5} is (H_4-H_5) and may or may not be applied to partially offset the work of compression. Finally the useful refrigeration of the cycle is $Q_{(6-5)}=(H_6-H_5)$ and the heat transfer in the recuperative exchanger is $Q_{(1-6)}=(H_1-H_6)$.

The net work input to the cycle, W_n, (assuming the expander work output is used to reduce the compression work) is

$$W_n=(H_2-H_1)-(H_4-H_5) \tag{2.74}$$

The heat extracted in refrigeration is

$$Q_c=(H_6-H_5) \tag{2.75}$$

Therefore the coefficient of performance is

$$\mathrm{COP} = \frac{Q_c}{W_i} = \frac{(H_6 - H_5)}{(H_2 - H_1) - (H_4 - H_5)} \tag{2.76}$$

COMPRESSORS AND EXPANDERS

Compression and expansion processes have so far been assumed to be isothermal or isentropic. In practice they are neither of these but occur in a manner that is characteristic of the machine used following what is termed generally a *polytropic* process.

Compressors

Figure 2.36 illustrates on the P–V and T–S planes, the different processes for *compression*.

i. Process (1–2) is an isothermal compression represented by $PV =$ const.
ii. Process (1–4) is isentropic represented by $pV^{\gamma} = $ const. γ is the ratio of specific heats of the gas, $\gamma = C_p/C_v$.
iii. Process (1–3) is polytropic between an isothermal and an isentropic process. The compression can be represented by $pV^n = $ const, where $(1 \leq n \leq \gamma)$.
iv. Process (1–5) is polytropic represented by the equation $pV^m =$ const, where $m \geq \gamma$.

The ideal reference cycle for reciprocating compressors is the cycle with isothermal compression. For rotary systems the isentropic compression

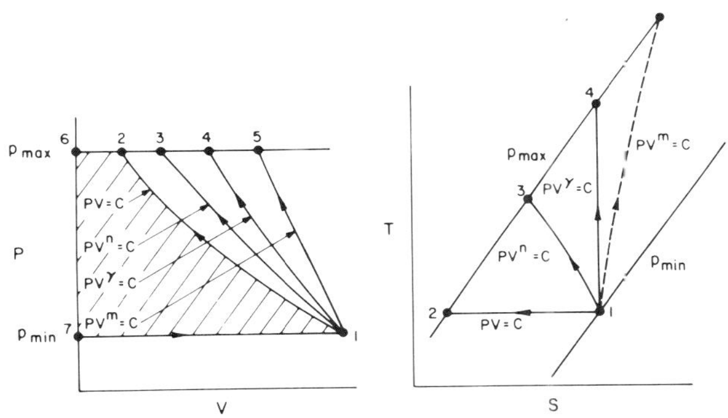

Fig. 2.36. Compression processes on the P–V and T–S planes.

cycle is the ideal reference cycle. In a rotary machine at high speed the gas particles are moving so fast that there is simply no time for cooling to take place. In the reciprocating compressor the gas passes through the machine at a slower pace and so some cooling is possible. To achieve isothermal compression would require either infinite rates of heat transfer or an impracticably low speed of operation. In practice reciprocating compressors follow the intermediate polytropic path, process (1–3) on Fig. 2.36. Rotary compressors follow the polytropic process (1–5).

The *isothermal efficiency* for a compressor (usually reciprocating) is defined as

$$\eta_{(\text{iso})} = \frac{\text{isothermal compression work}}{\text{actual compression work}} = \frac{\text{area 1–2–6–7}}{\text{area 1–3–6–7}} = \frac{H_2 - H_2}{H_3 - H_1} \tag{2.77}$$

The *isentropic efficiency* for a compressor (usually rotary but sometimes also used for reciprocating) is defined as

$$\eta_{(\text{isen})} = \frac{\text{adiabatic compression work}}{\text{actual compression work}} = \frac{\text{area 1–4–6–7}}{\text{area 1–5–6–7}} = \frac{H_4 - H_1}{H_5 - H_1} \tag{2.78}$$

Areas are measured on the P–V work diagram. Enthalpies, H, can be calculated or simply read from the appropriate point on the T–S diagram. The compression work measured by areas on the P–V plane or changes in the enthalpy, is called the *indicated* work. This is considerably less than the work actually required to drive the compressor. Additional work is necessary to overcome mechanical and aerodynamic friction losses, pneumatic losses in charging [process (7–1) in Fig. 2.36] and discharging [process (6–2)] and for driving accessories.

The ratio of indicated work to actual shaft input is called the mechanical efficiency:

$$\eta_{(\text{mech})} = \text{indicated work/brake work} \tag{2.79}$$

This usually ranges from 60% to 80% for reciprocating engines and up to 90% for rotary systems.

Reciprocating compressors have crankshafts, with main bearings, connecting rods with big and small end bearings, pistons with piston guide rings and sealing rings. They may also be equipped with water cooling requiring some pump work, with forced lubrication requiring oil pump work and a fan for air cooling. The mechanical and hydraulic losses tend to be appreciable compared with the pneumatic effects arising from the relatively slow moving gas being compressed.

Rotary systems have only main shaft bearings and rotary shaft seals which consume little power. The principal losses are pneumatic arising from friction and windage effects of the high-speed rotation of the disks, drums, and blades within the compression space.

Frequently the isentropic efficiency is quoted for reciprocating compressors rather than isothermal efficiency. This was adopted by compressor manufacturers to put their machines in a better light since the ratio of the isentropic work to the actual shaft input is larger than the ratio of isothermal work to actual shaft input. Both are arbitrary ratios and the actual work required remains the same regardless of the efficiency used. When discussing compressor efficiencies it is important to establish if the efficiency being quoted is isentropic, isothermal or mechanical.

Yet another efficiency is much used, the *volumetric efficiency*, defined as

$$\eta_{(\text{vol})} = \frac{\text{volume of gas actually pumped}}{\text{displacement volume of compressor}} \tag{2.80}$$

The volume of gas passing through the compressor is always less than the piston displacement volume. The reduction is due to a number of causes illustrated by the work diagram shown in Fig. 2.37. This shows the ideal work diagram 1–2–3–4 superimposed on an actual compressor indicator diagram. The effects have been exaggerated for the sake of clarity. Due to pressure losses in the intake duct (perhaps including a filter) the pressure at the start of compression (5) is less than the theoretical value (1).

The mass of gas in the cylinder can be calculated as

$$m_c = \frac{p_5 V_5}{RT_5} \tag{2.81}$$

where m_c is the mass of gas in the cylinder, p_5, V_5, and T_5 are the pressure, volume, and temperature at point (5), and R is the characteristic gas

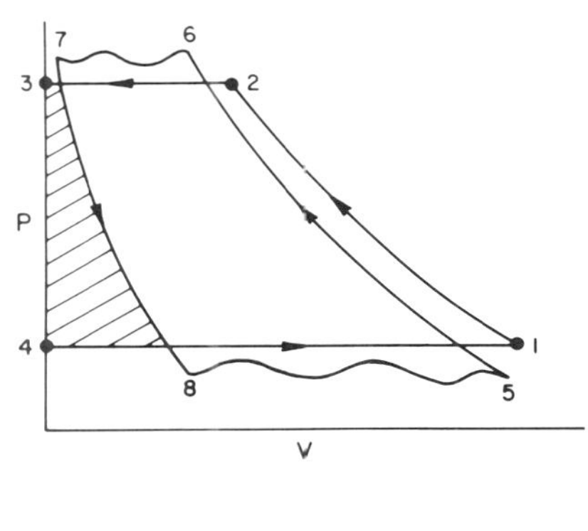

Fig. 2.37. Comparison of actual and ideal compressor work diagrams.

constant. Clearly a reduction in the pressure will mean a reduction in the mass of gas contained.

The actual compression process follows the path (5–6) rather than the ideal path (1–2). It is necessary to compress the gas to a higher final pressure (6) than theoretically required (2) so as to overcome pressure losses in the delivery valve and ductwork including the aftercooler. This has no effect on volumetric efficiency but requires extra work and reduces isothermal or isentropic efficiency.

The principal effect on volumetric efficiency arises from residual gas. It is not possible for all the compressed fluid to be discharged from the compressor cylinder. A small amount of residual fluid (2% to 5%) will remain in valve cavities and clearance spaces when the piston reaches top dead center position. Now as the piston descends the compressed residual gas expands from (7) to (8). No fresh charge will be inhaled to the cylinder until the pressure falls to (8) and the inlet opens to admit new gas. A sizable fraction of the piston stroke is consumed in this re-expansion of the compressed gas and is the biggest factor reducing the volumetric efficiency. Expansion of the clearance gas does work on the piston. Work input to the compressor is therefore proportionately decreased as the mass flow entering the compressor is decreased.

A reduction in inlet pressure or increase in inlet temperature will reduce the mass flow to the compressor. When comparing one compressor with another on the basis of volumetric efficiency or the "free air delivery rating," it is important to specify the same reference pressure and temperature. For comparative purposes it is best to reduce the available data to terms of mass rate of flow at the standard temperature and pressure of 0°C and 101.3 kPa.

Multistage Compression

When gases are compressed over a large pressure range it is advantageous to perform the compression in two or more stages with intercooling between stages and aftercooling the final stage. In Fig. 2.38 the *P–V* and *T–S* diagrams for two-stage compression are superimposed on those for a single stage.

The single-stage compression diagram is represented by process 1–11–6–10. The two-stage compression process is represented by the low-pressure stage, process 1–2–8–9, and the high-pressure stage, process 3–4–6–7. At the intermediate pressure the fluid is assumed to be cooled to the original temperature. Process 1–3–5 is the isotherm for the temperature at state 1.

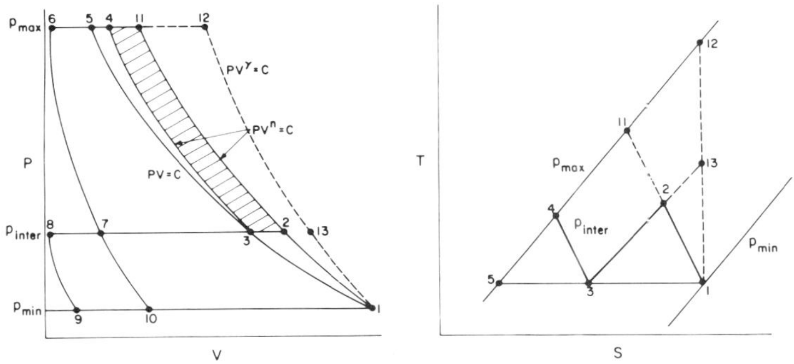

Fig. 2.38. Multistage compression.

The advantage of multistage compression is twofold. It increases the volumetric efficiency and also the isothermal (or isentropic) efficiency. Increase in the volumetric efficiency comes about because in a two-stage machine the clearance gas expands only to (9) rather than the larger volume, (10), with single-stage compression. The low-pressure stage can admit a greater volume of new gas. The isothermal or isentropic efficiency increases because of the reduced compression work required with intercooling back to the initial temperature. The saving in work is indicated by the shaded area, 2–11–4–3 in Fig. 2.38.

The disadvantage of multistage compression is that a more complicated and expensive machine is required with more moving parts, bearings, pistons, seals, valves, and intermediate pressure heat exchangers and intercoolers. The benefits of additional stages of compression occur at a progressively diminishing rate. It is rarely worthwhile exceeding four stages even for the highest pressure ratios.

To a first approximation the optimum division of pressure staging for equal compression work in each stage is:

$$p_r = \sqrt[n]{p_{\max}/p_{\min}} \tag{2.82}$$

where p_r is the pressure ratio per stage, n is the number of stages, $p_{\max}$ is the maximum pressure (at end of compression), and $p_{\min}$ is the minimum pressure (at start of compression).

Expanders

Processes for expansion are shown in Fig. 2.39. From an initial high pressure the fluid may expand:

i. Isothermally, process (1–2) represented by $pV = \text{const.}$

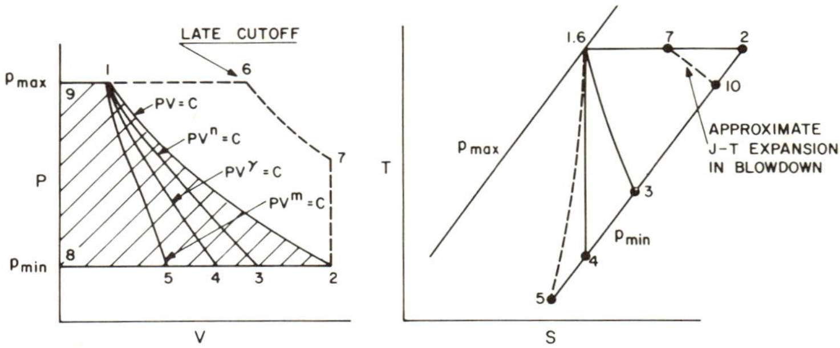

Fig. 2.39. Expansion processes on the *P–V* and *T–S* diagrams.

ii. Isentropically, process (1–4) represented by $pV^{\gamma} = \text{const}$, where γ is the ratio of specific heat at constant pressure and constant temperature ($\gamma = C_p/C_v$).
iii. Polytropically, process (1–3) represented by $pV^n = \text{const}$ where $1 \leq n \leq \gamma$.
iv. Polytropically, process (1–5) represented by the equation $pV^m =$ const where $m \geq \gamma$.

The process of expansion is never isothermal or isentropic but in both reciprocating or rotary machines lies somewhere between the two [process (1–3)]. It cannot be isothermal for this would require infinite rates of heat transfer between the gas and its surrounds. Similarly it cannot be isentropic for this would require perfect insulation. In a turbine aerodynamic friction causes a higher exit temperature, T_3, than with isentropic expansion. This frictional degradation causes an increase in the entropy of the fluid during expansion and is in accord with the second law of thermodynamics.

The expansion process (1–5) cannot be attained naturally for it requires the *withdrawal* of heat during expansion with consequent reduction of entropy. It would occur in an expander equipped to refrigerate the expansion process and is of little interest. Our main purpose is to use the expansion process to *generate* refrigeration rather than *consume* it.

In reciprocating expanders point 1 on the *P–V* diagram is called the *cutoff point*, where admission of working fluid is terminated and expansion begins. After expansion the exhaust valve opens and fluid is expelled from the cylinder.

The complete cycle for an expansion engine is the *P–V* diagram 1–2–8–9. The area of the *P–V* diagram is the work output of the engine. The refrigeration generated is

i. $dQ = dU + dW$ for a nonflow system (reciprocating expander);
ii. $dQ = dH$ for a flow system (rotary expander).

To increase the refrigeration the cutoff may be delayed to 6. The fluid expands to the end of the piston stroke at 7, the exhaust valve opens and the fluid "blows down" to low pressure 2. Delay in the cutoff point increases the size of the *P–V* work diagram and hence the refrigeration effect but at the expanse of increased use of compressed gas. The process of free expansion (7–2) in the exhaust blowdown occurs at a Joule–Thomson expansion (constant enthalpy) so the fluid is used less effectively than by controlled expansion in the engine. The refrigeration produced by the engine is *increased* but the refrigeration per unit mass flow is *reduced*.

An expander with variable cutoff can be used where a large refrigeration capacity is required for starting and cool-down purposes. Later the cutoff point can be adjusted to provide maintenance levels of refrigeration.

ISOTHERMAL VS. ISENTROPIC EXPANSION

An interesting point was discussed by Radebaugh (1977) in connection with the Siemens cycle and the use of isentropic or isothermal expansion. Consider the Siemens cycle shown in Fig. 2.40. It consists of isothermal compression (1–2), recuperative cooling at constant pressure (2–3), isentropic expansion (3–4), refrigeration (4–5), and recuperative heating at constant pressure (5–1). The heat extracted in the refrigeration process is the area 4–5–7–6 on the *T–S* diagram. Now consider an isothermal expansion (3–5) instead of the adiabatic expansion (3–4). With isothermal expansion the refrigeration effect is greater by the area 4–3–5. However, with adiabatic expansion the minimum temperature is much lower than

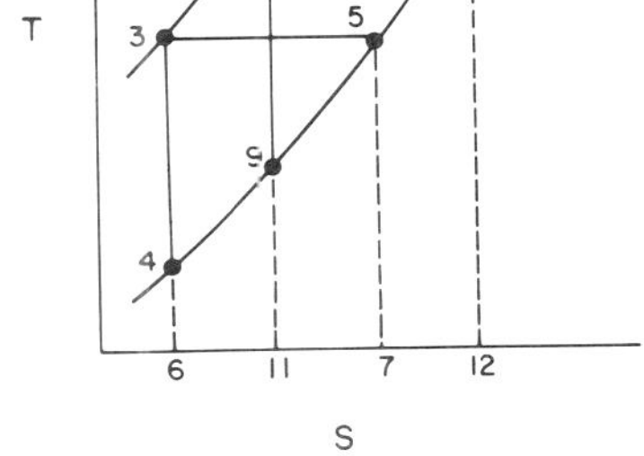

Fig. 2.40. Isothermal and isentropic expansion processes.

with isothermal expansion. In situations where the heat leak from surroundings dominates the internal heat generation (usually the case with small electronic or superconducting devices) it is advantageous to absorb heat over a range of temperatures. In these situations a better comparison might be to increase the temperature at the start of the adiabatic expansion to generate the same quantity of refrigeration (at variable temperature) with adiabatic expansion as with isothermal expansion. This modified adiabatic expansion is represented by the process (8–9). Refrigeration is available at variable temperatures in the process (9–10). The amount of refrigeration is area 9–10–12–11 and has the same area as the isothermal process 3–5–7–6. With refrigeration at variable temperatures the duty and hence the size of the recuperative heat exchanger is decreased since the temperature is only reduced to (8) rather than (3).

REFERENCES

Angrist, S. W. (1968). "Perpetual Motion Machines." *Sci. Am.* **218**, 114–22.

Köhler, J. W. L. (1965). "The Stirling Refrigeration Cycle." *Sci. Am.* **212**(4), 119–127.

Radebaugh, R. (1977). "Refrigeration Fundamentals: A View toward New Refrigeration Systems. Applications of Closed Cycle Cryocoolers to Small Superconducting Applications." NBS Special Pub. 508, Supt. of Doc. U.S. Govt. Printing Office, Washington, DC.

Rallis, C. J., Uurieli, I., and Berchowitz, D. M. (1977). A New Mathematical Model for Stirling Cycle Machines. Proc. 12th I.E.C.E.C., Paper No. 779254, Washington, D.C., August 28–September 2.

Reitlinger, J. (1876). "Über Kreisprozesse zwischen zwei Isothermen." *Z. Ost. Ing. Arch. Ver.*

Rogers, G. F. C., and Mayhew, Y. R. (1967). *Engineering Thermodynamics Work and Heat Transfer.* 2nd Edition. S.I. Units, Longmans Group Ltd., London.

Chapter 3

Stirling Cryocoolers

INTRODUCTION: HISTORICAL REVIEW

A Stirling engine operates on a closed thermodynamic regenerative cycle with the same working fluid repeatedly compressed and expanded at different temperature levels so there is a net conversion of heat to work or vice versa. It can be used as a cooling engine, prime mover, heat pump, or pressure generator.

As a prime mover, the Stirling engine has a long history dating back to 1817 and the invention of the first regenerative engine by the Reverend Robert Stirling. It is currently receiving an unprecedented level of research and development interest for applications ranging from miniature engines for artificial hearts to automotive engines and large stationary power generators using combustible wastes. Walker (1980) has recently reviewed this work in detail.

As a cooling machine, it extends back nearly as far. In 1834 John Herschel conceived the closed cycle regenerative cooling engine for making ice. The concept was not reduced to practice for 30 years. Alexander Kirk (1874) described a Stirling cooling engine in use for 10 years. Robert Stirling and his engineer brother James, did no work on Stirling engines as cooling machines. Nor did the hundreds of others busy developing the many variations of hot-air engines made in large quantities in the 19th Century in Britain, Europe, and North America.

The first Stirling cooling engines were made by Alexander Kirk, a Scottish engineer working at the oil works at Bathgate, Scotland. He constructed cooling engines for a variety of applications both in Great Britain and overseas. In addition, he established the cascade cycle now widely used for cryogenic and low-temperature gas liquefaction and processing. Collins (1958) credits him with the accolade of "scientific thinker as well as talented machine designer." Despite this promising base, the

Kirk machines were never made in larger numbers or used to achieve cryogenic temperatures.

The next significant development occurred with the start of the Philips cooling engine research about 1946. It followed a decade of effort already invested in small Stirling engine prime movers. The Philips program was the first time that substantial effort was invested simultaneously in Stirling engines for power and cooling applications. Although there were separate development teams, they were located in the same research building and cooperated well together with members common to both groups.

Köhler (1965), the leader of the cooler development group, has provided an intensely interesting account of the early program at Philips. He recalls that, in 1945, H. Rinia and F. K. du Pre operated a one horsepower engine as a refrigerator and obtained temperatures in the liquid air range of 80 K. Thereafter, research on the Stirling cooling engines was assigned to a separate group with Köhler in charge. Introducing him to the group Rinia observed:

> . . . our hot-air engines cooled marvelously, whereas when we built machines expressly designed as refrigerators they were unsatisfactory.

Köhler writes:

> I designed my first models along the lines of the hot-air engines and as a result they never failed as refrigerators.

It was in the early 1950s that Stirling engines were recognized as a specific generic class of machine. Up to then they had been generally classified as "hot-air engines" with the name of a particular researcher attached. This was no longer appropriate when hydrogen or helium supplanted air as the working fluid in high performance prime movers then under development at Philips in Eindhoven.

Since that time "Stirling engine" has been widely accepted as the generic title for closed cycle regenerative thermal systems whether used as prime movers or cooling engines. This is preferred to the practice of some who refer to the Kirk cycle when describing Stirling engines working as refrigerators. It is worth separating the two principal classes of regenerative machines into

i. Stirling engines (without valves),
ii. Ericsson engines (with valves),

depending on whether the internal fluid flows are controlled by volume changes or valves.

Development of Stirling cryocoolers was brilliantly successful. Commercial versions were in production by 1953; a remarkable achievement compared with the parallel development of Stirling engine prime movers where development continued over the past 40 years with no commercial product so far. There are now clear signs this sustained effort will be rewarded by successful commercial products (domestic heat pumps) in the near future.

The manufacture of Stirling cryocoolers has been profitable for the Philips Company. The first engine shown in Figs. 3.1 and 3.2 was an air liquefier capable of producing 5 or 6 liters an hour within 15 minutes of being switched on. It required no special expertise to install and maintain, was small and sufficiently low priced to be within the means of any university physics, chemistry, or engineering laboratory. In the quarter century since its introduction, several thousand models of this engine and derivatives have been made.

It is available in single-cylinder or four-cylinder versions (see Fig. 3.48) and may be equipped with a fractionating column to separate oxygen from liquid air and produce liquid nitrogen. For many applications, this is attractive because of the reduced safety hazard. A version of the engine

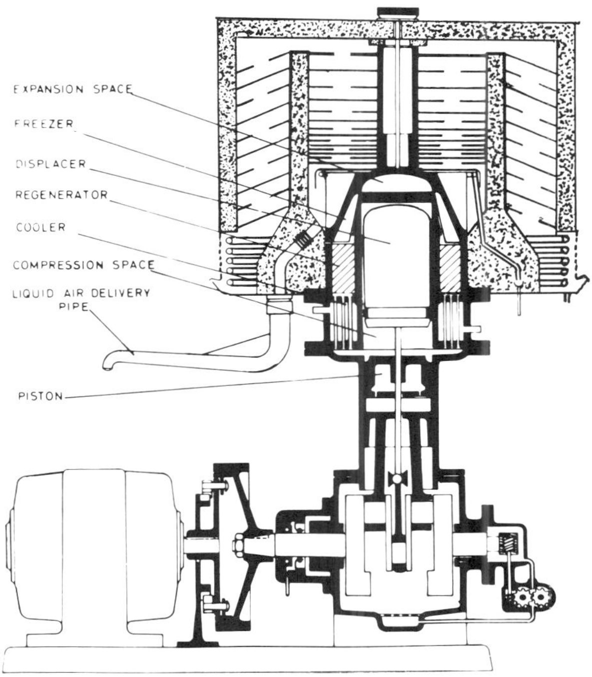

Fig. 3.1. Early Philips Stirling air liquefier.

Fig. 3.2. Early Philips Stirling air liquefier.

with double expansion was also developed which can achieve temperatures of 20 K and has been used for research and hydrogen, neon, and helium liquefaction.

Derivatives of the early engine have been incorporated in many cryogenic systems involving cooling or gas liquefaction and reliquefaction. These include standard cabinets with cold air recirculation for metallurgical processing, freeze drying, food processing, and a whole range of research applications requiring refrigeration at precisely controlled temperatures in the range from ambient to 20 K. It is used for reliquefaction of "flash" gas in liquid oxygen or other cryogenic reservoirs.

In addition to the basic Stirling cryocooler, much other work has been carried out in miniature engines and in the intermediate size coolers shown in Figs. 3.3 and 3.4. No significant application for the intermediate size

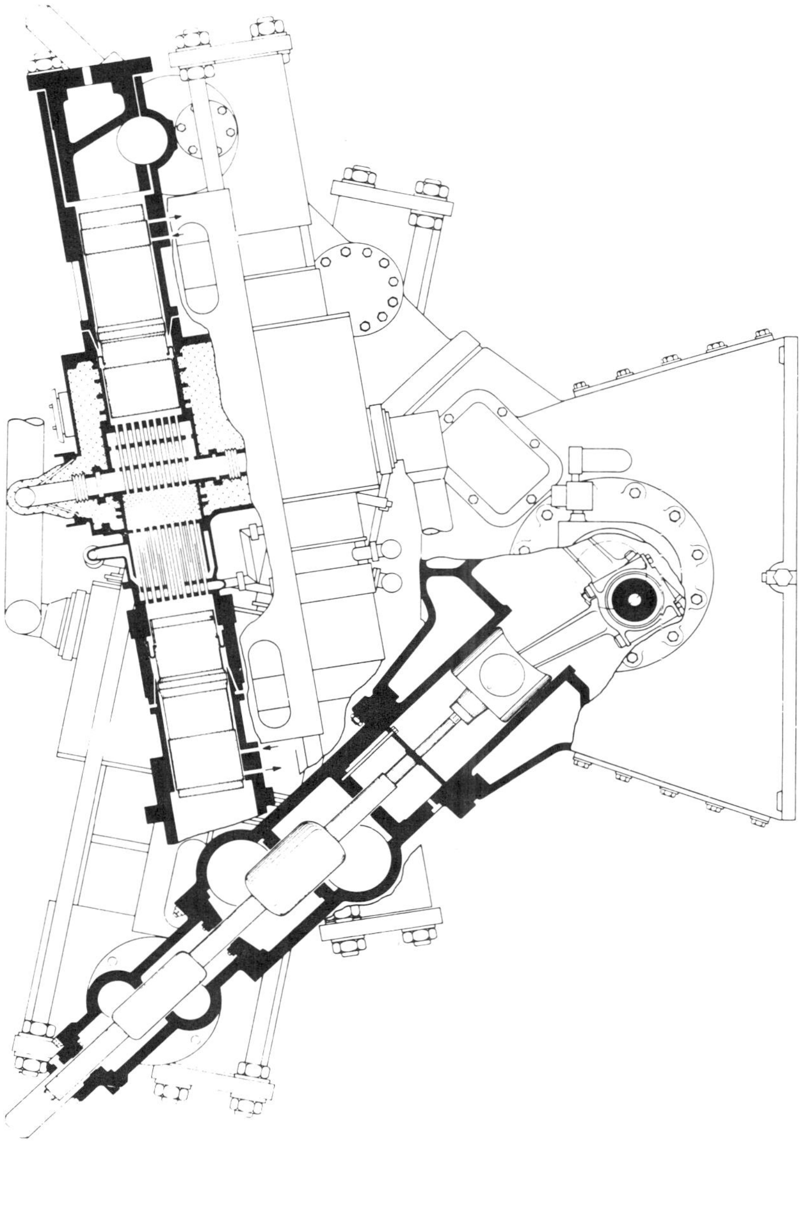

Fig. 3.3. Intermediate capacity Stirling cryocooler.

cooler has yet developed but will come with the growth of LNG transportation and storage systems and the eventual introduction of superconducting systems.

Military interest in infrared thermal imaging equipment for night vision and heat-seeking missile guidance focused research attention on miniature cryocoolers in the 1960s. This is now sustained by continued military interest, by spacecraft instrument applications, and by incipient applications of superconducting electronic and electrical systems on a broad front.

The principal military interest in miniature cryocoolers was manifest in the United States. Policies of the U.S. Government led Philips to concentrate their development and production of miniature, militarily supported cryocoolers at the associate company North American Philips. The research facilities of the company are located in Briarcliff Manor, New York and the production facilities were at Ashton, Rhode Island. In the early 1970s, the production facilities were sold to a competitor, Cryogenic Technology Incorporated (CTI), of Waltham, Massachusetts. CTI also distributes in the United States the larger Philips cryogenic equipment made in Eindhoven.

The research laboratory continues at Briarcliff Manor principally engaged in research and prototype development of Stirling cryocoolers and

Fig. 3.4. Intermediate capacity Stirling cryocooler.

prime movers supported by various agencies of the United States government. There is a close relationship between the Philips Company of Eindhoven and Philips Laboratories, although the companies are independent. Philips Laboratories also has an uneasy cooperative, but yet competitive, association with CTI and, perhaps more significantly for the future, much common interest with the Magnavox Corp., the United States affiliate of Eindhoven-based Philips Company.

The Philips Laboratory has assembled a remarkably competent resourceful and innovative scientific engineering team under the leadership of A. Daniels and F. K. du Pre. In the Bibliography many reports and papers originating from this team will be found under the authorship of Daniels, du Pre, Pitcher, Lehrfeld, Balas, Lindale, and Storace.

An excellent comprehensive summary of recent work on miniature Stirling engines for cryogenic cooling can be found in the unclassified report

Fig. 3.5. Miniature Stirling cryocooler with rhombic drive (after Lindale, 1978).

by Lindale (1978). The development is described of six long-life low-power cooling engines for gamma ray detectors in a satellite experiment. A photograph of one of the miniature cryogenic cooling engines is shown in Fig. 3.5 and in cross section in Fig. 3.6.

Research on Stirling engines continues at Eindhoven and has resulted in the development and recent commercial introduction of an ultracompact 80 K cooler operating as a free piston Stirling engine with a linear motor drive. The military version of this unit is shown in Fig. 3.7 (Haarhuis, 1978; de Jonge, 1979).

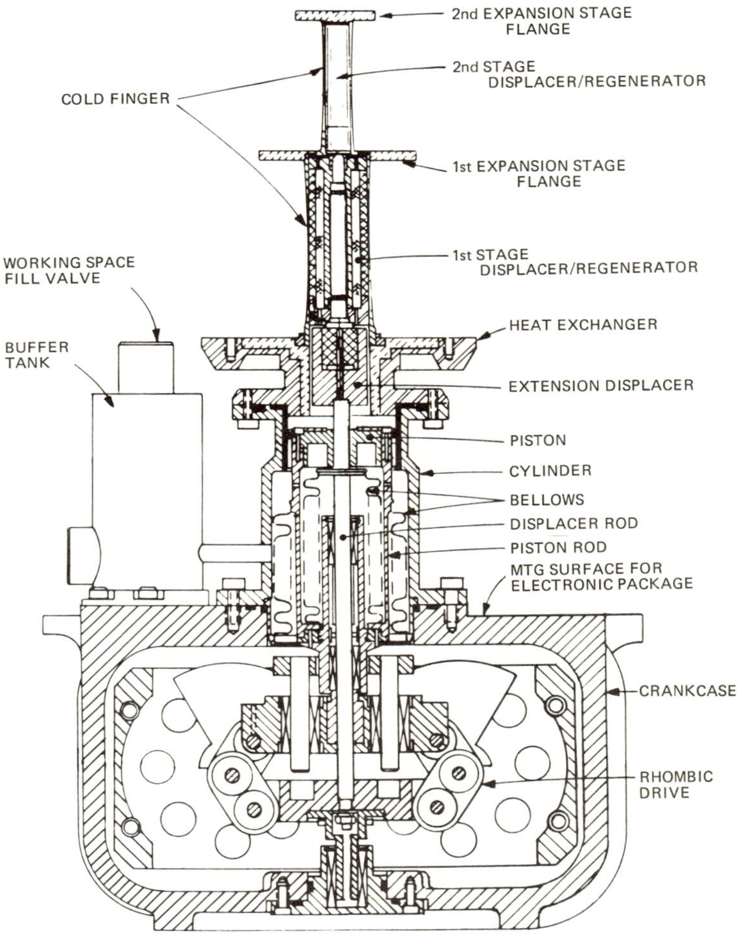

Fig. 3.6. Cross section of miniature Stirling cryocooler with rhombic drive (after Lindale, 1978).

Fig. 3.7. Ultracompact free-piston Stirling cryocooler with linear motor drive (Philips Research Labs).

The availability of research funds induced other companies to work on Stirling cryocoolers. In the mid 1960s the Malaker Corporation, High Bridge, New Jersey was active in miniature cryocooler development. With the unfortunate failure of the company, the Malaker cryocoolers have disappeared from the scene.

About the same time, Hughes Aircraft Co., Culver City, California, developed small Stirling cryocoolers for heat-seaking infrared missile guidance. Thousands of these engines have been made. The Martin-Marietta Corp., Orlando, Florida, Texas Instrument Co., Dallas, Texas, and CTI, Waltham, Massachusetts are other U.S. defense contractors with significant research and development experience in miniature Stirling cryocoolers. Much of their work has been sponsored by the U.S. Army Night Vision Laboratory, Fort Belvoir for forward-looking infrared (FLIR) coolers for night vision equipment. Figure 3.2 is an integral Stirling FLIR cooler in current production. Other work has been done on split-Stirling systems.

Significant research programs directed to the engineering development and eventual production of Stirling cooling engines have been variously reported from both Japanese and Russian sources. Details of these developments are given in Chapters 13 and 14.

CLASSIFICATION

The essential elements of a Stirling engine are two spaces, at different temperatures, whose volumes can be varied cyclically and which are connected by ducts containing heat exchangers. These simple elements can be combined in a surprisingly wide range of mechanical arrangements. Many of these have been identified (by Finkelstein, 1959) with the name of the originator. Many variations used in the 19th Century have been adopted or reinvented for application in contemporary developments and new arrangements are being devised.

Figure 3.8 is a family tree of fundamental characteristics on which the following classification scheme is based. The key identifying feature of any *regenerative* system is whether or not valves are used to control the flow of the working fluid. If valves are used, the machine is an Ericsson engine. If valves are not used the machine is a Stirling engine.

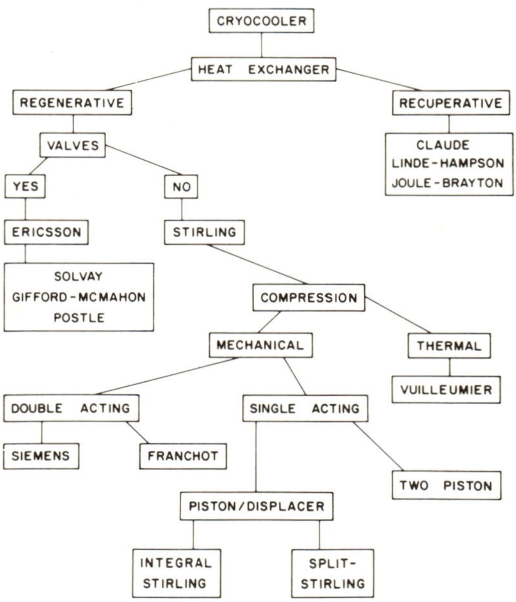

Fig. 3.8. Classification of cryocoolers.

The use of valves for flow regulation has the advantage of increased flexibility in flow control and timing with the possibility of virtually unrestricted pressure ratios (p_{max}/p_{min}). Valves add to the mechanical complexity of the system and provide sources of noise and additional points of wear so the prospect for long life with low maintenance is reduced.

Stirling cryocoolers, machines operating on a regenerative cycle without valves, can be classified into two principal groups, namely, machines with mechanically induced compression and machines with thermally induced compression. In this review, machines with thermal compression are treated separately as Vuilleumier Engines in Chapter 4.

Machines with mechanical compressors may be further subdivided into two families identified as single-acting and double-acting engines.

Single-Acting Stirling Engines

A single-acting Stirling engine is an ensemble of expansion space, compression space, and associated heat exchangers in one or two cylinders with *two* reciprocating elements, one of which must be a piston. The other may be a piston or a displacer.* Every ensemble constitutes a complete system operating independently of other single acting systems coupled on a common crackshaft or other kinematic mechanisms.

All existing arrangements of single-acting Stirling engines may be subdivided into two groups as shown in Figure 3.9; two-piston machines and piston/displacer machines.

A further distinction can be made in the piston/displacer group between (a) machines where the piston and displacer operate in a single cylinder called an integral-Stirling machine, and (b) machines where separate cylinders are provided for the piston (compressors) and the displacer (expander), an arrangement called the split-Stirling machine.

In all these machines the reciprocating elements, the pistons or displacers, may or may not be connected to a kinematic drive mechanism (cranks and connecting rods, Scotch yoke or rhombic drive mechanism). Machines with a kinematic drive mechanism are called *disciplined piston* (and displacer) engines. Machines with no kinematic drive mechanism are called *free piston* engines. Another variation has one of the reciprocating elements (usually the piston) coupled to a kinematic mechanism and the

* There are important differences between a piston and a displacer. A piston sustains a high pressure difference but no appreciable temperature difference between the upper and lower faces. A piston must, therefore, have a good fluid seal to prevent leaks and act as a structural element with high-pressure forces acting on it. A displacer is an element with a low pressure difference and a high temperaure difference between the upper and lower faces. It can, therefore, be a lightweight structural element with a minimal seal requirement but designed with long thermal conduction paths to minimize heat leaks.

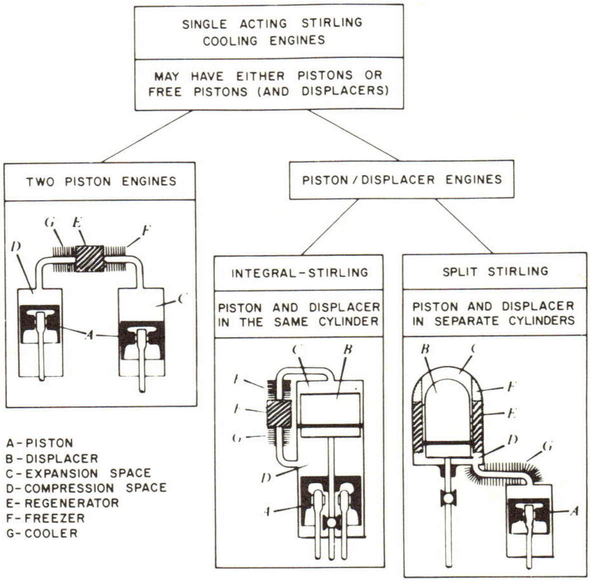

Fig. 3.9. Classification of single-acting Stirling cryocoolers.

other operating as free piston (or displacer). This has been called the hybrid crank controlled (or disciplined) piston/free displacer engine. It is widely used for split-Stirling coolers and can also be used for integral-Stirling machines (see Walker, 1973).

A distinction also has to be made between machines with a separate stationary regenerator and those where the regenerator is incorporated in the displacer and moves with it.

Design variants for integral-Stirling engines are shown in Fig. 3.10 and for split-Stirling engines in Fig. 3.11. Design variants for two-piston machines are shown in Fig. 3.12. The originator of a particular configuration is given on these figures where it is known. In most cases, the system was conceived for use as a prime mover and many of these cases have never been made as cooling engines. This does not preclude future development of new and different arrangements.

The spectrum of possibilities for two-piston arrangements is very wide. It includes the conventional piston/cylinder systems shown in Fig. 3.12 and many others employing metal bellows and diaphragms, rotary assemblies and free-piston devices where motion is induced by fluidic forces alone, or in combination with mechanical springs. The only two-piston machine available commercially (from Philips) is the intermediate capacity cooler shown in Fig. 3.4. This interesting machine utilizes a Joy crompressor

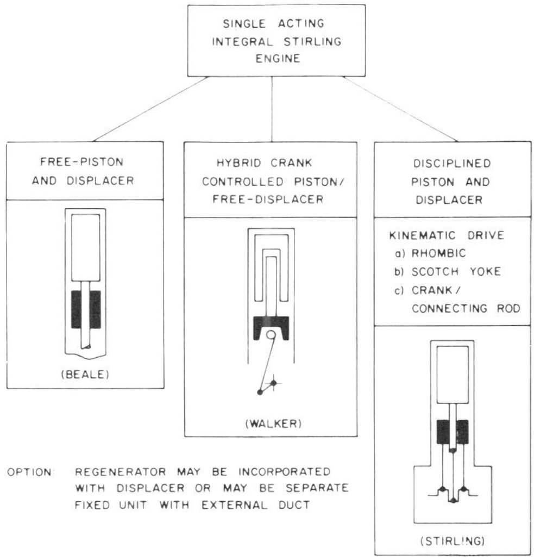

Fig. 3.10. Design variants of integral Stirling cryocoolers.

crankcase and pistons to generate the hydraulic drive for the two opposed pistons of the upper, central Stirling engine cooler. It is a difficult machine to classify. The Stirling system pistons are activated by fluidic forces only, pneumatic on one side, hydraulic on the other, to make it a free piston

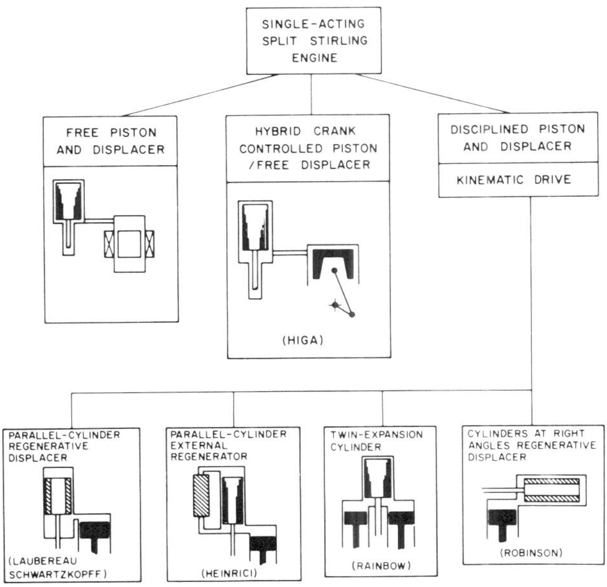

Fig. 3.11. Design variants of split-Stirling cryocoolers.

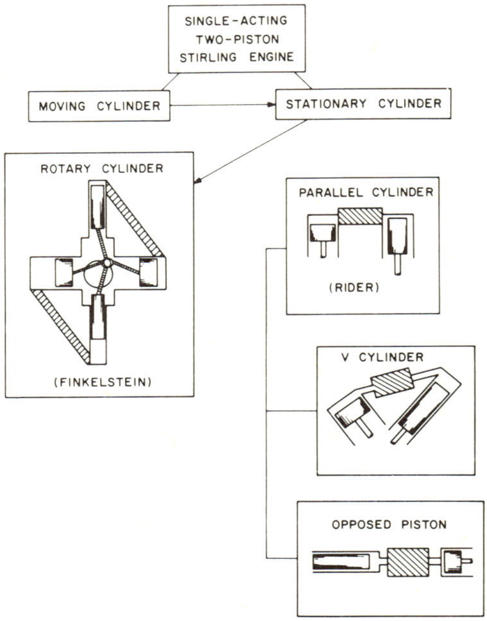

Fig. 3.12. Design variants of two-piston Stirling cryocoolers.

engine. Yet, it has a kinematic vee cylinder crank connecting rod mechanism through which the input work is supplied to the engine to generate the hydraulic forces inducing the Stirling piston motion.

Regenerative Stirling engines using bellows are not uncommon in the power systems field. The artificial heart engines of the McDonnel Douglas/Richland Energy Laboratory and the Thermo-Electron Corporation incorporate a variety of bellows and diaphragms. The radio-isotope/propane-fueled power generator developed by Cooke-Yarborough at the U.K. Harwell Atomic Energy Establishment uses a diaphragm (Walker, 1980).

Double-Acting Stirling Engines

Double-acting Stirling engines, shown in Fig. 3.13, are ensembles of multiple cylinders arranged so the expansion space of one cylinder is connected through associated heat exchangers to the compression space in an adjacent cylinder. There is, therefore, the same number of Stirling engine systems as there are cylinders, and only one reciprocating element per cylinder. It may be called a piston-displacer, for it has both a large

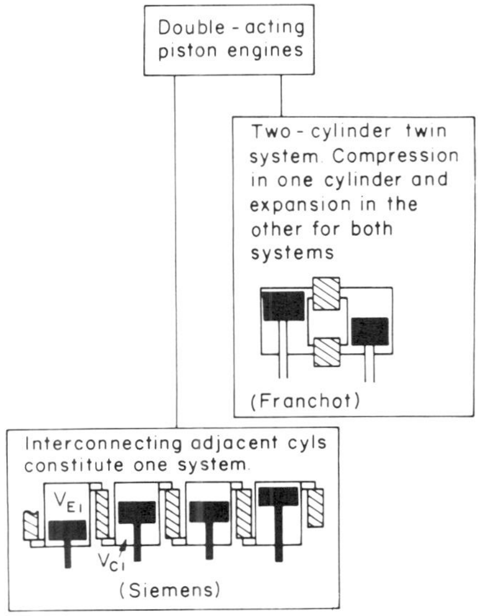

Fig. 3.13. Classification of double-acting Stirling engines.

pressure and a large temperature difference between the top and bottom faces.

The great advantage of double-acting engines is that the number of reciprocating elements is *half* the number required in multiple arrangements of single-acting engines. This can lead to major simplifications of the kinematic drive arrangements and to reduced costs.

The principal disadvantage is the limited flexibility in design, and to a lesser extent, operation. Moreover, for prototype development, it is necessary to proceed with the entire multiple cylinder engine rather than an experimental single-cylinder unit which can then be reproduced in multiple cylinder versions. Nevertheless, the elimination of half the moving parts is so significant an economy, that the major interest for prime mover Stirling engines greater than 15 kW is concentrated on double-acting arrangements.

Double-acting engines were invented in the 19th Century. Babcock (1885) attributes the invention of the two-cylinder, twin system to the Frenchman, Charles Louis Franchot in 1853. It was reinvented a century later by Finkelstein (1959) for the prototype unit of a domestic refrigerator and was later incorporated by him in the concept of a multiple-cylinder free-piston unit for power generation.

The general form of double-acting engines with three or more cylinders is ascribed by Babcock to Sir William Siemens who, in 1863, conceived

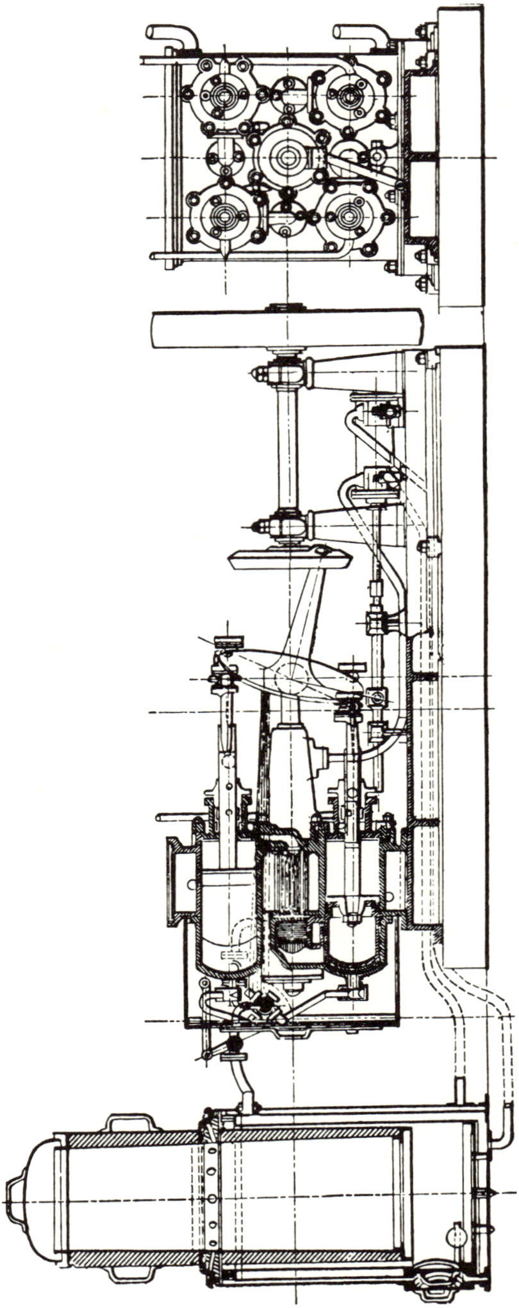

Fig. 3.14. Four-cylinder double-acting Stirling engine invented by Sir William Siemens in 1863 (after Babcock, 1885).

the four-cylinder engine shown in Fig. 3.14. The engine had four cylinders each containing a single piston and which were coupled in a square arrangement through the associated heat exchangers and regenerators. It was activated by a wobble-plate drive mechanism.

So far as is known, the Siemens engine was not taken beyond the conceptual stage until it was reinvented, complete with wobble-plate, 80 years later by van Weenan in course of the Philips program. Following some early success at Philips, the Siemens engine was abandoned in favor of the single-acting "rhombic drive" engine. Later, in 1965, one of the Philips licensees, General Motors, in the United States, resumed work on double-acting engines and by 1970, Philips and the other licensees had concentrated their efforts on double-acting engines for automotive engine applications. The Philips unit, shown in Fig. 3.15, is a four-cylinder version with a swash-plate drive* and exists in prototype 150- and 75-kW forms. Until recently (1978), the engine was under development for automotive application by Philips and the U.S. Ford Motor Co. in a 7-year $180 million program funded by the U.S. Department of Energy. Ford abandoned its development effort on Stirling engines in 1978.

An alternative kinematic mechanism has been adopted by the Philips licensee United Stirling of Sweden using twin cranks and connecting rods as shown in Fig. 3.16. Another version with the cylinders arranged in a shallow vee form was earlier favored. This used a single crankshaft but the inclination of the cylinders increases both production and maintenance costs. The parallel cylinder version is the unit intended for preproduction evaluation in 40-, 75-, and 150-kW versions for automotive use, in mining applications, for boats and stationary power generation (see Walker, 1980). The United Stirling engine is the power unit for development of a Stirling automotive unit in a program sponsored by the U.S. Department of Energy. Mechanical Technology Inc. of Latham, New York and American Motors are participating in the program.

Double-acting engines have not been used for cryogenic refrigerating purposes. This is due entirely to the absence of need for cryogenic coolers of intermediate or large capacity. Significant change is now at hand with the development of relatively large-scale superconducting systems and liquid natural gas transportation and reliquefaction systems. These developments will require new cryocooler concepts. Double-acting Stirling cryocoolers of intermediate and large-scale capacity will likely be prominent. Already one major U.S. corporation is known to be actively developing a

* A distinction between a wobble-plate and a swash-plate should be made but is sometimes difficult to appreciate. Maki *et al* (1971) define a swash plate as an inclined plate rigidly attached to a rotating shaft. A wobble plate is not rigidly attached to a rotating shaft but is mounted so that the plate nutates as the shaft rotates.

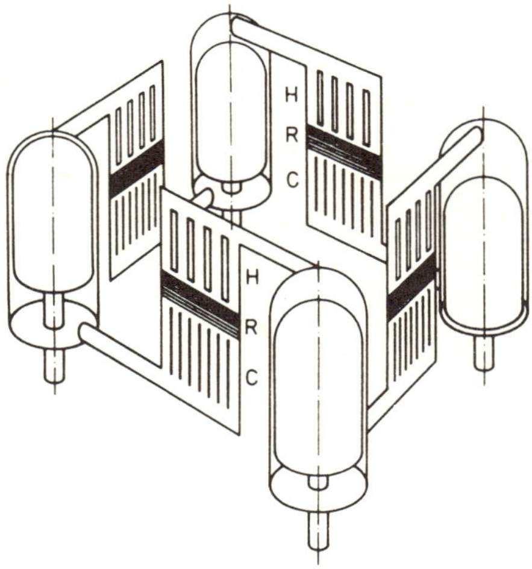

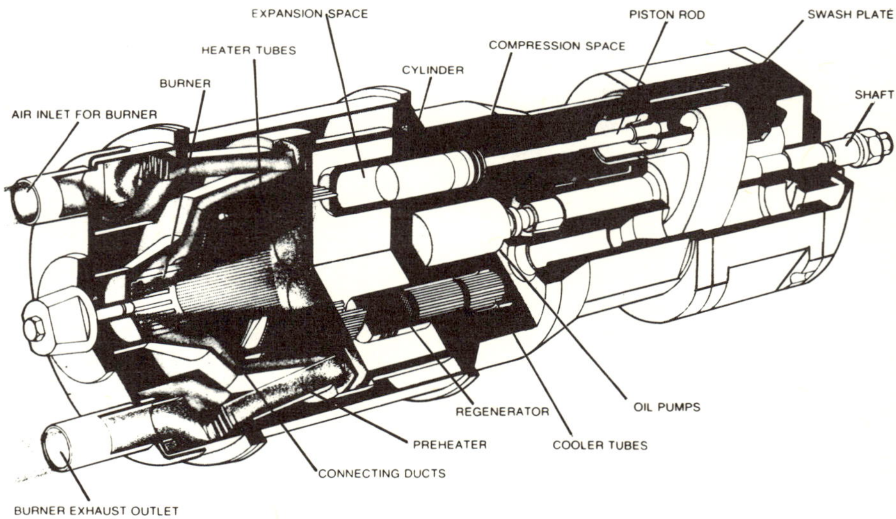

Fig. 3.15. Philips double-acting Stirling engine with swash-plate drive (after Walker, 1979).

double-acting free piston Stirling cooling engine along lines enunciated by Finkelstein (1978).

The first known application of double-acting cooling engines for refrigeration was a Franchot twin-cylinder engine reinvented in England by Finkelstein (1959). A practical form of engine was made and investigated as the potential replacement for domestic refrigeration units. Minor funds were provided and no significant attempts were implemented to optimize

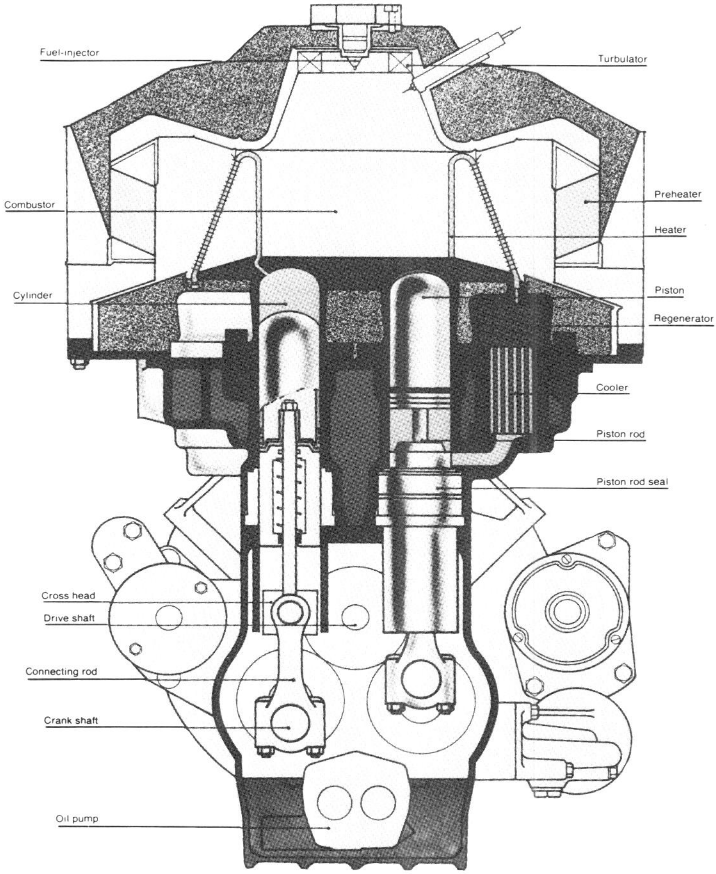

Fig. 3.16. United Stirling double-acting Stirling engine with crank-connecting rod drive.

the design or incorporate the most suitable materials. Nevertheless, the unit functioned well with a coefficient of performance that was somewhat less than the conventional vapor compression unit in the normal domestic refrigeration temperature range. At lower temperatures, the performance of the Finkelstein unit was superior to the vapor compression machine. The performance characteristics crossed in favor of the Stirling unit at about −20°C.

So far as is known, the earliest conceptual application of the Siemens engine with swash-plate drive as a cooling engine was in an outline design by Walker (1965), for a leading British vehicle components manufacturer. This unit was intended as the air conditioning cooler for an armored fighting vehicle. The design was compact with a mass and volume half the conventional vapor compression unit. Later on, other designs were evaluated for railcar air conditioning systems and for cockpit cooling of fighter aircraft and crop-spraying aircraft. In all cases, promising designs emerged but did not proceed to prototype development.

The growth of interest and capability in single-acting free-piston Stirling engines in the 1970s resulted in the development of concepts for multiple-cylinder double-acting free-piston Stirling engine coolers and prime movers. One engineering prototype development along the lines proposed by Finkelstein (1978) is under way and further hardware development is expected in the future.

One version of a three-cylinder free-piston Stirling cooling engine of intermediate capacity is shown in Fig. 3.17. Three cylinders are used to take advantage of the 120° phase difference in conventional 3-phase electric power supplies to energize the linear motor drive. Ingenious electrical design is required to devise a motor capable of accepting high-frequency (50 or 60 Hz) conventional power supplies and operating with a frequency half to one third this value.

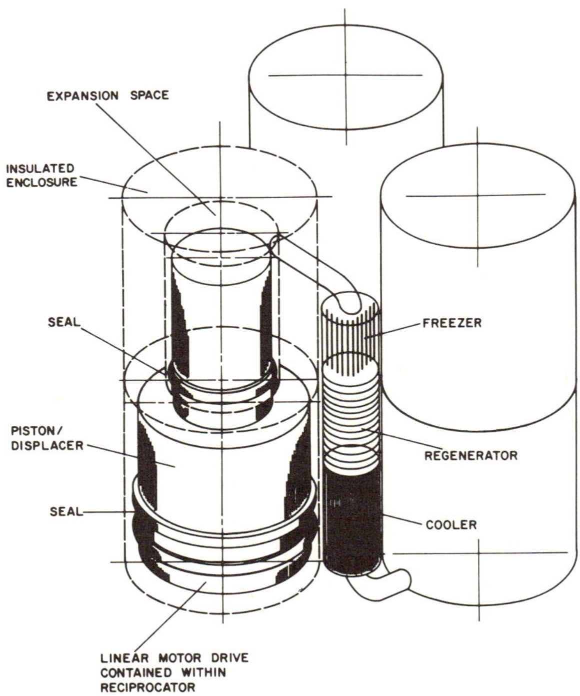

Fig. 3.17. Design concept for three-cylinder free-piston double-acting Stirling cryocooler.

The optimum swept volumes of the expansion and compression spaces of a cryocooler are markedly different. The ratio of swept volume $V_{(comp)}/V_{(exp)}$ is greater than unity and increases as the temperature difference between the space becomes greater. At cryogenic temperatures, the difference is as much as five times. To accommodate this difference in swept volume, various possibilities are illustrated in Fig. 3.18. In case (a), a stepped piston is used, with the diameter of the expansion piston being smaller than the compression piston. The annular space above the compression space can be utilized as a pneumatic spring to reduce the loading on the piston drive mechanism during the downward compression stroke. An alternative arrangement shown in Fig. 3.18(b) has a double-ended, stepped piston with central linear motor actuation to drive the piston. This would be one cylinder of a multiple-cylinder arrangement with the spaces all connected in appropriate fashion to ensure proper phasing between the volume variations in one system and another. Yet another variant is shown in Fig. 3.18c. The cylinder is equipped with a center post of substantial diameter projecting into the displacer interior. A pneumatic spring cavity is located below the center post. It is possible to engineer the system so the center post is equipped with linear gas bearings at the upper and lower ends to stabilize the location of the displacer in the cylinder. This accurate location allows the use of close tolerance seals between the displacer and cylinder enclosure so that rubbing seals are eliminated from the system. Attachment of the armature of a linear motor to the underside of the displacer and containment of the motor stator within the pressure cylinder

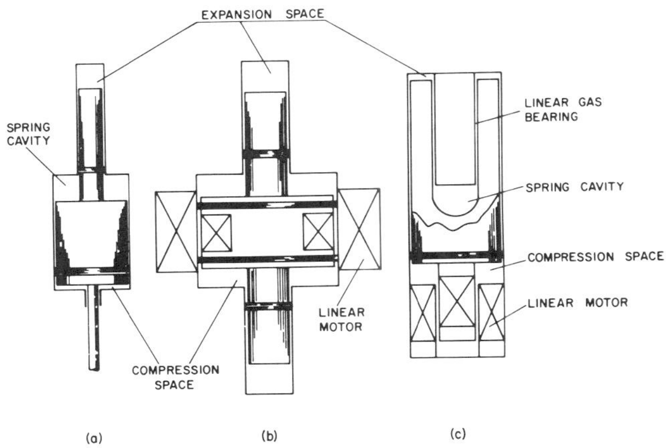

Fig. 3.18. Alternative design features to accommodate the difference in swept volumes in the expansion and compression spaces of double-acting Stirling cryocoolers.

completes the arrangement for one cylinder of an advanced multiple-cylinder free-piston double-acting Stirling system.

Multiple-Element Cooling Systems

Some military applications require multiple-element cryogenic spot cooling. For example, multiple infrared detectors are used in the forward and rear scanning missile detector systems of some manned bombers. Future computer applications for superconducting elements will likely involve multi-element cryogenic spot cooling. There are various ways to achieve this. Figure 3.19 shows a three-cylinder system. The two outer cylinders are identical and contain displacers *A* and *B* which may be activated by a kinematic drive or by fluid forces only. The central cylinder contains a piston which is double-acting. The underside of the piston is connected to the left-hand displacer cylinder. The top side of the piston is connected to the right-hand displacer cylinder.

Such an arrangement can be recognized simply as two single-acting Stirling systems with a common double-acting compressor. This was first applied as early as 1827 by the brothers Stirling, Robert and James, in a 25 horsepower Stirling engine used in a Dundee foundry. It ran at 45 revolutions per minute and was successful but suffered repeated failures of the displacer cylinder hot ends due to the poor materials then available.

An alternative arrangement, shown in Fig. 3.20, has multiple expansion cylinders supplied from a common (single-acting) compressor cylinder. This was identified by Walker and Burn (1976) as the Rainbow B configuration. The Rainbow A configuration with a split compression space and a single expander was adopted by a British engine designer, Horace Rainbow of Bristol, in the design of a Stirling boat engine. Cryogenic cooling engines of this type have been made by CTI, Waltham, Massachusetts for the U.S.

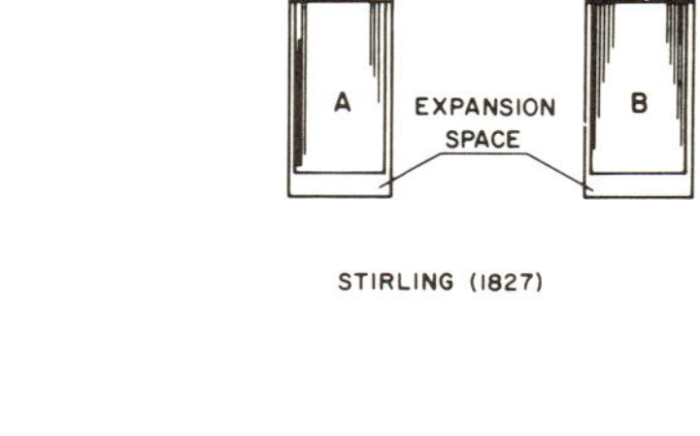

Fig. 3.19. Two single-acting Stirling systems with common compressor cylinder.

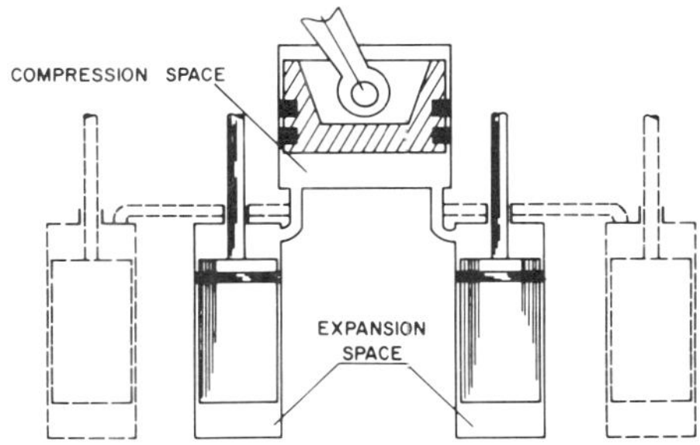

Fig. 3.20. Multielement cryogenic spot cooling.

Air Force (Chellis *et al.*, 1972). Up to 12 cooling heads may be supplied from a single compressor with lead distances (length of duct between compressor and expander) up to 60 in. In split-Stirling machines of this type, the displacers are not coupled to a kinematic drive but are activated only by fluidic forces, a pneumatic spring on one side and a variable pressure due to the compressor on the other.

PISTON AND DISPLACER MOTION

In Chapter 2, the ideal Stirling cycle was observed to be a thermodynamic cycle of two isothermal and two constant volume regenerative processes having the pressure-volume and temperature-entropy diagrams shown in Fig. 2.9.

The elements of a Stirling engine consist of two spaces of variable volume at different temperatures, coupled together through the regenerative heat exchanger and the freezer and cooler heat exchangers. These elements can be arranged in a wide variety of configurations and can be broadly classified into single- and double-acting engines. The single-acting machines are either two-piston or piston-displacer systems. These latter can be further divided into integral- or split-Stirling engines. Now it is of interest to consider the motion of reciprocating elements in all these variations and show the ideal cycle is the same, regardless of the configuration.

Two-Piston Stirling Engine

Consider first the two-piston Stirling engine shown in Fig. 2.9. This was used in Chapter 2 to describe the operation in detail of an ideal Stirling engine. The system consists of a cold expansion space at temperature T_E

and a compression space at the ambient temperature T_C. The regenerative heat exchanger is interposed between the two spaces with supplementary recuperative heat exchangers, the "freezer" on the cold side and the "cooler" on the ambient temperature side.

The cycle consists of four processes:

(1–2) Isothermal compression in the compression space at ambient temperature. Heat is rejected to the surroundings of the compression space.

(2–3) Constant volume regenerative cooling where heat is transferred from the working fluid to the regenerative matrix. The reduction in temperature at constant volume causes a reduction in the pressure.

(3–4) Isothermal expansion in the expansion space at the refrigeration temperature T_E. Heat is absorbed from the surroundings of the expansion space.

(4–1) Constant volume regenerative heating where heat is transferred from the regenerative matrix to the working fluid. The increase in temperature at constant volume causes an increase in the pressure.

The disposition of the two pistons in the expansion and compression spaces at the four terminal points of the cycle is shown in Fig. 2.9. The time-displacement diagram for the two pistons is also shown. It is of interest to note that volume variations in the expansion space *lead* those in the compression space, i.e., the expansion space volume is zero in process (1–2), whereas the compression space volume does not become zero until later [in process (3–4)].

Integral Stirling Engine

Now consider the integral Stirling machine shown in Fig. 3.21. This consists of a displacer and a piston operating in a single cylinder. The space *above* the displacer is the expansion space at temperature T_E. The space *between* the piston and displacer is the compression space at ambient temperature T_C. The two variable volumes are connected by a duct containing the regenerator and the two recuperative exchangers, the freezer above the regenerator and the cooler below the regenerator.

The cycle of operation of the engine is exactly the same as with the two-piston machine above. The four processes are (1–2) isothermal compression, (2–3) regenerative cooling at constant volume, (3–4) isothermal expansion, and (4–1) regenerative heating at constant volume. The disposition of the reciprocating elements at the four terminal points of the

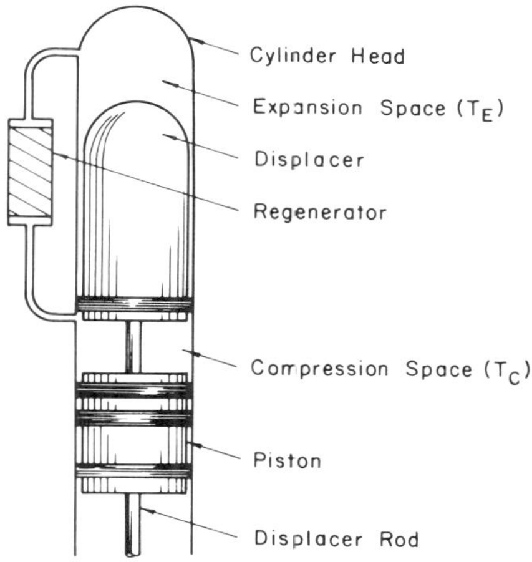

Fig. 3.21. Integral Stirling engine.

cycle and the time-displacement diagram for the reciprocating elements are shown in Fig. 3.22.

Split Stirling Engine

A similar set of disposition and time-displacement diagrams is drawn in Fig. 3.23 for the split-Stirling arrangement. In this case, the displacer and piston operate in separate cylinders. The cycle and the four processes are exactly the same as described above. In this arrangement, the compression space is divided between the compressor cylinder and the displacer cylinder. This is frequently overlooked in discussions of split Stirling systems where the displacer cylinder is often referred to as the "expander" and the piston cylinder as the "compressor."

For all the above engines, the *P–V* and *T–S* planes shown in Fig. 2.9 are applicable with the numbers 1–2–3–4 corresponding in all diagrams.

Double-Acting Stirling Engine

The double-acting Stirling engine is a little more complicated to understand. Consider the four-cylinder double-acting Stirling engine arrangement shown in Fig. 3.24. The four cylinders (a), (b), (c), and (d) each contain a single reciprocating element, a piston-displacer. The space above each element is the expansion space at low temperature T_E and the space below is the compresion space at ambient temperature T_C. The expansion space of one cylinder is connected to the compression space of the adjacent

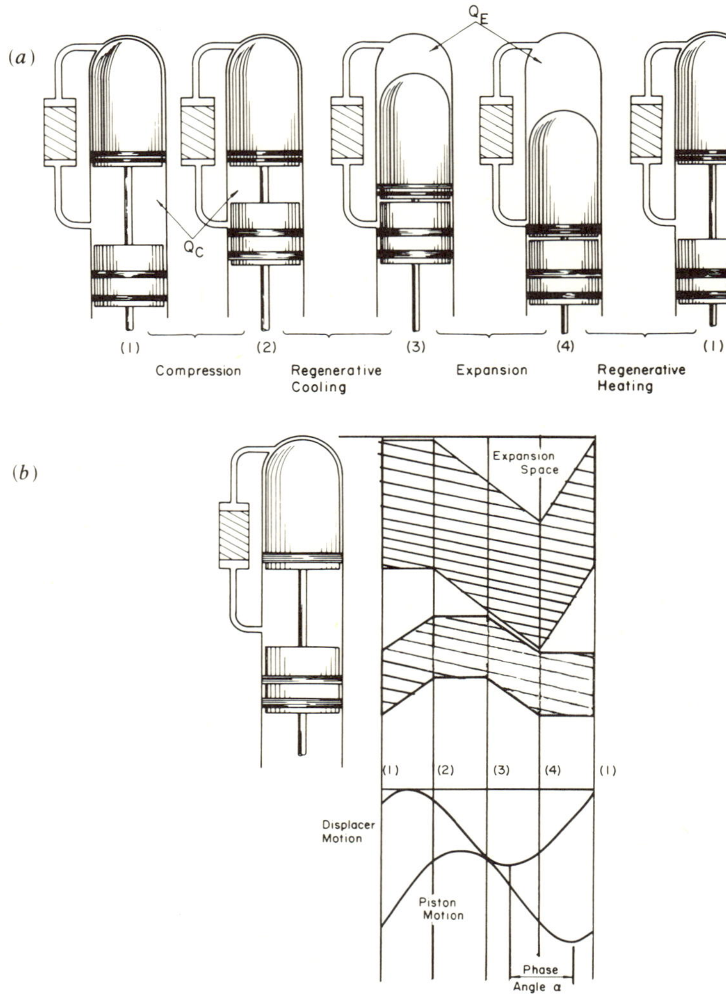

Fig. 3.22. (a) Piston and displacer disposition, and (b) displacement-time diagrams for the integral Stirling engine operating on the ideal Stirling cycle.

cylinder through the appropriate heat exchangers so that four separate Stirling engine systems are formed, A, B, C, and D.

System A combines the compression space of cylinder (a), $(V_C)_a$ with the expansion space of cylinder (d), $(V_E)_d$. Similarly system B is made up of $(V_C)_b$ and $(V_E)_a$ and so on. There are of course four work diagrams, one for each system and the numbers at the terminal points do not correspond

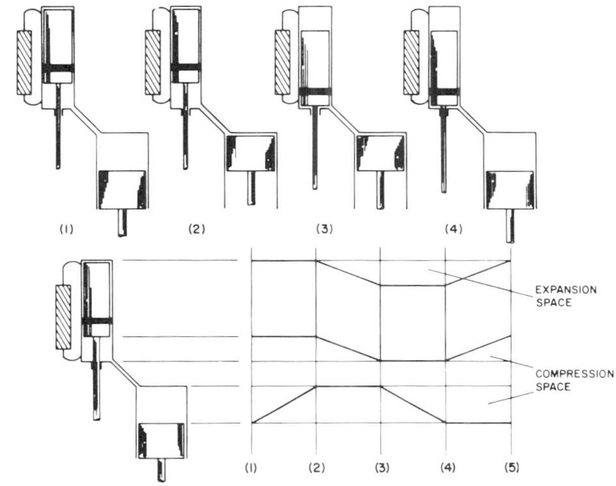

Fig. 3.23. Piston and displacer disposition and the displacement-time diagrams for the split-Stirling engine operating on the ideal Stirling cycle.

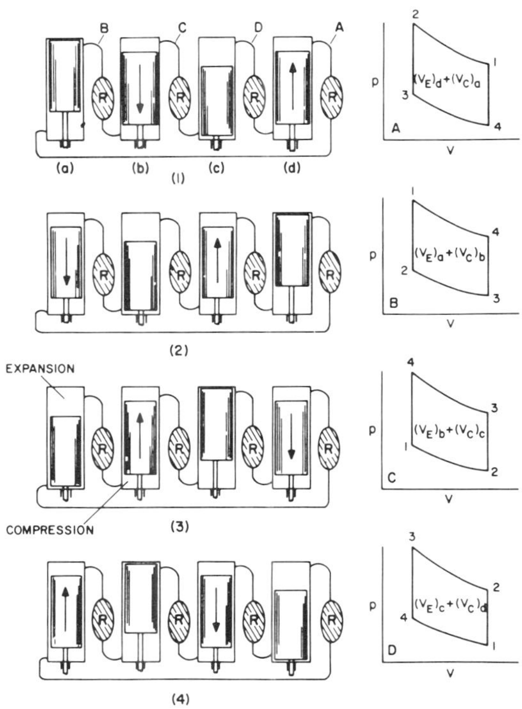

Fig. 3.24. Disposition of the reciprocating elements of the four-cylinder double-acting Stirling engine.

with those given on Fig. 2.9, except for system A. Therefore, the four P–V diagrams for the double-acting engine are also included in Fig. 3.24.

At station 1 the reciprocator in cylinder (a) is at the top dead center position. The reciprocator in cylinder (b) is at the midstroke and descending. In cylinder (c) the reciprocator is at the bottom dead center position and in cylinder (d) the reciprocator is at midstroke and ascending. There is, therefore, a phase difference of 90° between the reciprocator positions in adjacent cylinders. The phase difference depends on the number of cylinders, i.e., $\alpha = 360/n$, where α is the phase difference and n is the number of cylinders. Here n can be 3, 4, 5, or 6 for satisfactory operation.

At station 2 all the pistons have moved by one half stroke but the phase relationship has been preserved. In cylinder (a) the reciprocator has descended to midstroke, in cylinder (b) it has descended to the bottom dead center position, in (c) it has ascended to the midstroke position, and in (d) it has ascended to the top dead center position.

Consider what has happened in this process (1–2). System A, made up of $(V_C)_a$ and $(V_E)_d$, has decreased in volume to accomplish the isothermal compression 1–2 shown on the P–V diagram for system A. Simultaneously system B made up of $(V_C)_b$ and $(V_E)_a$ maintained the total system volume constant but pushed the working fluid from $(V_C)_b$ to $(V_E)_a$, that is the constant volume regenerative cooling process (1–2) shown in the P–V diagram for system B. While this was going on, system C was experiencing an isothermal expansion and system D was experiencing a constant volume regenerative heating process. If the pistons are now allowed to progress another half stroke to the disposition shown in station 3, the analysis of events will show the process (2–3), shown on the four P–V diagrams, will occur. Similarly, we can continue through the processes (3–4) and the final process (4–1).

To facilitate understanding, the time–displacement diagrams for the four Stirling systems have been drawn in Fig. 3.25. These diagrams combine the appropriate expansion and compression spaces constituting each system. Thus, the upper diagram for system A, combines the motion of the reciprocator in cylinder (d) (top half of the diagram) with the motion of the reciprocator in cylinder (a) (bottom half of the diagram). The shaded area in the top half represents the volume variation in the expansion space of system A (located in cylinder (d) *above* the piston/displacer). The shaded area in the lower half of the figure represents the volume variation of the compression space of system A [located in cylinder (a) *below* the piston/displacer]. At any instant, the sum of the two shaded areas represents the total "live" working space in system A. The constant volume of the connecting ducts containing the regenerator and associated heat exchangers must be added to obtain the total working space. The same pattern of

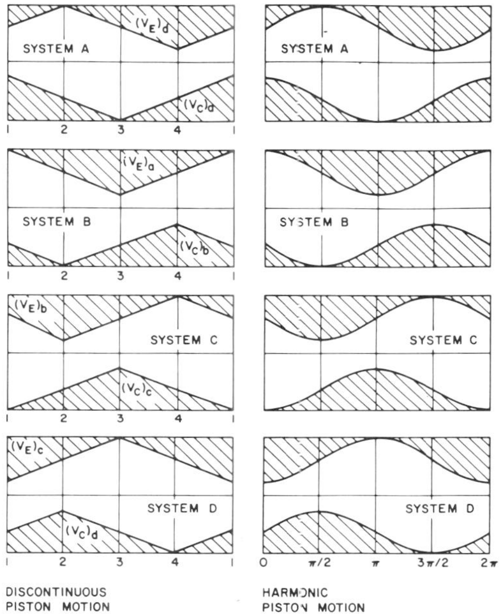

Fig. 3.25. Time-displacement diagrams for the reciprocating elements of the four-cylinder double-acting Stirling engine.

presentation is repeated for the displacement-time diagrams for systems *B*, *C*, and *D* included in Fig. 3.25.

Newcomers to the field will be perplexed by the complicated interplay of events in double-acting engines. Let them be assured the perplexity is shared by this author and others. If the subject of their interest is intermediate or large-scale refrigerating capacity, the closest study of double-acting systems is warranted. The Koh-i-noor of large-capacity regenerative cryocoolers has yet to be found and most surely lies in the double-acting free piston field.

PRACTICAL REGENERATIVE CYCLE

So far it has been assumed the Stirling engine works on the ideal Stirling cycle of two isothermal and two constant-volume regenerative processes. Now it is time to consider the difficulties arising if attempts are made to make cooling engines work on the ideal cycle. Having assessed that problem, we can define a more realistic model with which to compare the performance of actual engines.

Mass Distribution in Ideal Cycle

The ideal Stirling cycle assumes much that is beyond the bounds of practical possibility. At the four terminal points of the cycle, all the working fluid is assumed to be concentrated in either the compression or expansion space at temperature T_C or T_E. This would require the regenerative matrix and recuperative freezer and cooler heat exchangers to have zero void volumes. Since fluid must pass through the heat exchangers, they have some flow passages and a finite void volume. Therefore, it is never possible to have all the working fluid instantaneously at the same state (except of course at the start when the whole machine is at ambient temperature). Some working fluid will always be in the void spaces of the heat exchangers at intermediate temperatures.

The void volume will reduce the volume compression ratio (V_{max}/V_{min}) with the consequence of a reduced pressure compression ratio (p_{max}/p_{min}) and reduction in the pressure–volume (work) diagram for the cycle.

Harmonic Piston Motion

The ideal cycle demands the reciprocating elements move with discontinuous motion. It is not impossible to achieve the required motion but fairly elaborate kinematic or electrohydraulic mechanisms are required. It is easier to produce the simple harmonic motion customarily found in reciprocating engines with cranks and connecting rods or pneumatic springs.

A reasonable approximation to the ideal piston motion can be attained with harmonic motion. This is illustrated by the simple harmonic curves given along with the discontinuous time–displacement curves in Figs. 3.22 and 3.25.

Realistic Work Diagrams

The change in piston motion from discontinuous to continuous harmonic has profound effects on the operation of Stirling engines. With harmonic motion, the work diagram assumes the kidney shape figure shown in Fig. 3.26 rather than the familiar four-cornered figure.

The working fluid is distributed between the various spaces and three work diagrams may be drawn, one for the expansion space, one for the compression space, and one for the total working space, including the "dead space," the total volume of the internal voids in the heat exchangers and the duct connecting the two "live" spaces. Figure 3.27 shows a typical set of work diagrams. The diagram for the expansion space is traversed in a clockwise direction, whereas the diagrams for the compression and

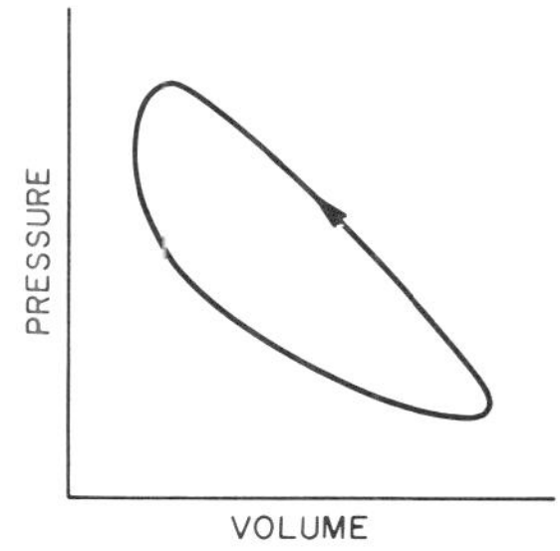

Fig. 3.26. Work diagram for a Stirling engine with harmonic piston motion.

expansion spaces are traversed in the reverse, anticlockwise direction. This is because work is produced by gas expanding in the expansion space, whereas work must be done on the gas in the compression space.

The area of the work diagrams is the work done *by* or *on* the gas. The *difference* in area of the pressure–volume diagrams for the compression and expansion spaces is the net work of the cycle. In the case of a refrigerator, the compression space diagram is larger than for the expansion space and work must be supplied to the cycle. The amount required is equal to the area of the diagram for the total working space or the difference in area of the compressor and expansion space diagrams. With a prime mover, the expansion space diagram is larger than the compression space diagram. As before, the difference in the area of the diagrams is equal to the area for the total working space, but in this case, the total working space diagram is traversed in the clockwise direction and represents the net cycle *output* of the engine.

When compression and expansion occur isothermally, the areas of the work diagrams also represent the quantities of heat transfer *to* the expansion cylinder (and freezer) and *from* the compression space (and cooler).

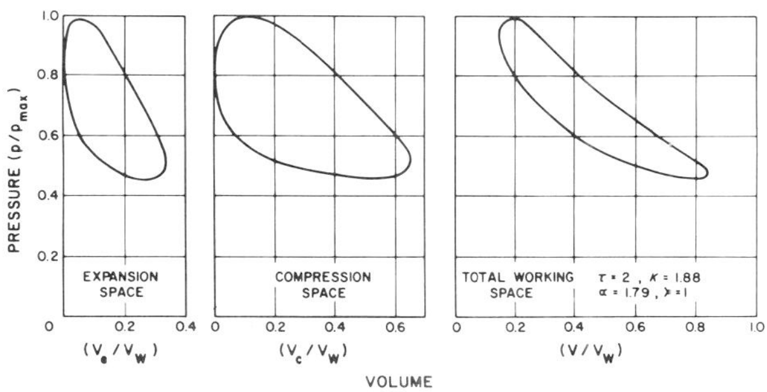

Fig. 3.27. Typical work diagrams for the expansion, compression, and total working spaces of a Stirling cryocooler.

The validity of the equality $Q = \oint p\,dv$ with isothermal conditions and ideal gas is widely accepted for a closed system as a straightforward application of the first law of thermodynamics enunciated for a closed system: $dQ = du + dw$. It is less familiar for the nonclosed, variable mass, unsteady flow situations such as exist here. However, by means of thermodynamic argument too involved for inclusion here, it can be shown that the equality is obtainable just as in the simpler, closed system provided the conditions of ideal gas behavior and *constant* temperature conditions are met.

Temperature–Entropy Diagrams

The introduction of continuous harmonic piston motion and finite void volume in the heat exchangers, eliminates conditions inherent in the ideal cycle that all the working fluid is instantaneously at the same state and proceeds as a lumped mass around the cycle. With continuous harmonic piston motion, the fluid is distributed about the machine in various spaces at various temperatures.

Any particular element of mass inhabits a region of the engine and when the engine is operating, the element will not move entirely from the compression to expansion space, but will simply commute to and from, say, the expansion space to the freezer. Another particle may inhabit the regenerator and cooler and so on. It is not possible to draw meaningful temperature–entropy diagrams for all the working fluid. Rather, individual diagrams for particles at different regions of the engine can be constructed. The individual envelopes drawn on the T–S plane can then be wholly enclosed in boundaries that probably represent some kind of temperature–entropy diagram.

Such representations have only limited use and are rarely drawn in practice. Areas of these composite diagrams do not represent simple heat transfer as in the idealized cycle.

Schmidt Cycle

The case discussed above with harmonic piston motion and isothermal compression and expansion was analyzed by Gustav Schmidt (1871). It is convenient to identify this case as the Schmidt cycle. The analysis is presented in detail later in this chapter.

Nonisothermal Compression and Expansion

In practice, the condition of isothermality can never be achieved. It would require either infinite rates of heat transfer in the compression and expansion spaces or the engine running at very low speeds. In real engines running at realistic speed (15 to 50 Hz), conditions in the cylinders are closer to adiabatic (no heat transfer) than isothermal (infinite heat transfer). They may be close to isothermal in the freezer and cooler since these components are designed to maximize heat transfer.

The departure from isothermal conditions in the compression and expansion spaces causes a marked redistribution of working fluid in the machine resulting in a decrease of refrigerating capacity and an increase in the input work, thereby reducing the coefficient of performance. The reduction may be as much as one half the ideal Carnot value.

Analysis of the Stirling engine with adiabatic processes is somewhat more complicated than the isothermal Schmidt cycle. The theory has been developed, and will be discussed later. For reference purposes, the adiabatic case is called the Finkelstein adiabatic cycle.

Aerodynamic Friction Loss

Another important effect in real engines is the loss of refrigerating capacity due to aerodynamic pressure losses. The effect is illustrated in Fig. 3.28. The upper diagram shows the pressure–time diagrams for the compression and expansion spaces. Differences in these two curves are the pressure losses arising from flow through the regenerator, freezer, and cooler heat exchangers. The effect of pressure drop on refrigerating capacity can be gauged by reference to the work diagrams for the expansion and compression spaces. The shaded areas are the reduction in area of the

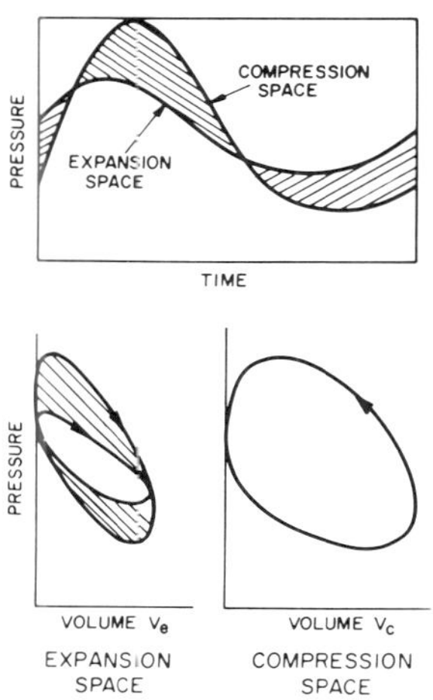

Fig. 3.28. Regenerator pressure drop and the effect on expansion cylinder work.

expansion space diagram due to regenerator pressure drop. The refrigerating capacity is decreased approximately by the same proportion as the area of the work diagram is decreased.

The regenerator pressure drop has a direct and immediate effect on the refrigerating capacity and should be kept to the minimum. The pressure drop is a function of the square of the velocity of the fluid. For minimum pressure drop, the velocity should be low. This can be achieved with a large, open, regenerative matrix and large fluid ducts in the freezer and cooler passages. Unfortunately, this increases the dead space relative to the "live" swept volumes and has deleterious effects on refrigerating capacity. The dead space must be as small as possible. This dichotomy between the smallest possible pressure drop and the lowest possible dead space provides a challenge to ingenious mechanical designers.

Heat Exchanger Thermal Potential

The function of the cooler and the freezer heat exchangers is to maximize heat transfer to or from the working fluid. In both cases, a temperature difference between the working fluid and the heat exchanger walls is necessary to establish the energy transfer.

In the cooler, energy is transferred by convection *from* the working fluid to the walls of the cooler, hence by conduction through the walls of the cooler and is carried off by convective heat transfer from the walls of the cooler to water or air at ambient temperature. Water is a much more effective coolant (by an order of magnitude) than air. If water is used in a closed cooling circuit, a circulating pump and a separate air-cooled convective recuperative heat exchanger (in automotive terminology, the "radiator") will be necessary and this addition will make the system large, heavy, and expensive.

Direct air cooling is attractive, but to improve effectiveness, a large finned area is required to increase the area for heat transfer. A fan may be necessary to blow cooling air through the fin complex.

The temperature difference between the average temperature of the working fluid and the coolant may be assumed to be somewhere in the range of 20 to 50 °C. To assure heat flow in the right direction, the gas must be hotter than the coolant.

Similar considerations apply at the low-temperature region of the machine. When a machine is operating in steady state, the average temperature of the working fluid in the cold region will be that temperature at which the refrigeration available exactly balances the thermal load. The thermal load is made up of several constituents:

i. The external refrigeration effect in cooling or condensing a steady flow of gas or liquid circulating through the freezer at constant temperature and pressure,
ii. Various heat "leaks" by conduction, convection, and radiation from the surroundings at higher temperature.

Careful attention to design, the use of structural materials having a low thermal conductivity and high performance thermal insulating materials can reduce the heat leaks to a surprising degree. Nevertheless, whatever measures are taken, some residual heat leak must remain. Indeed, in many cryogenic coolers, particularly the miniature and microminiature engines used for electronic and infrared sensing applications, the external refrigerating component is minimal, a nominal fraction of a watt. The sole objective of the machine is to achieve and maintain a cryogenic temperature at the cold finger. The entire refrigerating effect is dissipated by external heat leaks to the cold region. The electronic device or infrared sensor to be cooled may be mounted directly on the cold finger and connected by fine metallic conductors to the electronics associated with it. In most cases, the presence of the sensor has little effect on the temperature achieved in the cold region of the engine.

The temperature difference between the *average* temperature of the gas in the cold region and the surroundings of the cold region, might be 2 to 20°C. For energy flow in the right direction, the gas must be cooler than the cylinder surrounds.

To assess the effects of these temperature potentials, consider a cryocooler producing refrigeration at 80 K and supplied with coolant at ambient temperature, 290 K. The ideal coefficient of performance of a refrigerator operating between these temperatures is

$$\mathrm{COP}_{(\mathrm{Carnot})} = \frac{T_E}{T_C - T_E} = \frac{80}{280-90} = 0.38$$

Now include the heat exchanger thermal potentials of, say, 50 K in the cooler and 20 K in the freezer. The thermodynamic temperature range of operation is, therefore, increased to $(290+50)=340$ K and $(80-20)=60$ K. The coefficient of performance under these conditions is, therefore,

$$\mathrm{COP}_{(\mathrm{Carnot})} = \frac{(T_E - \Delta T_E)}{(T_C + \Delta T_C) - (T_E - \Delta T_E)} = \frac{60}{340-60} = 0.214$$

The effect of the temperature potentials is evidently profound. In the case chosen, the COP, with the admittedly exaggerated temperature potentials, decreased to a value only 56% of the COP calculated ignoring the heat exchanger thermal potentials.

Regenerator Contamination

The importance of the regenerator effectiveness to the ideal Stirling cycle was demonstrated in Chapter 2 (see Fig. 2.21). Remarkably effective regenerators ($\varepsilon > 0.95$) are routinely used in Stirling cryogenic coolers with matrices of finely divided materials in the form of wires or balls. This finely divided material is also an excellent filter and effectively "strains out" any contaminants contained in the working fluid passing through. The contaminants include lubricating oil leaking from the crank case, wear debris from dry rubbing piston rings or seals, and condensed liquids or solids of water and carbon dioxide from the working fluid. Blockage of regenerator flow passages increases the fluid friction in the matrix (i.e., a larger pressure drop) with all the unfortunate consequences to refrigerating capacity discussed above.

Regenerator Thermal Saturation

Another important phenomenon associated with the regenerator is "thermal saturation." In moving through the matrix from one end to the other, there is a progressive decrease in the temperature of the matrix although at any point the temperature of the matrix material is assumed to remain constant with ideal regeneration. This assumption implies the thermal heat capacity (mass times specific heat) of the matrix is infinitely large compared with the gas passing through it.

When the heat capacity of gas becomes significant compared with the matrix, there is increasing departure from ideality and the regenerator approaches the state of thermal saturation. This can arise in two ways; the first is by increasing the frequency of operation or the pressure of the working fluid (and hence the density). To a first approximation, the refrigerating capacity is a linear function of the machine speed or the pressure of the working fluid. Increasing either or both increases the refrigerating capacity but at a progressively diminishing rate until a maximum is attained. Further increase in speed or density actually decreases the refrigerating output. The reduced rate of increase arises from the combined action of increased pressure drop and thermal heat capacity of the gas (incipient thermal saturation of the matrix).

The second cause of thermal saturation becomes significant at low temperature levels (below 20 K). A decrease in temperature results in *decreased* specific heat of *solids and liquids* but *increased* specific heat of gases. At very low temperatures, helium is the only gas not liquefied and so must be used as the working fluid for cryocoolers. Helium has a high specific heat compared with many other gases and this already high value increases sharply at low temperatures. At the same time, the specific heat of conventional matrix materials (copper and bronze wire, etc.) decreases so that thermal saturation prevents effective operation of the matrix.

Change of the matrix material to lead produces an improvement. Lead is not available in the form of very fine wires but can be obtained as fine powder or "shot." Further improvement can be gained by the use of "rare earth" materials (europium sulfide) with advantageous discontinuities in the specific heat temperature characteristic. Eventually, though, thermal saturation of the matrix prevails as the helium working fluid nears the liquefaction temperature. The ultimate temperature achievable is rarely less than 6 K. Zimmerman (see Chapter 1) has achieved 3.5 K in a special expansion machine with a "regenerative annulus," a pressure of less than 1 atmosphere of the helium working fluid and precooling with liquid helium to 10 K.

THEORETICAL ANALYSIS OF STIRLING ENGINES

Introduction

Theoretical analyses of Stirling engines acting as prime movers or cooling engines have been developed with varying degrees of sophistication. The most simple analysis is that for the ideal Stirling cycle, the thermodynamic cycle, comprised of two isothermal and two constant-volume regenerative processes. As we have seen, this involves such gross idealization as to be suitable only for the most elementary design calculations.

A more realistic analysis devised by Gustav Schmidt (1871) has become the classical analysis of the cycle and is believed to give a reasonable approximation of machine performance. The analysis is highly idealized, and in practice, the indicated performance will likely be no better than 50% of the Schmidt cycle performance and often a good deal less. This is because the Schmidt cycle assumes compression and expansion to be isothermal. In practical engines running at 15 Hz or more, the processes are more nearly adiabatic. This apparently minor difference causes a redistribution of the mass working fluid and has a critical effect on engine performance.

Following introduction of the Schmidt analysis in 1871, nearly a century was to elapse before Finkelstein (1960) devised the generalized analysis of engines with processes of compression and expansion other than isothermal. The Schmidt cycle with isothermal compression and expansion then became a special case of the generalized Finkelstein analysis. Another special case is the cycle with adiabatic processes in the engine cylinder (the

Finkelstein adiabatic cycle). The generalized theory permits consideration of other cases of *limited* heat transfer in the engine cylinders.

Further development of the adiabatic model allowing for secondary effects, primarily aerodynamic flow losses, and thermal effects, is possible to make the adiabatic model the preferred level of analysis for routine performance predictions. The analysis is sufficiently complex to require the use of digital computers, but only to a moderate extent, involving little computer time and reasonable expenditure.

More advanced theoretical models for regenerative engine analysis were devised by Finkelstein (1975) culminating in a nodal analysis program installed on a commercial computer network and available for general use on payment of a royalty fee. In the closing years of the 1970s substantial efforts were devoted to Stirling engine simulation by other workers and a variety of advanced engine simulation programs have become available.

A comprehensive discussion of the generalized Finkelstein analysis and subsequent nodal analysis programs is beyond the scope of this work. Here, with particular regard to cooling engines, we shall

i. summarize equations for the ideal Stirling cycle,
ii. present in reasonably comprehensive detail, the Schmidt isothermal analysis,
iii. outline the Finkelstein adiabatic analysis,
iv. compare and comment briefly on the advanced level analyses for Stirling engines making reference to the source documents for those who wish to investigate further.

Ideal Stirling Cycle

Equations for analysis of the ideal Stirling cycle are summarized below. The reader is advised that the gross idealization of the Stirling cycle precludes the use of these equations for anything other than the most elementary investigations.

The principal deficiency of the ideal cycle is that *all* the working fluid is assumed to be, instantaneously, at the same condition in either the compression or expansion spaces. This implies the internal void volume of the regenerator is zero and requires the pistons or other reciprocating elements to move with a discontinuous motion. The processes of expansion and compression are assumed to be isothermal and the effects of imperfect regeneration and aerodynamic pressure drops are not considered.

A more complete treatment of the ideal cycle, based on the Rallis regenerative reference cycle, is given in Chapter 2. With reference to Fig. 2.6 and the discussion of the ideal cycle given in Chapter 2, the equations for the ideal cycle may be summarized as follows:

Required Data.

i. Some reference temperature and pressure, or volume, say, conditions, at state 1.
ii. Temperature ratio $\tau = T_C/T_E$.
iii. Volume ratio $r = V_{max}/V_{min}$.

For unit mass of working fluid, assumed to be a perfect gas, then $V_1 = RT_1/p_1$, from the characteristic gas equation.

The following data are required for each of the four cycle processes:

(*a*) *Isothermal Compression Process* (*1–2*). In this process, heat is abstracted from the working fluid and rejected from the cycle at the maximum cycle temperature T_C. Work is done on the working fluid equal in magnitude to the heat rejected from the cycle. There is no change in internal energy, and there is a decrease in entropy.

$$p_2 = p_1V_1/V_2 = p_1r; \qquad T_2 = T_1 = T_C$$

Heat transfer $(Q) =$ work done $(W) = P_1V_1 \ln(1/r) = RT_1 \ln(1/r)$. Change in entropy $(S_2 - S_1) = R \ln(1/r)$.

(*b*) *Constant-Volume Regenerative Cooling Transfer Process* (*2–3*). In this process heat is transferred to the regenerative matrix from the working fluid, decreasing the temperature from T_C to T_E. No work is done, and there is a decrease in the entropy, and internal energy, of the working fluid.

$$p_3 = p_2T_3/T_2 = p_2/\tau, \qquad V_3 = V_2$$

Heat transfer $(Q) = C_v(T_3 - T_2)$. Work done $(W) = 0$. Change in entropy $(S_3 - S_2) = C_v \ln(1/\tau)$.

(*c*) *Isothermal Expansion Process* (*3–4*) In this process, heat is supplied to the cycle at a low temperature T_E, during expansion of the working fluid. Work is done, by the working fluid, equal in magnitude to the heat supplied. There is no change in the internal energy, but an increase in the entropy of the working fluid:

$$p_4 = p_3V_3/V_4 = p_3(1/r), \qquad T_4 = T_3 = T_E$$

Heat transfer $(Q) =$ work done $(W) = p_3V_3 \ln r = RT_3 \ln r$. Change in entropy $(S_4 - S_3) = R \ln r$.

(*d*) *Constant-Volume Regenerative Heating Transfer Process* (*4–1*). In this process, heat is transferred to the working fluid from the regenerative matrix increasing the temperature of the working fluid from T_E to T_C. No work is done, and there is an increase in the internal energy and entropy of the working fluid.

$$p_1 = p_4 T_4 / T_1 = p_1 \tau, \qquad V_1 = V_4$$

Heat transfer $(Q) = C_v(T_1 - T_4)$. Change in entropy $(S_1 - S_4) = C_v \ln \tau$.

In the regenerative processes, the heat transferred to the matrix from the working fluid in process (2–3) is restored to the working fluid from the matrix in process (4–1). There is no net gain or loss of heat by the working fluid or the matrix. Therefore

$$\text{the total heat extracted (at } T_E) = RT_3 \ln r$$

$$\text{the total heat rejected (at } T_C) = RT_1 \ln r$$

and

$$\begin{aligned} \text{coefficient of performance} &= \frac{RT_3 \ln r}{RT_1 \ln r - RT_3 \ln r} \\ &= \frac{T_3}{T_1 - T_3} = \frac{T_E}{T_C - T_E} \\ &= 1/(\tau - 1) \end{aligned}$$

This value corresponds to the Carnot value between the same temperature limits.

The Schmidt Cycle

The classical analysis of the operation of Stirling engines is due to Schmidt (1861). The theory provides for harmonic motion of the reciprocating elements, but retains the major assumptions of isothermal compression and expansion and of perfect regeneration. It, thus, remains highly idealized, but is certainly more realistic than the ideal Stirling cycle. Provided a reasonable level of caution is exercised in interpretation, the predictions of the Schmidt theory can be a useful tool for engine design.

Principal Assumptions of the Schmidt Cycle.

1. The regenerative process is perfect.
2. The instantaneous pressure is the same throughout the system.
3. The working fluid obeys the characteristic gas equation, $pV = RT$.
4. There is no leakage, and the mass of working fluid remains constant.
5. The volume variations in the working space occur sinusoidally.
6. There are no temperature gradients in the heat-exchangers.
7. The cylinder wall, and piston, temperatures are constant.
8. There is perfect mixing of the cylinder contents.

9. The temperature of the working fluid in the ancillary spaces is constant.
10. The speed of the machine is constant.
11. Steady state conditions are established.

Nomenclature.*

A = a factor $(\tau^2 + 2\tau\kappa \cos\alpha + \kappa^2)^{1/2}$.
B = a factor $(\tau + \kappa + 2S)$.
K = const.
M = total mass of working fluid.
N = machine speed.
p = instantaneous cycle-pressure.
p_{max} = maximum cycle-pressure.
p_{mean} = mean cycle-pressure.
p_{min} = minimum cycle-pressure.
P = engine output.
$P_{mass} = P/RT_C$ dimensionless power parameter based on the mass of working fluid.
$P_{max} = P/(p_{max} V_T)$, dimensionless power parameter, based on the maximum cycle-pressure and combined swept volume.
Q = heat transferred to the working fluid in the expansion space, the heat extracted.
$Q_{mass} = Q/RT_C$ the dimensionless cooling parameter, based on the mass of working fluid.
$Q_{max} = Q/(p_{max} V_T)$, the dimensionless heat, based on the maximum cycle pressure.
R = characteristic gas constant of the working fluid.
$S = (2X\tau)/(\tau + 1)$, reduced dead volume.
T_C = temperature of the working fluid in the compression space, generally assumed to be 300 K.
T_D = temperature of the working fluid in the dead space.
T_E = temperature of the working fluid in the expansion space.
V_C = swept volume in the compression space.
V_E = swept volume in the expansion space.
V_D = total internal volume of heat-exchangers, volume of regenerator, and associated ducts and ports.
$V_T = (V_C + V_E) = (1 + \kappa)V_E$, combined swept volume.
$V_W = \frac{1}{2}V_E(1 + \cos\phi) + \frac{1}{2}V_C[1 + \cos(\phi - \alpha)] + V_D$, volume of total working space.
$V_{W\,max}$ = maximum volume of total working space.

* *Note:* Lower case subscripts indicate instantaneous values of temperature, pressure, volume, and mass. Upper case subscripts indicate maximum (or constant) values. Thus: E or e refers to expansion space, C or c refers to compression space, D or d refers to dead space.

$X = V_D/V_E$, dead-volume ratio.
α = angle by which volume variations in the expansion space lead those in the compression space (in fractions of π radians, or degrees).
$\delta = (\tau^2+\kappa^2+2\tau\kappa \cos\alpha)^{1/2}/(\tau+\kappa+2S)$.
$\theta = \tan^{-1}[(\kappa \sin\alpha)/(\tau+\kappa\cos\alpha)]$.
$\kappa = V_C/V_E$, swept-volume ratio.
$\tau = T_C/T_E$, temperature ratio.
ϕ = crank angle.

Basic Equations. Volume of expansion space:

$$V_e = \tfrac{1}{2}V_E(1+\cos\phi). \tag{3.1}$$

Volume of compression space:

$$V_c = \tfrac{1}{2}V_C[1+\cos(\phi-\alpha)] \tag{3.2}$$

$$= \tfrac{1}{2}\kappa V_E[1+\cos(\phi-\alpha)] \tag{3.3}$$

Volume of dead space, being that constant volume of the total working space not included in the volumes of the expansion or compression space:

$$V_D = XV_E \tag{3.4}$$

Mass of working fluid in expansion space:

$$M_e = (p_eV_e)/(RT_e)$$

Mass of working fluid in compression space:

$$M_c = (p_cV_c)/(RT_c)$$

Mass of working fluid in dead space:

$$M_d = (p_dV_d)/(RT_d)$$

Since the total mass of the working fluid remains constant

$$M_T = (p_eV_e)/(RT_e)+(p_cV_c)/(RT_c)+(p_dV_d)/(RT_d) = (KV_E)/(2RT_c) \tag{3.5}$$

If the instantaneous pressure is the same throughout the system, and equal to p, say, and if T_e and T_c are constant at T_E and T_C then, substituting for the volumes, eliminating R, and rearranging

$$K/p = (T_C/T_E)(1+\cos\phi)+\kappa[1+\cos(\phi-\alpha)]+(2V_DT_C)/(V_ET_D) \tag{3.6}$$

If the temperature variation in the dead space is linear in the axial direction, then the mean temperature

$$T_D = T_C+\tfrac{1}{2}(T_E-T_C) = (1+T_E/T_C)(T_C/2)$$

and, since $T_C/T_E = \tau$ then, from Eq. (3.6)

$$K/p = \tau(1+\cos\phi)+\kappa[1+\cos(\phi-\alpha)]+2S \tag{3.7}$$

where S (the reduced dead volume) $= 2X\tau/(\tau+1)$.

To simplify Eq. (3.7) consider first

$$y = x\cos\phi + z\sin\phi \tag{3.8}$$

then $y=\sqrt{r^2}\cos(\phi-\beta)$, where $\tan\beta = z/x$, $z = r\sin\beta$, and $x = r\cos\beta$ since

$$\begin{aligned}\sqrt{r^2}\cos(\phi-\beta) &= \sqrt{r^2}(\cos\phi\cos\beta+\sin\phi\sin\beta)\\ &= r\cos\phi\cos\beta + r\sin\phi\sin\beta\\ &= x\cos\phi + z\sin\phi\end{aligned}$$

Equation (3.8) is similar in form to Eq. (3.7) therefore, by analogy

$$K/p = [(\tau+\kappa\cos\alpha)^2+(\kappa\sin\alpha)^2]^{1/2}\cos(\phi-\theta)+\tau+\kappa+2S$$

$$= (\tau^2+2\tau\kappa\cos\alpha+\kappa^2)^{1/2}\cos(\phi-\theta)+\tau+\kappa+2S \tag{3.9}$$

where

$$\tan\theta = (\kappa\sin\alpha)/(\tau+\kappa\cos\alpha)$$

Let

$$A = (\tau^2+2\tau\kappa\cos\alpha+\kappa^2)^{1/2}, \qquad B = \tau+\kappa+2S, \qquad \text{and} \quad \delta = A/B$$

then

$$K/p = A\cos(\phi-\theta)+B$$

and

$$p = K/\{B[1+\delta\cos(\phi-\theta)]\}$$

The instantaneous pressure p is

(a) a minimum, when $\phi=\theta$, i.e., $(\phi-\theta)=0$,

(b) a maximum, when $\phi=(\theta+\pi)$, i.e. $(\phi-\theta)=\pi$.

Therefore, $p_{\min} = K/[B(1+\delta)]$, and $p_{\max} = K/[B(1-\delta)]$.

Thus

$$p = p_{\max}(1-\delta)/[1+\delta\cos(\phi-\theta)] \tag{3.10a}$$

$$= p_{\min}(1+\delta)/[1+\delta\cos(\phi-\theta)] \tag{3.10b}$$

and the pressure ratio $p_r = p_{\max}/p_{\min} = (1+\delta)/(1-\delta)$. (3.11)

Mean Cycle Pressure. The mean cycle pressure is given by

$$p_{\text{mean}} = (\tfrac{1}{2}\pi) \int_0^{2\pi} p\, d(\phi - \theta)$$

$$= (\tfrac{1}{2}\pi) \int_0^{2\pi} \{p_{\max}(1-\delta)/[1+\delta \cos(\phi - \theta)]\}\, d(\phi - \theta) \tag{3.12}$$

which can be resolved to

$$p_{\text{mean}} = p_{\max}[(1-\delta)/(1+\delta)]^{1/2} \tag{3.13}$$

Heat Transferred and Work Done. Since the processes of expansion and compression take place isothermally the heat transferred Q is equal to the work done P, therefore

$$Q = P = \int p\, dV$$

If $V = \frac{1}{2}V_E(1 + \cos\phi)$,

$$dV = -\tfrac{1}{2}V_E \sin\phi\, d\phi \tag{3.14}$$

and, if

$$p = p_{\text{mean}}[1 - \Delta \cos(\phi - \theta)] \tag{3.15}$$

approximately, where

$$\Delta = 2\delta/[1 + (1-\delta^2)^{1/2}]$$

then

$$Q = -\tfrac{1}{2} \int_0^{2\pi} \{p_{\text{mean}} V_E[1 - \Delta \cos(\phi - \theta)] \sin\phi\}\, d\phi$$

$$= -\tfrac{1}{2}p_{\text{mean}} V_E \int_0^{2\pi} [\sin\phi - \Delta(\cos\phi \cos\theta \sin\phi + \sin\theta \sin^2\phi)]\, d\phi$$

$$= -\tfrac{1}{2}p_{\text{mean}} V_E \{-\cos\phi - \Delta[-\cos\theta \cdot \tfrac{1}{2}\cos 2\phi + \sin\theta(\tfrac{1}{2}\phi - \tfrac{1}{4}\sin^2\phi)\}]_0^{2\pi}$$

$$= -\tfrac{1}{2}p_{mean} V_E \left[-\Delta \sin\theta \frac{\phi}{2}\right]_0^{2\pi}$$

$$= -\tfrac{1}{2}\pi p_{\text{mean}} V_E \Delta \sin\theta \tag{3.16}$$

Expansion Space. The variation in volume in the expansion space follows the equation

$$V_e = \tfrac{1}{2}V_E(1 + \cos\phi)$$

which conforms to the required equation [Eq. (3.14)], therefore the heat transferred in the expansion space, from Eq. (3.16), is given by

$$Q = \pi p_{\text{mean}} V_E \delta \sin\theta / [1 + (1 - \delta^2)^{1/2}] \tag{3.17}$$

There is no need to invoke the approximate expression for p in Eq. (3.15). Instead, Eq. (3.10b) is a suitable starting point, and can be combined with (3.14) to calculate the heat transfer per cycle, Q, as follows:*

$$Q = \oint p\,dV = \oint \frac{-p_{\min}(1+\delta)}{1+\delta\cos(\phi-\theta)} \cdot \frac{1}{2} V_E \sin\phi\, d\phi$$

$$= \frac{-V_E p_{min}(1+\delta)}{2} I$$

where

$$I = \oint \frac{\sin\phi\, d\phi}{1+\delta\cos(\phi-\theta)}$$

let $\phi - \theta = \psi$, then $\psi/2 = t$: then $d\phi \; x \; d\psi = 2dt/(1+t^2)$ and

$$\delta I = \oint \left\{ \frac{\delta \sin\psi \cos\theta}{1+\delta\cos\psi} + \frac{\delta\cos\theta + \sin\theta}{1+\delta\cos\psi} - \frac{\sin\theta}{1+\delta\cos\psi} \right\} d\psi$$

$$\delta I = [-\cos\theta \log(1+\delta\cos\psi)]_{\psi=0}^{2\pi} + \sin\theta[\psi]_{\psi=0}^{2\pi} - \sin\theta \oint \frac{2\,dt}{1+t^2+\delta(1-r^2)}$$

$$= 2\pi\sin\theta - 2\sin\theta \oint \frac{dt}{(1+\delta)+(1-\delta)t^2}$$

$$= 2\pi\sin\theta - \frac{2\sin\theta}{1-\delta} \oint \left(\frac{1-\delta}{1-\delta}\right)^{1/2} \oint \frac{[(1+\delta)/(1-\delta)]^{1/2}\,dt}{\{[(1+\delta)/(1-\delta)]^{1/2}\}^2 + t^2}$$

$$= 2\pi\sin\theta - \frac{2\sin\theta}{(1-\delta^2)^{1/2}} \left| \tan^{-1}\{t/[(1+\delta)/(1-\delta)]^{1/2}\} \right|_{\psi=0}^{2\pi}$$

In evaluating the definite integral note that $t = 0$ at both limits; $\tan^{-1}(0) = 0, \pi$ etc. It is clear from (a), if $\delta \to 0$, that the cyclic integral must take the value $-2\pi \sin\theta$ for its third term. Thus

$$\delta I = 2\pi\sin\theta - 2\pi\sin\theta/(1-\delta^2)^{1/2} = 2\pi\sin\theta[1 - 1/(1-\delta^2)^{1/2}]$$

and

$$Q = \frac{-V_E p_{\min}(1+\delta)}{2} \cdot \frac{2\pi\sin\theta}{\theta} [(1-1)/(1-\delta^2)^{1/2}]$$

* I am indebted to Dr. A. J. Oxley of the United Kingdom Rutherford Laboratory for the alternative derivation given here—G. W.

Substituting for p_{min} gives

$$Q = V_E p_{mean} \delta\pi \sin\theta/[(1-\delta^2)^{1/2}+1]$$

which is identical to Eq. (3.17).

Compression Space. The variation in volume of the compression space follows the equation

$$V_c = \tfrac{1}{2}\kappa V_E[1+\cos(\phi-\alpha)] \tag{3.18}$$

and, by a process similar to that above, we can obtain expressions for the pressure and volume in the required form, so that heat transferred in the compression space is given by

$$Q_c = [\pi p_{mean} V_E \kappa\delta \sin(\theta-\alpha)]/[1+(1-\delta^2)^{1/2}] \tag{3.19}$$

Dividing Eq. (3.19) by Eq. (3.17),

$$\begin{aligned} Q_c/Q &= [\kappa \sin(\theta-\alpha)]/\sin\theta \\ &= \kappa(\sin\theta\cos\alpha - \cos\theta\sin\alpha)/\sin\theta \\ &= \kappa[(\cos\alpha - \sin\alpha)/\tan\theta] \end{aligned}$$

but $\tan\theta = \kappa\sin\alpha/(\tau+\kappa\cos\alpha)$ and, therefore, $Q_c/Q = -\tau$.

The heat transferred in the expansion space is of opposite sign to the heat transferred in the compression space, and is numerically different by the temperature ratio τ. By analogy, the work done in the two spaces has the same relationship, $P_C = -\tau P_E$, and the net power is $P = P_E + P_C = (1-\tau)Q$.

In the case of the machine acting as a prime mover $T_E > T_C$, i.e., $\tau < 1$, and the thermal efficiency

$$\begin{aligned} \eta &= (\text{heat supplied} - \text{heat rejected})/(\text{heat supplied}) \\ &= (Q-\tau Q)/Q = 1-\tau = (T_E - T_C)/T_E \end{aligned}$$

This corresponds to the Carnot efficiency.

When the machine acts as a refrigerator, $T_C > T_E$, i.e., $\tau > 1$, and

$$\begin{aligned} \text{coefficient of performance} &= \text{heat extracted/work done} \\ &= Q/(Q-Q_c) = 1/(1-\tau) \\ &= T_E/(T_E - T_C) \end{aligned}$$

For a heat pump, $T_C > T_E$, i.e., $\tau > 1$, and

$$\begin{aligned} \text{coefficient of performance} &= \text{heat rejected/work done} \\ &= Q_c/(Q-Q_c) = \tau/(1-\tau) = T_C/(T_E - T_C) \end{aligned}$$

This corresponds to the inverse thermal efficiency.

Mass Distribution in the Machine. From the characteristic gas equation,

$$M = pV/RT$$

where

$$p = p_{\text{mean}}(1-\delta^2)^{1/2}/[1+\delta\cos(\phi-\theta)]$$

(*a*) Expansion Space $V_e = \frac{1}{2}V_E(1+\cos\phi)$. The instantaneous mass of working fluid in the expansion space is given by

$$M_e = \frac{\frac{1}{2}V_E p_{\text{mean}}(1-\delta^2)^{1/2}(1+\cos\phi)}{[RT_E(1+\delta\cos(\phi-\theta)]} \tag{3.20}$$

The rate of change of mass of working fluid in the expansion space is

$$\frac{dM_e}{d\phi} = \frac{V_E p_{\text{mean}}(1-\delta^2)^{1/2}\{\delta[\sin(\phi-\theta)-\sin\theta]-\sin\phi\}}{2RT_E[1+\delta\cos(\phi-\theta)]^2}$$

(*b*) Compression Space $V_c = \frac{1}{2}\kappa V_E[1+\cos(\phi-\alpha)]$. The instantaneous mass of working fluid in the compression space is given by

$$M_c = \frac{\frac{1}{2}\{\kappa V_E p_{\text{mean}}(1-\delta^2)^{1/2}[1+\cos(\phi-\alpha)]\}}{2RT_c[1+\delta\cos(\phi-\theta)]} \tag{3.21}$$

The rate of change of mass of working fluid in the compression space is

$$\frac{dM_c}{d\phi} = \frac{\kappa V_E p_{\text{mean}}(1-\delta^2)^{1/2}\{\delta[\sin(\phi-\theta)+\sin(\alpha-\theta)-\sin(\theta-\alpha)]\}}{2RT_c[1+\delta\cos(\phi-\theta)]^2}$$

(*c*) Dead Space ($V_D = XV_E$, constant). The instantaneous mass of the working fluid in the dead space is given by

$$M_d = \frac{[XV_E p_{\text{mean}}(1-\delta^2)^{1/2}]}{RT_D[1+\delta\cos(\phi-\theta)]} \tag{3.22}$$

The rate of change of mass of working fluid in the dead space is

$$\frac{dM_d}{d\phi} = \frac{[XV_E p_{\text{mean}}(1-\delta^2)^{1/2}\delta\sin(\phi-\theta)]}{RT_D[1+\delta\cos(\phi-\theta)]^2}$$

Now $dM_c + dN_c + dM_d = 0$, so that the total mass of working fluid M_T is constant. Now

$$M_T = \frac{V_E p_{\text{mean}}(1-\delta^2)^{1/2}\{\tau(1+\cos\phi)+\kappa[1+\cos(\phi-\alpha)]+2S\}}{2RT_c[1+\delta\cos(\phi-\theta)]}$$

and when $\phi = 0$

$$M_T = \frac{V_E P_{\text{mean}}(1-\delta^2)^{1/2}[\tau+S+(\kappa/2)(1+\cos\alpha)]}{RT_C(1+\delta\cos\theta)} \tag{3.23}$$

Heat Extracted and Engine Output in Dimensionless Units. (a) The heat lifted per unit mass of working fluid, combining Eqs. (4.17) and (4.23) is given by

$$Q_{\text{mass}} = \frac{Q}{RT_C} = \frac{\pi\delta \sin\theta(1+\cos\theta)}{\{(1-\delta^2)^{1/2}[1+(1-\delta^2)^{1/2}][\tau+(\kappa/2)(1+\cos\alpha)+S]\}} \tag{3.24}$$

Similarly, the net engine-output per unit mass of working fluid is given by

$$P_{\text{mass}} = P/RT_C = (\tau-1)Q_{\text{mass}} \tag{3.25}$$

(b) Nondimensional expressions, in terms of characteristic pressures and volumes, may be devised as follows. The combined swept volume is given by

$$V_T = (V_E + V_C) = (1+\kappa)V_E$$

Combining this with Eqs. (3.1) and (3.13), then

$$\begin{aligned} Q_{\text{max}} &= Q/(P_{\text{max}}V_T) \\ &= [\pi(1-\delta)^{1/2}\delta \sin\theta]/\{(1+\kappa)(1+\delta)^{1/2}[1+(1-\delta^2)^{1/2}]\} \end{aligned} \tag{3.26}$$

and

$$P_{\text{max}} = (\tau-1)Q_{\text{max}} \tag{3.27}$$

The Finkelstein Adiabatic Cycle

Finkelstein (1960) devised a generalized thermodynamic analysis of Stirling engines in which the processes of compression and expansion were not confined to isothermal conditions. In this generalized analysis the processes of compression and expansion *in the engine cylinders* could be specified to occur anywhere between the two limiting cases: (a) isothermal (infinite rates of heat transfer between the working gas and the cylinder walls); (b) adiabatic (zero rates of heat transfer between the working gas and the cylinder wall).

The model assumed that, in the *freezer* and *cooler*, infinite heat transfer isothermal conditions prevailed so that fluid in the heat exchangers was *always* at the lower temperature T_E or the upper temperature T_C. The temperature of the working fluid in the cylinders varied during the cycle and could be greater or less than T_E (in the expansion space) or T_C (in the compression space).

Finkelstein's generalized analysis retained all the other significant assumptions of the Schmidt cycle so that it remained highly idealized.

Nevertheless, the possibility for nonisothermal processes represented the most significant theoretical development in nearly a century.

The limiting isothermal case of Finkelstein's theory corresponded exactly to the Schmidt cycle and the equations for this case reduced to a form identical with the Schmidt cycle equations given above. For the adiabatic limiting case, closed form solutions were not obtained nor for other intermediate cases, but the theory was readily amenable to numerical analysis by standard methods. Significant simplifications were introduced by the assumption of adiabatic conditions.

In his presentation, Finkelstein included only one set of numerical results. This referred to a cooling engine with the temperature $\tau = T_C/T_E = 2$. The coefficient of performance of 1.0 with isothermal processes was reduced to 0.543 with adiabatic processes. Similarly, Stoddart (1960) found that a Stirling engine prime mover having a Schmidt cycle (Carnot) efficiency of 50% with isothermal compression and expansion processes had an efficiency of only 34% with adiabatic processes.

Later, Khan (1962) investigated the effect of variation in the principal design parameters and obtained numerical results for a large number of different cases with adiabatic compression and expansion processes. Some of these results were summarized by Walker and Khan (1965).

In the adiabatic cycle, the thermal efficiency and coefficient of performance become functions, not only of temperature (as in the isothermal cycle) but also the swept volume ratio, κ, the phase angle, α, and the dead volume ratio, X. The power and heat extracted are functions of all these parameters both in the isothermal and adiabatic cases. The results presented by Walker and Khan provided some indications of the effects on efficiency of design parameters other than temperature. One unexpected result was that increase in dead space increased the thermal efficiency, even though the actual power output declined.

More recently, Lee (1976), has reconsidered the Finkelstein adiabatic cycle for both cooling engines and prime movers. This included the preparation of a fast, well-behaved computer program in FORTRAN IV language. The program, containing extensive documentation to facilitate understanding is reproduced in entirety as an appendix to Lee's thesis.*

Figure 3.29, reproduced from Lee (1976), shows the cyclic temperature variation of the working fluid in an adiabatic cycle Stirling cooling engine, having temperatures of 100 K in the freezer (T_E), and 300 K in the cooler (T_C), so that $\tau = 3$. The phase angle $\alpha = 105°$, swept volume ratio $K = V_C/V_E = 1.0$, and the dead volume ratio $X = V_D/V_E = 1.0$. It is interesting

* Copies of the thesis by Lee may be obtained through the usual channels or by direct application to Dr. G. Walker, Department of Mechanical Engineering, University of Calgary, Alberta, Canada. Payment of a reproduction and handling charge will be required.

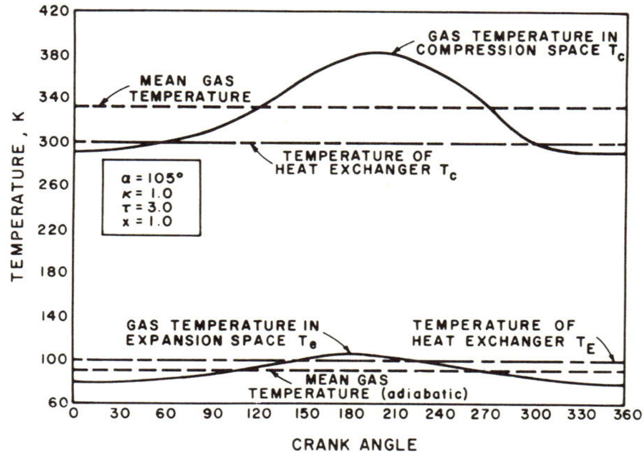

Fig. 3.29. Cyclic temperature variation in the compression and expansion space of a Stirling cryocooler operating on the Finkelstein adiabatic cycle (after Lee, 1976).

to note the temperature of fluid in the expansion space was less than the freezer temperature for much of the time. The mean temperature in the expansion space was 90 K, a difference of 10 K below the freezer temperature, T_E. In the compression space, the mean temperature of the fluid coincided more or less with the cooler temperature of 300 K.

Significant work using the adiabatic cycle approach was reported by Qvale and Smith (1969) and by Rios and Smith (1969). They consider a basic cycle with adiabatic compression and expansion and then separately assess the effects of irreversibilities. This approach allows for independent study of individual heat exchange components in a series of successive approximations that can be extended to the required degree of complexity.

Little has been published about theoretical work on Stirling engines at the Philips Company or their licensees. Mention is made in the various papers about the extensive use of computers and the ability to predict engine performance to one or two percent of test bed performance. Odd straws in the wind gathered over the years have led to the understanding that a family of computer programs have been developed by Philips. The ensemble of programs are thought to include some overall cycle simulation programs for the complete engine at various levels of sophistication and complexity, whereas others are available for the detail simulation and design of single components or subassemblies in the engine. The basis of the Philips thermodynamic analysis simulation program is thought to be similar to that developed by Smith, basically the Finkelstein adiabatic cycle with successive approximations and corrections. The adherence by the Philips Company over many years to similar design configurations has permitted the accumulation of extensive practical experience. On the basis of the experience, "fudge" factors of appropriate magnitude can be judiciously applied to the analytical results to provide close and realistic simulation.

Nodal Analysis

Nodal analysis of Stirling engines was pioneered by Finkelstein (1975) but other nodal analysis programs were prepared independently. Development was principally directed to prime movers, although the programs are equally applicable to cooling engines. In nodal analysis programs, the attempt was made to model the simultaneous energy and fluid flows occurring in the engine and thus simulate exactly the engine cycle and performance. This was achieved by writing and solving equations for the conservation of mass, momentum, and energy for particular nodes, cells, or elements of the engine. The equations were too complex for general analytical solution and were solved numerically in terms of small incremental time steps. The equations were invariably reduced to one-dimensional form with additional simplifications. A comparative discussion of the various programs has been given by Urieli (1979).

All the nodal analysis programs are basically similar in their general approach. The design of engine to be simulated must be known in exact detail to the extent that the mechanical arrangement, cylinder wall thickness and material, heat exchanger tube diameter, fin dimensions or matrix pore sizes, are all specified. This design is then broken down, as experience dictates, into a number of nodes, elements, or control volumes. Some operating conditions must be specified such as charge, pressure, and temperature of the energy source and sink.

Differential equations for the conservation of mass, momentum, and energy must be developed and generally converted to difference equations. Empirical formulas for the aerodynamic friction and heat transfer effects must be included as well as an equation of state for the working fluid. A mathematically stable method must then be found for numerical solution of the difference equations to resolve the pressure, temperature, and mass distribution in the engine at the end of a particular time step, given the conditions at the beginning of the step.

The usual procedure for solution is to start with some initial arbitrary assumed conditions and then proceed through several engine cycles until quasisteady state is achieved, when the instantaneous cyclic values of pressure, temperature, and mass distribution are not significantly different from the preceding cycle. The cyclic pressures and volumes are integrated to calculate work transfer, heat flows are estimated, and the overall thermal efficiency or coefficient of performance is computed. In most cases, an isothermal Schmidt cycle-type calculation is included at the beginning or end of the nodal analysis program.

The results are presented in a variety of formats. Typically the input data in terms of engine geometry and operating parameters are restated for comparison purposes. This is followed by tabulations at selected cyclic crank intervals of pressures, temperatures, velocities, mass content, and mass velocities at representation stations in the engine. Finally, considerations of the overall cycle are presented, usually in terms of energy flows, work terms, heat transferred, or ratios of these such as the thermal efficiency and coefficient of performance.

Most programs now incorporate automatic plot routines so that, if required, the data can be presented pictorially. Remarkable three-dimensional pictures can be produced in this way. Figures 3.30 and 3.31 were given by Schock (1978) and are typical of the interesting and informative representation achieved by computer plot routines. These particular diagrams show the variation in temperature with node position or volume as a function of the cyclic angle for a prime mover.

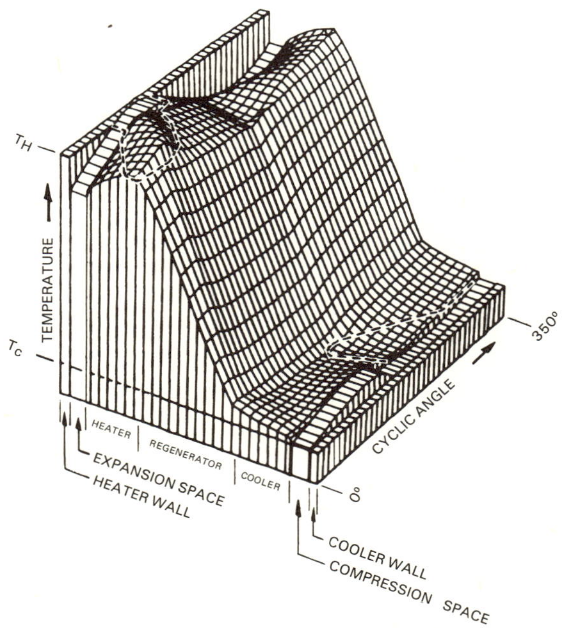

Fig. 3.30. Three-dimensional representation of the gas temperature versus node position and cyclic angle in a Stirling engine prime mover (after Schock, 1978).

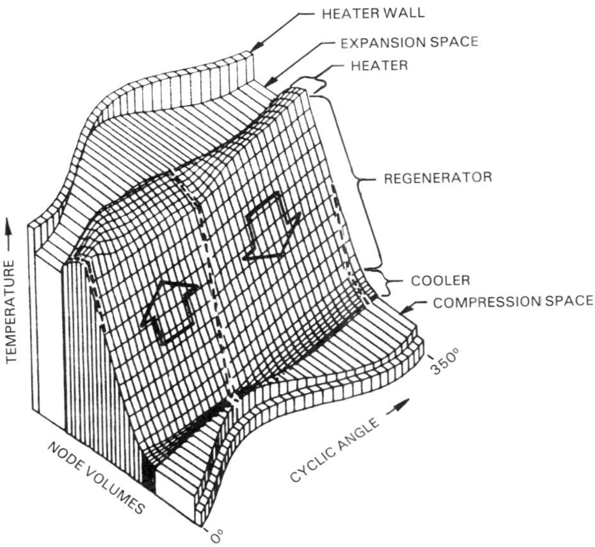

Fig. 3.31. Three-dimensional representation of the gas temperatures versus node volumes and cyclic angle in a Stirling engine prime mover (after Schock, 1978).

Finkelstein Nodal Analysis. After development of the adiabatic analysis, Finkelstein worked on increasingly sophisticated computer simulation of Stirling engines. This culminated in the late 1960s in adaptation of the NASA Thermal Analysis Program (TAP) to Stirling cycle simulation and the creation of the first nodal analysis program.

Its first practical application was at the University of Calgary in 1968/69 in support of the development of miniature cryocoolers for the British Ministry of Technology. Figure 3.32 shows a simplified cross section through the cooling engine. Figure 3.33 shows the two-dimensional representation of the engine used for the nodal analysis simulation. This work was briefly discussed by Finkelstein *et al.* (1970).

Following further refinement and development, Finkelstein (1975) presented the theoretical basis for the analysis and described the numeric differencing technique for solution of the partial differential equations obtained. The Finkelstein nodal analysis program is installed on a commercial computer network and available for general use on payment of a royalty fee. Potential users may obtain further details from Dr. T. Finkelstein.*

Urieli Nodal Analysis. Israel Urieli, working under the supervision of Professor Rallis at the University of Witwatersrand, provided the most

* TCA Inc., P.O. Box 643, Beverly Hills, California.

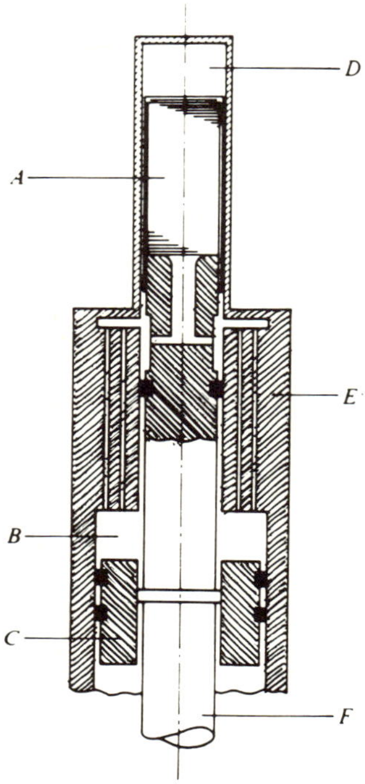

Fig. 3.32. Simplified cross section of Stirling cryocooler (after Finkelstein *et al.*, 1971).

complete discussion (Urieli, 1977) of Stirling engine simulation by nodal analysis. His thesis is recommended for all those professionally interested in the field of Stirling engine computer simulation. A good outline of the approach, more readily available than copies of the thesis, was provided by Urieli *et al.* (1977) and Urieli (1979). The thesis contains a complete program listing in FORTRAN of the Urieli program. The program is likely to be applicable to engines other than the model assumed for investigation by Urieli, but a very substantial effort would be required to install a fully operational version of the Urieli program.

Work on the nodal analysis field continues at the University of Witwatersrand under the direction of Professor Rallis with experimental work to validate the computer model and investigation of alternative theoretical models (Berchowitz *et al.*, 1977). Further information may be obtained from I. Urieli* and Professor C. Rallis.†

Sunpower Nodal Analysis. Gedeon (1978) outlined the techniques of Stirling engine simulation in use at Sunpower of Athens, Ohio, manufacturers of Beale free-piston Stirling engines. Numerical simulation is an

* Ormat Turbines, P.O. Box 68, Yavne, Israel.
† Department of Mechanical Engineering, University of Witwatersrand, South Africa.

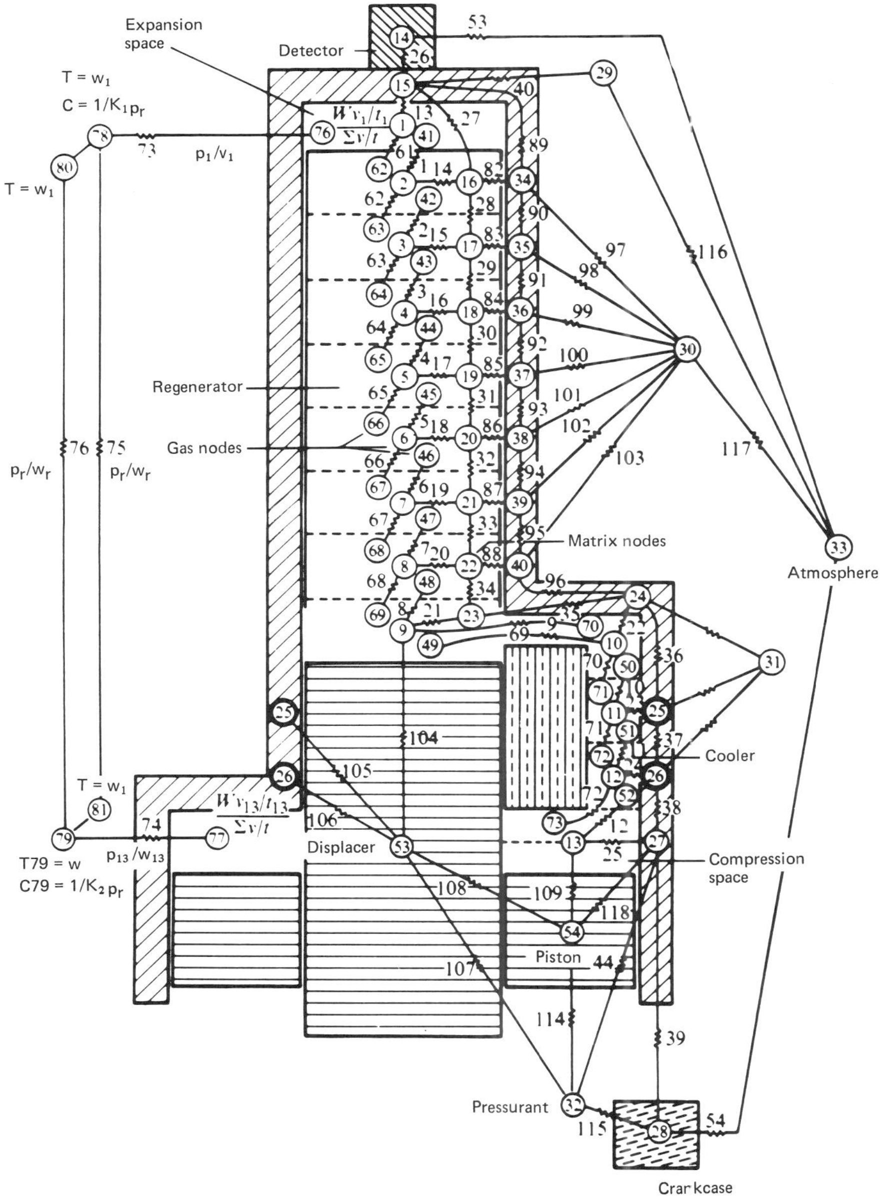

Fig. 3.33. Two-dimensional representation of Stirling cryocooler used for nodal analysis (after Finkelstein *et al.*, 1971).

integral part of the development approach at Sunpower. Engine refinement and program validation proceeds hand in hand with routine daily exchanges between the staff engaged in laboratory and theoretical work.

The Sunpower program is fast and fully validated. It is a complex of programs capable of cycle simulation at different levels of sophistication as selected by the program controller. The program is self-optimizing to the degree specified in the program input. In the field of Stirling engines, Sunpower has accumulated the greatest body of practical engine development and operating experience, supported by computer simulation, outside of Philips and their licensees. It is, therefore, gratifying that the Sunpower simulation programs are available for general use on a contract basis. Active moves are afoot (Beale, 1979) to install the program on a commercial computer network and organize educational courses in the use of the programs. Further details of Sunpower Stirling engine development may be gained from William Beale.*

Other Nodal Analyses. Schock (1978) described the Stirling Nodal Analysis Program (SNAP) prepared by Fairchild Industries under contract to the U.S. Department of Energy in support of a Beale free-piston Stirling engine development. The Schock program is closely similar to the Urieli program. It is well described in a draft report (Schock, 1978a). An abbreviated description may be found in the paper (Schock, 1978b) contained in the proceedings of the 1978 Intersociety Energy Conversion Engineering Conference. Details of the status of the program may be obtained from Alfred Schock.†

Roy Tew and others at the NASA Lewis Research Center, Cleveland, Ohio, have developed a Stirling engine nodal analysis program as part of the DOE/NASA Stirling Engine Automotive Program. The description of the Lewis program was given by Martini (1978). Experimental studies are being carried out at the Lewis Research Center (Cairelli *et al.*, 1977). The final answer to computer simulation and experimental validation will eventually be placed in the public domain by the team at NASA Lewis Research Center. Readers are urged to consult the current literature relating to the DOE/NASA Stirling Automotive Engine. Further details may be obtained on application to Robert Ragsdale.‡

Finegold and Vanderbrug (1977) outlined a nodal analysis program for Stirling engine simulation developed at the Jet Propulsion Labroaratory, Pasadena, California. The program is explained and completely listed in a 42 page appendix to the final report discussing Stirling engines for under-

* President, Sunpower Inc., Athens, Ohio.

† Energy Systems Department, Fairchild Space and Electronics Co., Germantown, Maryland.

‡ Stirling Engine Project Office, NASA Lewis Research Center, 21000 Brookpark Road, Cleveland, Ohio.

water applications. Present status and potential applications for the computer program are not known.

Dr. Allen Organ, formerly of Kings College, University of London, now at the School of Engineering, University of Cambridge, has recently published papers indicative of a powerful nodal analysis program in formation, but not yet fully developed. Readers are advised to contact Dr. Organ directly for up-to-date information.

Several papers have hinted at the existence and use of nodal analysis type programs at Philips and their licensees. No details have been given. It is understood that for general design work, adiabatic cycle programs are preferred over the more sophisticated nodal analysis programs.

Summary

Stirling engine simulation by nodal analysis is an expensive, time-consuming activity to be reserved for those professionally engaged in Stirling engine development or academics engaged in the training of engineers at the graduate level.

At present, no comprehensive evaluation is available for the various Stirling engine simulation programs described above. Martini (1978) attempted the preparation of a design manual including a comparison of the various theoretical procedures, but concentrated his efforts principally on isothermal type analyses with only passing reference to the more sophisticated nodal analyses. He later extended this work to higher-level simulation programs (1980).

The program of Urieli and Finegold–Vanderbrug are completely listed in the references cited above. However, neither of these programs has been extensively validated by experimental data. The programs by Schock and by Tew are neither listed nor have been validated experimentally. Neither program is available commercially. Finkelstein's program is available commercially on payment of moderate royalties, but no documentation of validity with reference to practical engines is available. The Sunpower program is available commercially and has been validated extensively during the development of Beale-type free-piston engines. The applicability of the program to engines with crank mechanisms (disciplined-piston engines) has also been established recently.

For those wishing to obtain a simulation or optimization of a Stirling engine concept, the choice is between Sunpower and Finkelstein. Another possiblity is that Philips could offer consultancy services to evaluate new concepts using their undoubted expertise and experience in the field.

It is simply out of the question to "dabble" in nodal analysis even using the existing programs. The few companies and individuals wishing to operate this high level of simulation must be prepared to invest substantial time and money to establish the program procedure and to accomplish the work. Obtaining results is easy. Understanding them and appreciating the implications is beyond the level of dilettanish efforts.

For those unable to indulge in nodal analysis, close study of the excellent manual by Martini (1978) and its successor is recommended. Martini offers a commercial Stirling engine simulation service using programs at various levels of sophistication. Further information may be obtained on application to Dr. William Martini, 2303 Harris, Richland, Washington.

The adiabatic cycle with the appropriate thermo/fluid corrections is suitable for most design and optimization of Stirling engines. The adiabatic cycle requires use of digital computers, but not to an unconscionable degree, and these are so readily available that their use is routine to most engineers. The adiabatic cycle is a good deal better than the isothermal case and requires only minor computing cost and time compared with nodal analysis. Readers are referred to the thesis by Lee (1976), for a program listing, and to the works of Rios, Qvale, and Smith referenced below. At yet earlier stages in design, the isothermal analyses are adequate. Martini (1978) has developed programs suitable for analysis using pocket calculators.

At the start of a project, it is usually sufficient to carry out simple Schmidt-cycle calculations. To attain the likely performance of a practical engine, one simply divides the coefficient of performance and heat extracted by 2 (if an optimist) or 3 (if a realist).

PRACTICAL DESIGN

Design Parameters

The principal independently chosen design parameters for Stirling cooling engines are as follows:

1. The temperature ratio $\tau = T_C/T_E$, the ratio of temperatures in the compression and expansion spaces.
2. The swept volume ratio $\kappa = V_C/V_E$, the ratio of swept volumes in the compression and expansion spaces.
3. The dead volume ratio $X = V_D/V_E$, total internal volume of heat exchangers and associated ducts expressed as a fraction of the swept volume in the expansion space.
4. The phase angle α by which volume variations in the expansion space *lead* those in the compression space.

5. The pressure of the working fluid expressed as the maximum or mean pressure, p_{max} or p_{mean}.
6. The speed of the engine N in cycles per unit time.
7. The bore and stroke of the reciprocating member in the expansion and compression spaces.
8. The number of Stirling engine systems in the total machine, i.e., single or multiple systems.
9. For multiple systems the unit may be a series of single-acting Stirling engines operating independently or a double-acting arrangement with all systems coupled.
10. The method of activating the piston and displacer action, i.e., free piston or kinematic drive.
11. The design configuration, integral Stirling, split Stirling, two-piston machines, fixed or moving regenerator, type of kinematic drive, crank-connecting rod, Scotch yoke, rhombic drive, swash-plate, wobble-plate, etc.

The Schmidt cycle is useful to explore the effects of variation in the first six parameters. The remainder are decisions in the bailiwick of the mechanical designer. Selections are made on the basis of familiarity, past experience, constraints imposed by the sponsor (including weight, volume, cost, vibration), cooling system, and material considerations; or progressive optimization involving sophisticated nodal analysis of alternative outline designs.

Optimization of Design Parameters

In the Schmidt analysis presented above, the heat extracted in the expansion space per cycle was found to be [Eq. (3.17)]

$$\frac{Q_E}{p_{max} V_T} = \frac{\pi(1-\delta)^{1/2}\delta \sin\theta}{(1+\kappa)(1+\delta)^{1/2}[1+(1-\delta^2)^{1/2}]}$$

where

$$\theta = \tan^{-1}[(\kappa \sin\alpha)/(\tau+\kappa+2S)]$$

$$\delta = (\tau^2+\kappa^2+2\tau\kappa\cos\alpha)^{1/2}/(\tau+\kappa+2S)$$

$$S = (2X\tau)/(\tau+1)$$

Further

$$Q_C = \tau Q_E$$

$$W = (\tau-1)Q_E$$

$$\mathrm{COP} = 1/(\tau-1)$$

From this it is easy to see the refrigerating capacity is a linear function of (a) the engine operating frequency N, (b) the maximum pressure of the working fluid p_{max}, and (c) the size of the engine expressed in terms of characteristic volume V_E or the combined swept volumes $V_T = (V_C + V_E)$. According to the Schmidt theory, to double the refrigerating capacity one simply doubles the frequency, the maximum cycle pressure, or the combined swept volumes, i.e., uses a bigger machine. Alternatively, a combination of increased speed, pressure, and size might be used.

In practice, the results are not as straightforward. Increase in the pressure and speed will certainly increase the refrigerating capacity, but the improvement occurs at a progressively diminishing rate because of increasing friction losses and thermal saturation of the matrix. Increase in speed or pressure to excess actually decreases the available refrigeration.

The effect on performance of the four principal design parameters (τ, κ, α, and χ) is less obvious. It is not clear which combinations should be used to achieve the optimum performance. This is important, for these parameters are determined at the design stage and, except for temperature, τ, cannot be readily changed without structural change to the engine.

The temperature ratio τ is fixed by the refrigeration duty (T_E) and the temperature of the cooling available (T_C). The refrigeration temperature T_E is the temperature achieved when the refrigeration load, including the various heat leaks to the engine, exactly balance the refrigerating effect generated. Lower temperatures can be achieved if necessary by increasing the pressure and speed of the engine, thereby increasing the available refrigerating effect.

Figure 3.34 shows the effect of variation in the expansion space temperature on the heat extracted, $(Q_E/p_{max}V_T)$, and the input work per cycle $(P/p_{max}V_T)$ calculated using the Schmidt theory for a case where all

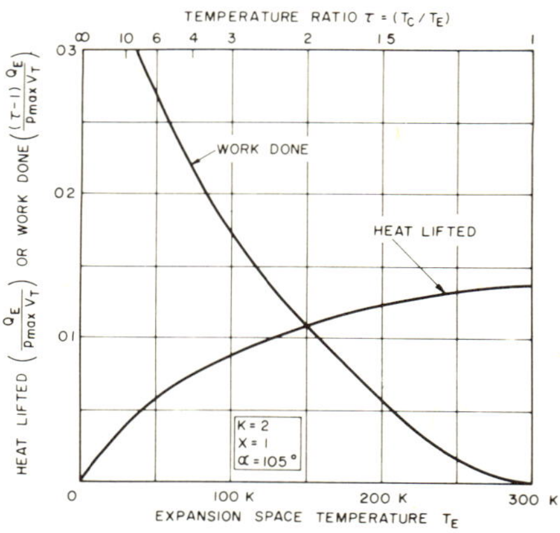

Fig. 3.34. Effect of variation in expansion space temperature T_E on heat extracted, input power and coefficient of performance with constant values of κ, X, and α.

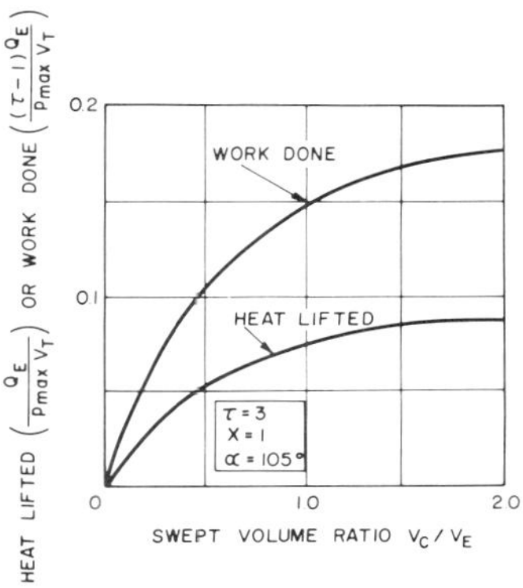

Fig. 3.35. Effect of variation in the swept volume ratio κ on the heat extracted $(Q_E/p_{\max} V_T)$ with the constant values of τ, X, and α.

the other design parameters were maintained constant, i.e., $X = 1, \alpha = 105°$, $\kappa = 2$ and with $T_C = 300$ K. As the expansion space temperature decreases, the heat lifted decreases, the input work increases and the coefficient of performance decreases. This is perfectly in line with our expectations from previous experience with the Carnot cycle. The Schmidt cycle has the same thermodynamic performance as the Carnot cycle since it has the same isothermal heat transfer to and from the cycle.

Figures 3.35, 3.36, and 3.37 show the effect on heat lifted $(Q_E/p_{\max} V_T)$ and the work done $(P/p_{\max} V_T)$ as a result of independent variation of one of the three parameters κ, α, and X with the other two maintained constant and with τ maintained at 3 ($T_E = 100$ K, $T_C = 300$ K) for all cases. The coefficient of performance is constant since it depends only on τ.

Figure 3.35 shows the effect of variation in the swept volume ratio k. For a given combination of τ, X, and α, there is a definite optimum value

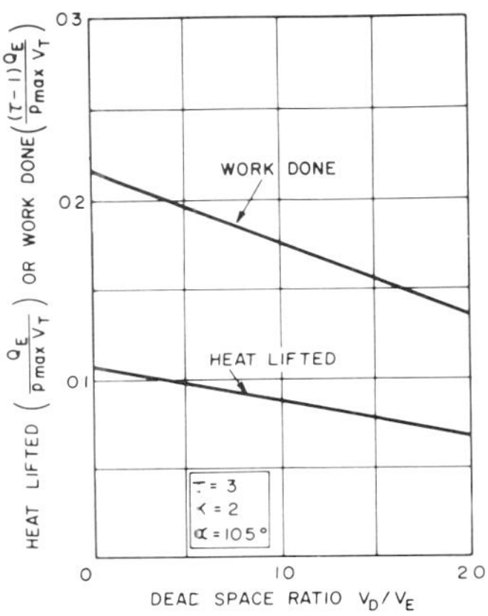

Fig. 3.36. Effect of variation in the dead space ratio X on the heat extracted $(Q_E/p_{\max} V_T)$ with constant values of τ, α, and κ.

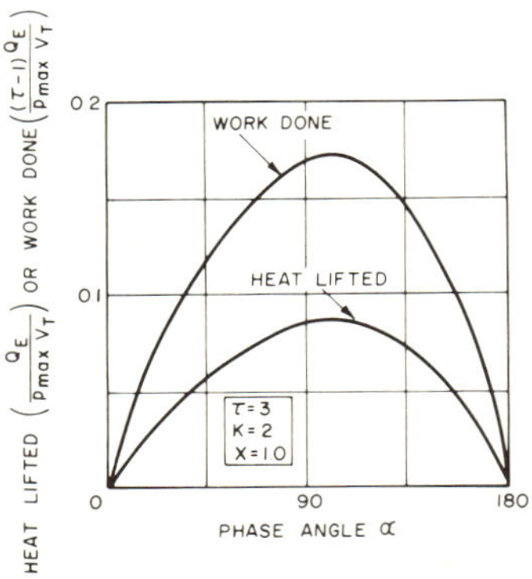

Fig. 3.37. Effect of variation in the phase angle α on the heat extracted $(Q_E/p_{max}V_T)$ with constant values of τ, X, and κ.

of the swept volume ratio V_C/V_E. Change in the temperature ratio τ produces a change in the refrigeration capacity and in the optimum swept volume ratio. Increase in the value of τ (decrease in the expansion space temperature) produce a *decrease* in the heat extracted and an increase in the optimum swept volume ratio.

Figure 3.36 shows the effect of variation in the dead space ratio, X. Increase in the dead space causes a decrease in the heat lifted so the dead space volume must be minimized for high refrigerating effects. In practical engines, the situation is more complicated with increased dead space. A positive improvement in refrigeration performance can result, provided the increased dead space is effectively utilized as increased void volume in the heat exchangers. This results in increased thermal capacity or bigger flow passages with reduced fluid friction losses. It is vital that the dead volume be limited to effective space and not wasted in oversized ducts, large ports, and excessive clearance space.

Theoretical deterioration of refrigerating capacity with increase in dead volume can be understood by considering the effect on the volume compression ratio (V_{max}/V_{min}). Increases in dead space decreases the volume compression ratio and the pressure compression ratio (p_{max}/p_{min}). Reduction in the pressure ratio results in decreased area of the expansion space work diagram and a smaller refrigerating capacity.

Figure 3.37 shows the effect of variation in the phase angle α. The refrigerating capacity is remarkably insensitive to changes in the phase angle from 60° to 120° of crank rotation, but does exhibit an optimum value around 90° at which the heat lifted is a maximum.

For any given combination of temperature ratio, τ and dead space, X, three-dimensional surfaces may be drawn as shown in Fig. 3.38. This illustrates the heat lifted $(Q_E/p_{max}V_T)$ as a function of swept volume ratio κ and the phase angle α. Different combinations of τ and X result in the

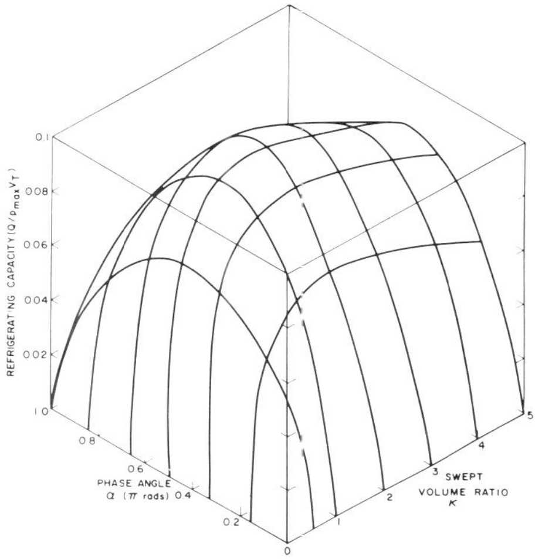

Fig. 3.38. Three-dimensional representation of refrigerating capacity ($Q_E/p_{\max}V_T$) as a function of swept volume ratio κ and phase angle α for a given combination of temperature ratio ($\tau = 3$) and dead volume ratio ($X = 1$).

generation of different surfaces, sitting one upon the other like onion skins. The apex of the surface in Fig. 3.38 is the maximum refrigeration capacity achievable with the given combination of τ and X. The values of κ and α corresponding the apex of the surface are the optimum values necessary to achieve the maximum refrigeration.

Consolidated Design Chart

Construction of surfaces such as Fig. 3.38 provides a visual demonstration of the interdependence of the refrigeration capacity on various design parameters. It would be tedious indeed to resolve optimum combinations of parameters by such diagrams. The advent of computers made the task of optimization straightforward. Optimal combinations for several hundred cases were determined (Walker, 1962) and consolidated design charts prepared for both prime movers and refrigerating machines. The chart for refrigerating machines is reproduced in Fig. 3.39. The chart for engines is reproduced elsewhere (Walker, 1980).

The design chart is in three sections, having a common abscissa of expansion space temperature T_E (T_C is assumed to be constant at 300 K). The corresponding scale for the temperature ratio τ is shown at the top of the diagram. The upper diagram shows the refrigerating capacity as a function of expansion space temperature T_E for different values of the dead space, X. The center diagram shows the optimum phase angle α (units of the α ordinate are π radians) as a function of expansion space temperature T_E for the same several dead space ratios. The lower diagrams shows the

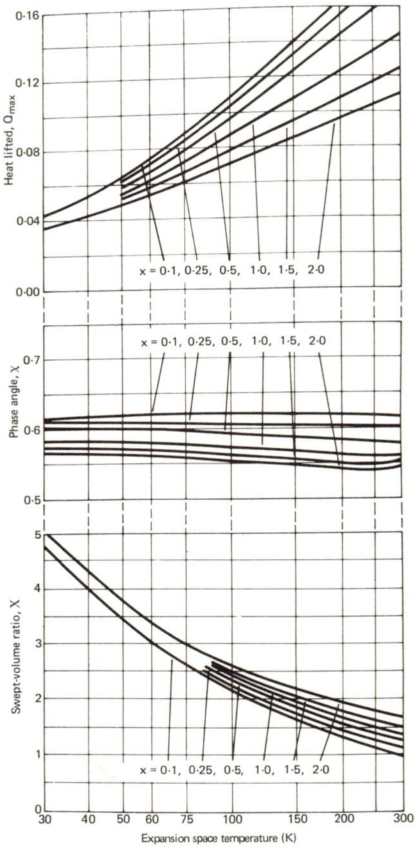

Fig. 3.39. Consolidated design chart for Stirling cryocoolers based on Schmidt theory.

optimum swept volume ratio κ as a function of expansion space temperature, T_E for the different values of the dead space, X.

The maximum refrigeration decreases linearly with expansion space temperature and is a strong function of the dead space ratio, X, falling progressively as X increases. The optimum phase angle is independent of both dead space and temperature effects. The optimum swept volume ratio κ is more or less independent of the dead space but is a strong function of the expansion space temperature. As the temperature decreases, the optimum swept volume increases at a progressively increasing rate from approximately unity near ambient temperature to 5 or so at cryogenic temperatures. This shows that at lower temperatures, a bigger compressor with increased power input is necessary to achieve a smaller refrigerating capacity.

To use the chart, one simply determines the expansion space temperature at which refrigeration is required. Through this point, a vertical line is drawn crossing the three diagrams. For different values of the dead space, X, one can read off values of the maximum refrigeration capacity, the optimum phase angle and swept volume ratio to achieve capacity. The actual refrigerating capacity will be 30% to 50% of the indicated capacity.

Alternative Basis for Optimization

The design chart, Fig. 3.39, was prepared for the optimum design in terms of heat lifted Q_E expressed in dimensionless units based on the maximum pressure (p_{max}) and the combined swept volume (V_T). This is the preferred basis for optimization because the maximum pressure is indicative of the weight of an engine and the combined swept volume (V_T) is indicative of the size of an engine. Thus, optimization generates an engine design having the maximum refrigeration capacity for a given size and weight.

There are alternative bases for optimization. The heat extracted can be expressed per unit mass of working fluid (Q_E/mRT) [Eq. (3.25)], where m is the mass of working fluid. Thermodynamic cycle calculations are often calculated on the basis of unit mass of working fluid so the parameter appears as a familiar and suitable basis for optimization.

However, when applied to Stirling engines, an optimization based on unit mass of working fluid results in an engine design where the working fluid is limited. The design is grossly distorted to achieve the maximum refrigerating effect *from a small mass of working fluid*. It is clearly not the proper basis for optimization. There are other bases, equally false, explored in some depth by Walker (1962).

Once the combination τ, X, α, and κ has been established, it is just as valid to use (Q_E/mR_T) as ($Q_E/p_{max}V_T$). They produce identical values of Q_E. The danger of a distorted design only arises when *optimization* studies are in progress.

Machine Design Procedure

The consolidated design chart was prepared using Schmidt theory based on an idealized thermodynamic cycle not attainable in practice. Use of the chart should be confined to the preliminary layout stages of design. It will provide a handy basis for starting the design and the calculation of approximate performance.

Various aspects of mechanical and thermal design are discussed in Chapter 9. If these guidelines are followed, a first estimate of actual indicated refrigerating capacity might be obtained by dividing the chart value $(Q_E/p_{\max}V_T)$ by 3 (this decrease largely arising from adiabatic rather than isothermal processes). The corresponding indicated input cycle work might be estimated by assuming $W = 1.2\tau(Q_E/p_{\max}V_T)$.

Now the various thermal losses can be estimated and can be debited against the initial estimate of the indicated refrigerating value to provide a first guess at the net available refrigerating capacity. For a machine operating in thermal balance, the equality of thermal losses to indicated refrigerating capacity will allow a preliminary estimate of the operational refrigerating temperature.

The indicated cycle work must be increased to allow for mechanical and fluid friction losses in the kinematic drive and rubbing friction of seals and guides on the pistons or crossheads. The shaft input is likely to be anywhere from 20% to 80% greater than the indicated cycle work depending on the size of the machine, the ingenuity and skill of the design engineer, and the quality of the final manufactured product.

Another important factor is the efficiency of the input power drive motor. Electric drive motors are used, both alternating and direct current, with and without brushes. In large sizes, the efficiency of these motors may approach 90%. In the smaller sizes used for low capacity coolers, the efficiency rarely exceeds 60% and is often substantially less.

Various design layouts and machine configurations can be quickly evaluated and preliminary performance estimates made. When a suitable design has evolved which conforms to the constraints imposed by the application or client, one can proceed to the next level of analysis.

The second stage of design would likely involve analysis according to the Finkelstein adiabatic cycle using digital computers, but at a relatively low level. Supplementary corrections for fluid flow and thermal losses can be made using the methods of Martini or Smith, Rios, Qvale, and Rea mentioned above. Various revisions to the outline design may be made at this stage to optimize the system according to the application.

Finally, a detail design may be prepared and analyzed using the nodal analysis programs discussed above. A relatively coarse network may be specified initially to minimize computer processing time with progressive elevation of the level of sophistication as the design approaches the ultimate stage.

This is an idealized procedure for machine design rarely followed in practice. Time and budget do not allow comprehensive exploration of design alternatives. Many machine designers have "blinkers" which inhibit their disinterested assessment of new and better ways to go. Sponsors rarely have an adequate basis of technical expertise to critically assess their contractors' work. Many times they naively assume that, even though the contractor is a recent entrant to the field, he will go straight to the optimum design. There is a great divide between computer and hardware specialists. Too often, computer simulation becomes an end in itself providing endless delight to those pursuing it but contributing little to the machine design. Designers cannot understand the acronymic sublanguage of the computer specialist and system analysts often have unrealistic expectations of engineering materials and production processes.

Nothing impresses a sponsor so much as working hardware. Frequently, contract renewals depend on the existence and demonstration of a prototype. Therefore, the project manager is anxious to start cutting metal at the earliest possible stage and is impatient with the design and simulation types who, in his terms, "frig about" unable to make up their minds. Eventually he insists the design be frozen at some arbitrary level and production started. Thereafter, various small changes are possible with all the attendant red tape and paperwork but substantial design modifications cannot be made. Frequently, man-years of effort are devoted to the engineering development of an inferior design or improperly evaluated concept.

TYPES OF STIRLING ENGINES

Multiple-Expansion

The possiblity of an engine with multiple-expansion spaces coupled to a single compressor was discussed above. The expansion process in all spaces occurs simultaneously and at the same pressure. In electrical engineering terms, this is "parallel expansion." There also exists the possibility of "series expansion" at different temperature levels in a double (triple or quadruple) expansion engine.

The principle of the double-expansion engine is illustrated in Fig. 3.40. This shows the cross section of an integral Stirling single-acting piston-displacer engine. The regenerator is contained within the displacer and moves with it. The displacer–regenerator is arranged in two sections having different diameters, with the top section substantially smaller than the lower section. The stepped displacer operates in a cylinder which has corresponding changes in diameter to fit the displacer.

Two expansion spaces are thereby created, namely, the customary expansion space above the displacer, and an intermediate or "extra" expansion space as shown in Fig. 3.40. The volume variations of the two expansion spaces are in phase with each other. The two spaces are mutually coupled through the porous regenerators and with the compression space

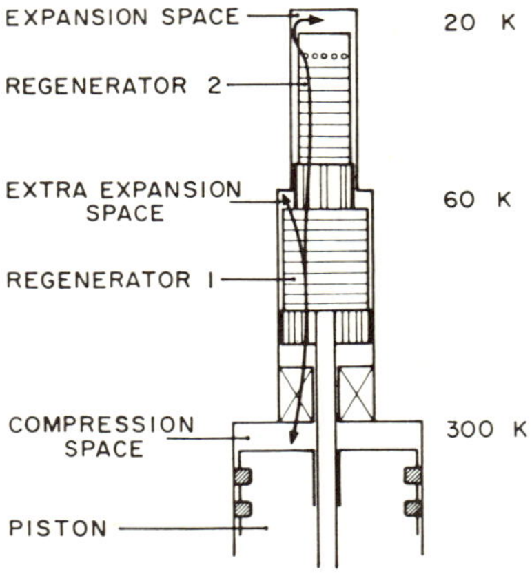

Figure 3.40. Double-expansion Stirling engine (after Lindale, 1978).

so the pressure in all spaces is the same, apart from slight differences due to regenerator pressure drop.

When the machine operates, it produces refrigeration at two temperature levels, shown in Fig. 3.40 as 60 K in the intermediate expansion space and 20 K in the upper expansion space. The generation of intermediate temperature refrigeration creates a "thermal buffer" to absorb heat leaks from ambient temperature sources and thereby protect the refrigeration generated at the lower temperature. The heat leak to the upper expansion space is now much reduced for the temperature potential between the buffer and low-temperature expansion space is only 60–20 K. The drain of low-temperature refrigeration is correspondingly reduced. Space limitations preclude a full discussion here of thermodynamic aspects of multiple-expansion engines, but the same principle of cascade refrigeration at progressively lower temperatures is widely used in large-scale recuperative gaseous liquefaction systems.

The pioneer work in double-expansion Stirling engines was carried out at Philips in the early 1960s (Prast, 1963) using a modified air liquefier of the type shown in fig. 3.1 with additional expansion space. With this machine, temperatures of 20 K were readily achieved. It has been incorporated in a variety of hydrogen and helium liquefiers and other low-temperature refrigeration systems.

Later on, Daniels and du Pre (1971) described a miniature triple-expansion Stirling engine which attained a temperature of 7.8 K. A diagrammatic cross section of the engine is shown in Fig. 3.41 and the multi-element displacer in Fig. 3.42. Figure 3.43 is a view of the prototype unit with the actual cylinder head. The tangle of tubes and wires around the cylinder head are part of the apparatus for measuring the expansion

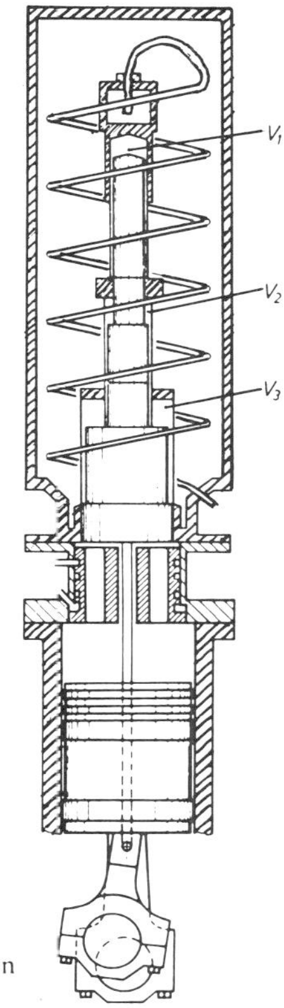

Fig. 3.41. Diagrammatic cross section of a miniature triple-expansion Stirling engine (after Daniels and du Pre, 1971).

space temperature using a hydrogen vapor-pressure technique with copper constantan thermocouples for the intermediate temperatures.

The cool-down characteristic of the engine is shown in Fig. 3.44. This shows the temperatures of the three expansion spaces as a function of time. The engine was relatively slow to cool down (about 1 hr) because of the low pressure and speed. The relatively large refrigerating capacity of the first stage enabled it to pull down the temperature of is expansion space very quickly (about 20 min to reach 60 K). The smaller second stage lagged behind in temperature quite considerably but was still substantially ahead of the small third stage most of the time. Eventually in "hare and tortoise" fashion the third stage passed both the other spaces to reach the lowest temperature. Part of the reason for the slow cool-down was the relatively large mass at the low-temperature end associated with the instrumentation. A production machine would be cleaner and have a much reduced cool-down period.

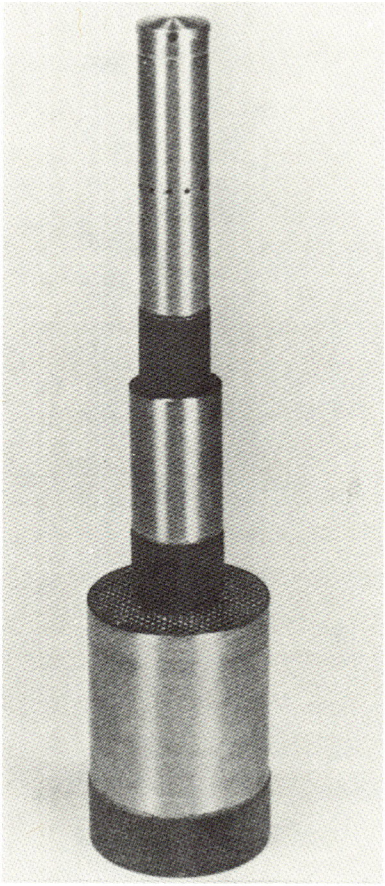

Fig. 3.42. Multielement regenerative displacer of the triple-expansion Stirling engine (after Daniels and du Pre, 1971).

Multiple-expansion systems have been used with other regenerative engines, i.e., Vuilleumier and Gifford–McMahon machines, with much the same result. Zimmerman and Radebaugh (1977) described a multiple-expansion engine, shown in Fig. 1.3, constructed of cheap, readily available materials that may be the forerunner of a low-cost, maintenance-free, cryocooler for widespread use in superconducting electronic applications.

Free-Piston

This already long chapter could be greatly extended by proper consideration of free-piston Stirling engines. In these machines, the pistons

Fig. 3.43. Laboratory prototype of miniature triple-expansion Stirling engine (after Daniels and du Pre, 1971).

and displacers are not coupled through a kinematic chain (crank and connecting rod) but are activated solely by fluidic forces acting on them.

Free-piston Stirling engines are in rapid development as prime movers and, to a lesser extent, as cryocoolers (Haarhuis, 1978; de Jonge, 1979). A comprehensive discourse on Stirling free-piston engines is contained in

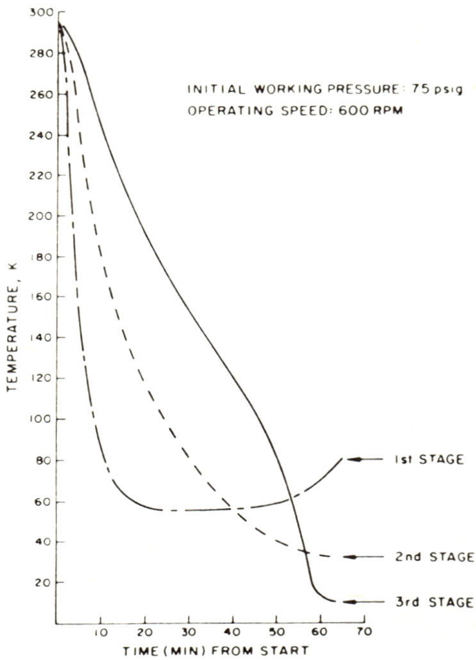

Fig. 3.44. Cooldown characteristics of a miniature triple-expansion Stirling engine (after Daniels and du Pre, 1971).

Chapter 11 (contributed by William Beale) in *Stirling Engines* (Walker, 1980). An increasing number of technical papers about Stirling free-piston engines are appearing in the technical literature. Enough material exists for a monograph on the topic planned to succeed this present work.

One type of free-piston engine, called the Beale free-piston Stirling engine, after the originator William Beale of Sunpower, Athens, Ohio, is shown in Fig. 3.45. The engine consists essentially of a lightweight displacer and a heavy, massive, piston contained in a single cylinder in the arrangement classified above as a single-acting integral Stirling system. The expansion space is above the displacer and the compression space is between the piston and displacer. The two spaces are coupled through a regenerator.

In a cryocooler, work input is necessary to drive the engine. This may come from a pulsating hydraulic actuator or, as in Fig. 3.45, from a linear motor providing an electromagnetic force on the piston.

The space below the piston, called the "bounce space," acts as a pneumatic spring and contains the same gaseous working fluid used in the Stirling engine above the piston. Application of an electromagnetic force on the piston by the linear motor will cause the piston to descend, thereby compressing gas in the bounce space. Release of electromagnetic force will allow the compressed gas in the bounce space to return to the piston.

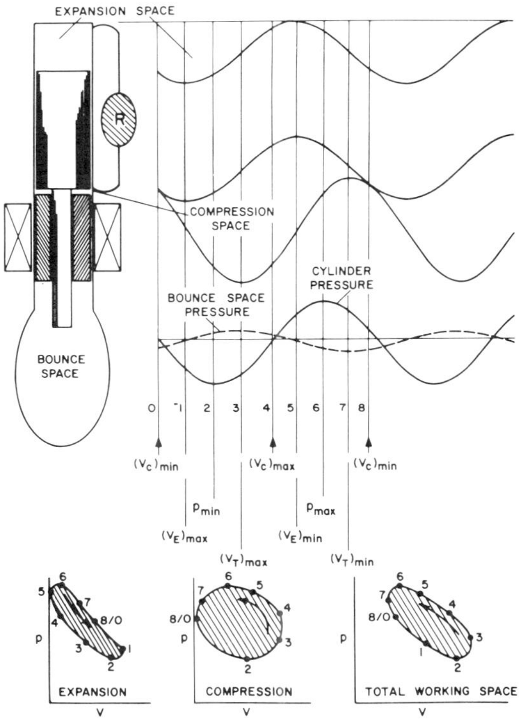

Fig. 3.45. Beale free-piston integral Stirling engine.

Application of the force at the resonant natural frequency of the piston/bounce space will cause the piston to oscillate with simple harmonic motion in the same way a cork bobs up and down in water or a weight suspended from a spring.

Oscillations of the piston in the cylinder will cause the pressure to vary in harmonic fasion. If the diplacer were not included, the space above the piston woudd act simply as another bounce space. The pressure variation above and below the piston would be exactly out of phase by one half cycle (180°) and vary at the same frequency as the piston oscillation.

In order to produce refrigeration, it is necessary to include the displacer and regenerative heat exchanger as shown in Fig. 3.45. The displacer is equipped with a large-diameter displacer rod passing right through the piston and subject to the pressure of fluid in the bounce space across the whole area of the displacer rod. An equivalent area at the top of the displacer is subject to the pressure of the working fluid in the expansion space. The net fluidic force acting on the displacer is equal to the difference between the pressure in the bounce space and the working space.

The *mass ratio* of piston to displacer and the *area ratio* of displacer rod area to cylinder cross section area are critical to successful operation. When these are in proper proportion, the displacer will be found to oscillate in the cylinder as shown in Fig. 3.45, at the same frequency as the piston but with a phase difference about 80° in advance of the piston. The motion of the displacer in the cylinder causes the working fluid to move back and forth between the expansion and compression spaces. The pressure is decreasing when the displacer is closest to the piston. Thus, there is a concentration of fluid in the expansion space and heat is abstracted from the surroundings of the expansion space at low temperatures. Similarly, the fluid is concentrated in the compression space when the pressure is increasing; the displacer is widely separated from the piston and the piston is rising on the compression stroke.

Figure 3.45 contains displacement–time curves for the piston and displacer and pressure–time curves for the working space and bound space. Typical pressure–volume diagrams are also included for the expansion, compression, and total working spaces. On Fig. 3.45, the numbers 0 through 8 correspond to critical points of the cycle of operation in accordance with the following key:

0—Minimum compression space volume $V_{C(\min)}$. The piston and displacer are in contact.
1—Maximum expansion space volume $V_{E(\max)}$. The displacer is at the bottom of its stroke.
2—Minimum pressure in the working space $p_{\min}$.
3—Maximum system volume $V_{(\max)}$. The piston is at the bottom of its stroke and the pressure in the bounce space is a maximum.
4—Maximum compression space volume $V_{C(\max)}$. The displacer and piston are at the point of maximum separation.
5—Minimum compression space volume $V_{E(\min)}$. The displacer is at the top of its stroke.
6—Maximum pressure in the working space, $p_{\max}$.
7—Minimum system volume $V_{(\min)}$. The piston is at the top of its stroke and the pressure in the bounce space is a minimum.
8—Minimum compression space volume $V_{C(\min)}$. This is a return to the same condition as 0 and the completion of the cycle.

Many alternative arrangements and configurations are possible. Some of these are briefly considered by Beale in the quoted reference (Walker, 1980).

This free-piston family also includes the resonant liquid-piston engines exemplified by the Harwell Fluidyne engine (West, 1971). A number of research papers on these engines were presented at the 14th Intersociety Energy Conversion Engineering Conference, Boston, August 1979. So far as is known, no practical work with liquid-piston Stirling cryocoolers has been attempted, but any number of diverting concepts exist which sooner or later will be the subject of investigation.

A liquid-piston engine overcomes the problem of piston seals and their dry-rubbing friction. Furthermore, it is interesting to reflect on the possibilities for cryocoolers where the expansion space piston is actually condensed working fluid. Successive condensation and vaporization of the working fluid in the cold space would allow substantial transfers of heat and large changes in pressure with minimal actual fluid movement. The prospect of a two-phase two-component reactive liquid-piston Stirling cryocooler is highly diverting. The condensed fraction of the unreacted two-component mixture would be the cold end piston with dissociated vapor working fluid and condensed liquid of perhaps one fraction as the ambient temperature piston. No fluids with the proper combination of properties come immediately to mind and no exploratory work along this avenue is known to be in progress.

Free-Displacer Split-Stirling

An important class of cryocooler much favored for infrared sensor and electronic applications is the split Stirling engine with a free displacer in the expander element. A cryocooler of this type manufactured by CTI is shown in Fig. 3.46. It consists of a compressor element and an expander element coupled by a single long tube. The compressor piston could be a free piston device energized to operate at resonant frequency with pneumatic or mechanical springs, but in all known machines, the compressor piston is activated by a conventional crank connecting rod mechanism driven by an electric motor.

A diagrammatic cross section of the expander is shown in Fig. 3.47. The expander contains a regenerative displacer with a displacer rod added at the ambient temperature end extending through the seal into the bounce space. The cold end of the expander cylinder is the expansion space and the ambient temperature end contains part of the compression space. The other part of the compression space is in the compressor above the piston.

When the compressor operates, the pressure of the working fluid p_c fluctuates periodically at the same frequency as the compressor. The pressure in the bounce space remains substantially constant at the mean cylinder pressure ($p_b = p_{\text{mean}}$). There is, therefore, a net fluidic force acting on the displacer equal to the difference in pressure, $\Delta p = p_c - p_b$, times the area of the the displacer rod.

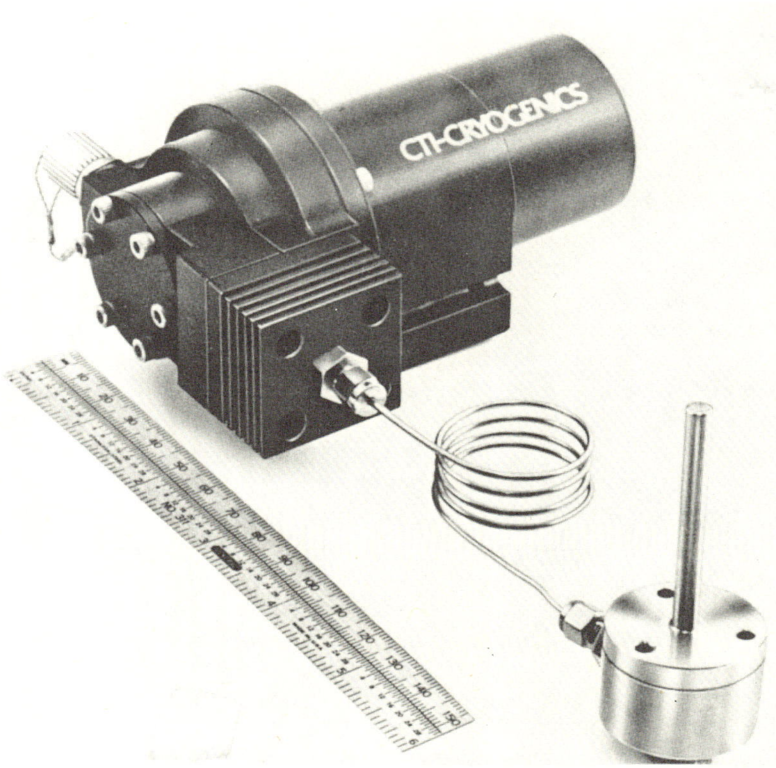

Fig. 3.46. Free displacer split-Stirling cryocooler, Model No. CM-3 80 K manufactured by Cryogenic Technology Inc., Waltham, Massachusetts.

Assume first that the displacer has a low mass (therefore a low inertia) and is completely unrestrained to motion in the cylinder within the confines of the permissible stroke. The net fluidic force acting on the displacer will cause the displacer to shuttle back and forth in the cylinder as shown in the *P–V* plane, Fig. 3.47b, for both the compression and expansion spaces in the cylinder. If the displacer had no mass, the motion would be the straight line *A–B–C*. However, the displacer must have some mass and so will follow the curved path *A–E–B–F–C*. It will return along the same path and consequently there will be no refrigeration effect, for the area of the work diagram on the return stroke will be exactly the same as on the outward stroke. The net area of cycle work will be zero.

Now let the displacer be partially restrained by the sliding seals at the warm end of the displacer, separting the working space and bounce space. Static friction will prevent the displacer moving until a sufficient fluidic

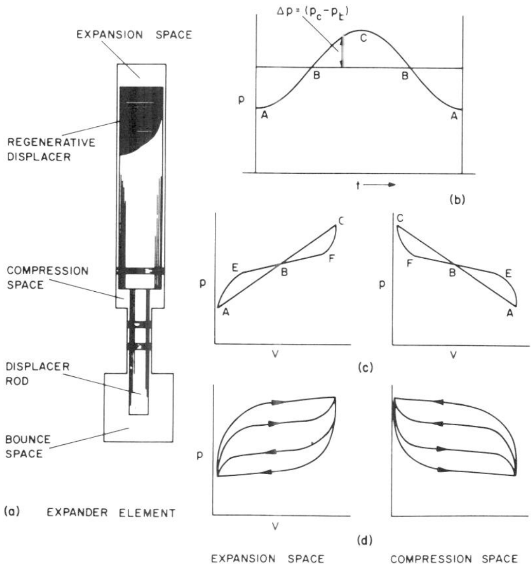

Fig. 3.47. Diagrammatic cross section of split-Stirling expander element, including the pressure–time and work diagrams for the compression and expansion spaces of the expander element.

force is acting to overcome the friction. The displacer will then move, but dynamic friction will require some fluidic forces to be acting on the displacer. Therefore, the line of motion followed on the *P–V* plane on the outstroke will be different from that followed on the instroke and a definite network diagram will be created. Diagrams arising from both a light and heavy seal restraint are shown by Fig. 3.47d. Increase in the restraint increases the separation of the outstroke and instroke paths, thereby increasing the cycle work and the refrigerating effect.

This ingenious free-displacer split-Stirling concept was originated by Walter Higa of the California Jet Propulsion Laboratory (Higa, 1965) and, independently shortly thereafter, by Fred Chellis at Cryogenic Technology Inc., Waltham, Massachusetts and Stuart Horn of the U.S. Army Night Vision Laboratory, Fort Belvoir, Virginia.

The principal difficulty is the need to maintain the seal friction restraint constant throughout the life of the unit. It is difficult to reproduce and maintain constant the friction conditions of presently available dry rubbing materials in a helium atmosphere. One solution (Walker, 1978) is to incorporate a small auxiliary compressor feeding a separate duct to the bounce space to provide a reproducible well-regulated fluidic drive force on the regenerative displacer.

Horn *et al.* (1973) have presented an excellent theoretical analysis of the free-displacer split-Stirling engine operation. No validation of the analysis by comparison with experimental data has been published in the open literature.

STIRLING CRYOCOOLERS OF INTERMEDIATE CAPACITY

The great majority of Stirling cooling engines in use are small systems of low capacity used for local gas liquefaction or miniature electronic instrument or infrared thermal sensors. Larger machines have been constructed, but presently there is no large demand for intermediate capacity cooling systems. Future developments anticipated for the relatively near term may change the situation. One development is the on-board reliquefaction of vaporized liquid natural gas for marine LNG tankers. As the "energy crisis" becomes more acute the marine transport of LNG is increasing dramatically. It is possible to drive the propulsion engines of the ship with the vaporized gas but it is difficult to arrange the rate of boil-off to correspond exactly with the rate of consumption and some reliquefaction may be required to eliminate the emission of combustible vapors. Similar considerations apply to land-based peak-saving LNG storage systems or tidewater terminal facilities.

Another area requiring cryocoolers of intermediate capacity is foreseen for future superconducting applications. These include the superconducting magnets of ship propulsion engines, magnetohydrodynamic power generators, magnetic levitation of high-speed rapid transit systems, atomic physics particle accelerators, and the magnetic plasma containment systems of fusion power reactors. Superconducting electric cable transmission (likely in conjunction with a liquid hydrogen/liquid natural gas cryogenic pipeline) is another area where major requirements for intermediate cryogenic cooling systems are required.

The future for these machines is, therefore, bright and it is timely to look briefly at presently available machines and to speculate on future development.

EXISTING INTERMEDIATE CAPACITY CRYOCOOLERS

A four-cylinder version of the original Philips Stirling gas liquefier is shown in Fig. 3.48. The unit can be equipped with a condensation pressure head operating at up to 20 atmospheres pressures (Model No. PPG-400) or with a cooling pressure head (Model No. PPS-401), able to provide cooling at any temperature and pressure in the range from 60 to 300 K and 1 to 30 atmospheres.

The refrigeration characteristics of the four cylinder engine are shown in Fig. 3.49. This shows the refrigeration capacity and power requirement as a function of the condensing temperature T_E for normal conditions of operation. The speed of the engine is 1450 revolutions per minute (24 Hz)

Fig. 3.48. Four-cylinder, single-acting integral-Stirling cryocooler, Philips Model No. PPG-400.

and the mean pressure of the helium working fluid is 25 atmospheres. The engine cool-down time is 10 min to attain rated capacity from a warm start and a duty cycle of up to 300 hr of continuous operation is prescribed. The unit requires 3 m^3 hr of cooling water at ambient temperatures. The weight of the system including the bedplate and drive motor is 1000 kg, excluding the concrete foundation. The installation envelope is 1 m high $\times$ 0.75 m wide $\times$ 1.8 m long, again excluding the concrete foundation which adds 0.9 m to the height.

A machine of large capacity, again developed and manufactured by Philips (Model No. PPG–2500) was shown earlier in Figs. 3.3 and 3.4. The development of this machine has been described in detail by Dros (1965). It is an interesting and unusual machine consisting of four single-acting Stirling engine systems of the two-piston type in an opposed configuration.

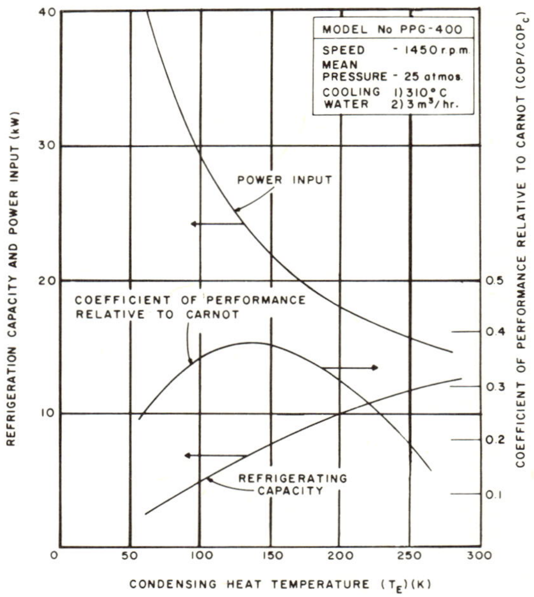

Fig. 3.49. Refrigeration characteristics of the four-cylinder, single-acting integral Stirling cryocooler, Philips Model No. PPG-400.

The unit embodies the hermetic rolling seal developed by Philips for their Stirling prime-mover engines. It is driven by an electric motor supplying shaft work to a vee-piston crank and connecting rod assembly (actually a Joy air compressor unit). The pistons activate a hydraulic fluid connection which operates on the compression and expansion space pistons causing them to oscillate in the proper sequence. Massive external tie rods, easily seen in Fig. 3.3, are used to contain the cylinder end caps. When the machine is operating, these elastic members can be seen to be visibly stretching and relaxing. It takes some getting used to this motion as normal for machine operation.

The refrigeration characteristics for the engine, presented by Dros, are reproduced in Fig. 3.50. The characteristics show the refrigeration capacity (designated P_E), the input power (designated P_M) and the coefficient of performance relative to the Carnot value (designated as η/η_c) as a function of the cold space temperature, T_E. The discontinuity in the characteristics, an increase in the slope of the refrigeration capacity curve and the conversion to constant input power, P_M, arises as a consequence of the control system. When the power input to the drive motor approaches the rated limit, 134 kW, working fluid is released from the cylinder to reduce the pressure and so maintain the power input constant as the expansion space temperature decreases further. Reduction in pressure of the working

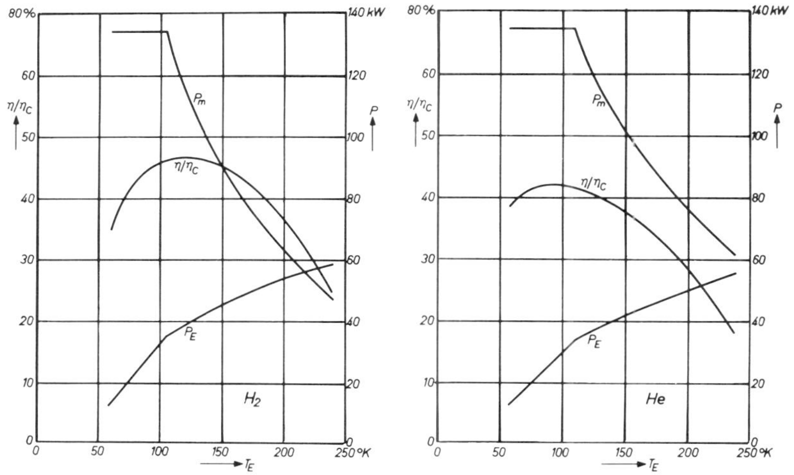

Fig. 3.50. Refrigeration characteristics of the four-cylinder, single-acting opposed-piston Stirling cryocooler, Philips Model No. PPG-2500.

fluid decreases the refrigeration capacity of the engine as shown by the change in slope of the capacity characteristic.

Dros presented refrigeration characteristics for the engine operating with both hydrogen and helium as the working fluid. It can be shown (Walker, 1980) that the ratio of heat transfer to pumping loss is a function of the molecular weight of the transfer fluid. The *highest* ratios are achieved with fluids having the *lowest* molecular weight. For this reason, hydrogen is the preferred working fluid for Stirling engines. Helium is the preferred alternative in many cases on grounds of safety, for helium is inert and hydrogen is very flammable. The superiority of hydrogen at the higher expansion space temperature can be clearly seen by comparing the characteristics presented in Fig. 3.50 for hydrogen (left-hand diagram) and helium (right-hand diagram). At 100 K, the coefficient of performance relative to the Carnot value is about 46% for hydrogen and about 42% for helium.

At lower temperatures, hydrogen loses the advantage over helium because helium deviates less from ideal gas behavior than hydrogen as it approaches the saturated vapor state. Hydrogen liquefies at higher tempertures than helium (20 K compared with 4 K at 1 atmosphere) so for low-temperature cryocoolers, helium is the only possible fluid in operation when a single-phase fluid desired.

The model PPG-2500 cooling engine operates at 730 revolutions per minute at a mean pressure of 40 atmospheres. It requires 20 m^3 per hour of cooling water at ambient temperatures. The weight of the engine including the drive motor is 6500 kg excluding the concrete foundation. Overall dimensions excluding the foundation are 1.95 m high, 1.6 m wide, and 2.97 m long.

The Werkspoor Cryocooler

Another Stirling cooling engine of intermediate capacity is shown in Fig. 3.51. This was developed in the 1950s and 60s by the Dutch marine and general engineering company Werkspoor N.V. of Amsterdam. It was marketed in the United States by the CVI Corporation of Columbus, Ohio as the principal refrigeration component for small-scale liquefaction equipment for natural gas. Vogelhuber and Parish (1968) outlined an integrated LNG system incorporating the Werkspoor engine and Joule–Thomson expanders capable of liquefying up to 2000 mscfd of natural gas.

A commercial brochure isssued by CVI entitled "Technical Description of an 80 MCFD LNG Liquefaction Facility," March 1968, contained a reproduction of an unfamiliar reference* that is the most complete description of the engine known. In Fig. 3.51, it can be seen that the engine was a single-acting integral Stirling machine with a rhombic drive mechanism for activating the piston and displacer in the proper phase relationship. The unit was driven by an electric motor. The engine and drive motor were skid-mounted on a bedplate with separate control console. With a rhombic drive, perfect dynamic balance could be readily achieved so that substantial foundations were not required.

Most descriptions of the engine feature a single cylinder version but include mention of the possiblity of two-, three-, and four-cylinder machines. It is not known if multiple cylinder versions were actually constructed.

The *VMF Review* article included the following technical description of the single-cylinder version with electric drive motor:

Total mass of skid mounted unit: 2000 kg,
Mean pressure of hydrogen working fluid at full load: 35 atm,
Cooling water consumption: 9 m^3 hr,
Cooling water temperature increase: 10°C,
Cooling water pressure drop: 1.5 atm,
Cooling water pump power input: +0.6 kW,
Refrigeration temperature range: +75 to 180 K,
Installation envelope: 1.9 m high, 1.03 m wide, 2.6 m long,
Control console dimensions: 0.98 m high, 0.95 m wide, 0.76 m long.

* "The Werkspoor Cold Gas Refrigerator," Anon. *VMF Review*, **III**(23), 125–131 (date unknown).

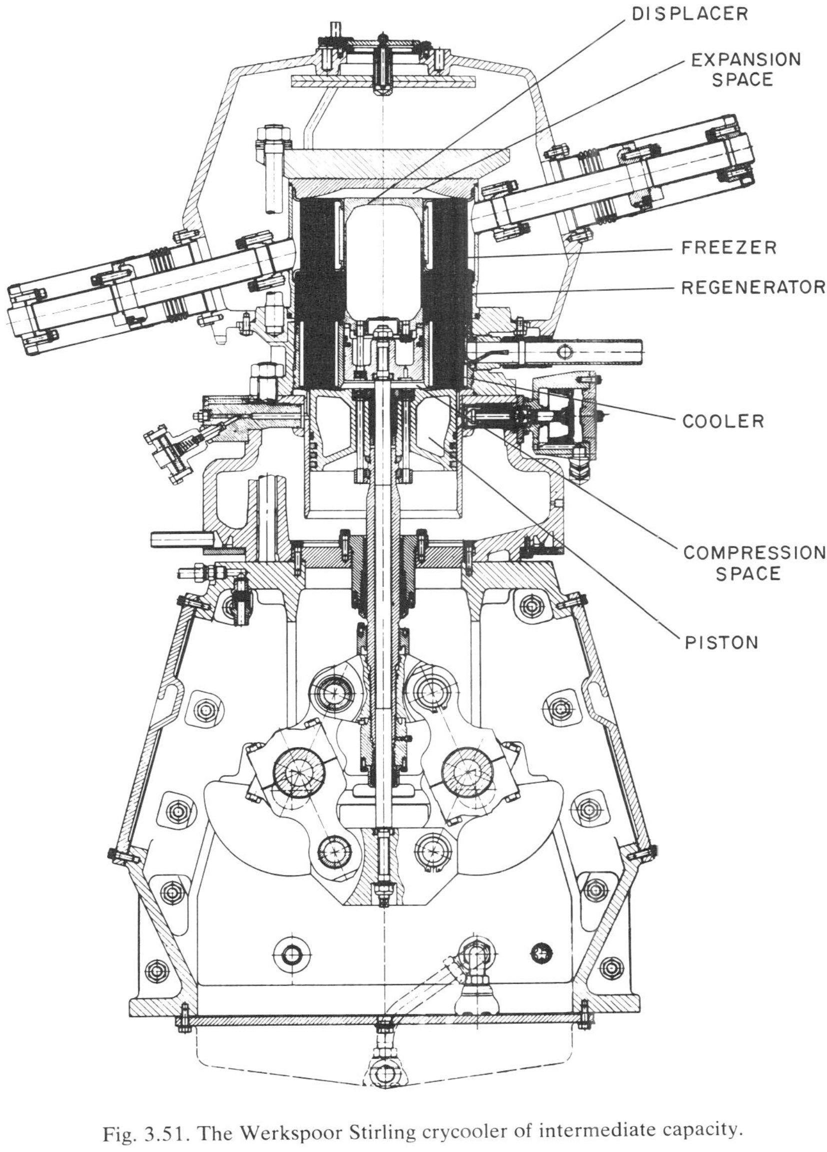

Fig. 3.51. The Werkspoor Stirling crycooler of intermediate capacity.

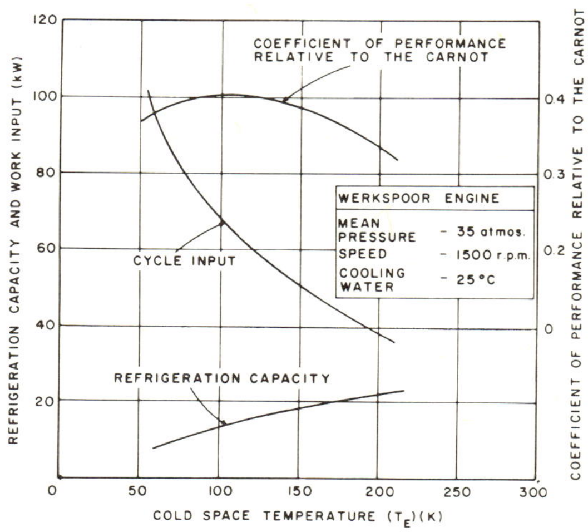

Fig. 3.52. Refrigeration characteristics of the single-cylinder rhombic drive, integral Stirling Werkspoor cryocooler.

The refrigerating characteristics of the Werkspoor single-cylinder Stirling cooling engine are shown in Fig. 3.52. The mean pressure was 35 atm and the cooling water outlet temperature was 25°C. The peak value of the ratio of actual coefficient of performance to the Carnot value was a little over 0.4. The peak value for the Dros hydraulic opposed piston engine was given as 0.47 (see Fig. 3.50). The reasons for this significant difference are not known, and so far as is known, have never been compared in the literature before. Perhaps they arise as a result of the difference in speed (730 rpm for the Philips engine, 1500 rpm for the Werkspoor machine). It is hard to believe that a properly designed rhombic drive engine would be so very much less efficient than the hydraulic opposed piston engine. Perhaps considerable potential for development remained inherent in the Werkspoor engine.

Other interesting characteristics included in the *VMF Review* article, reproduced in Fig. 3.53, show the effect on refrigerating capacity, Q_E, and shaft input power, P, as a function of the mean pressure of the working fluid, p_{mean}. The characteristics were measured at constant condensing and cooling water outlet temperatures of 75 K and 25°C, respectively. The engine speed was 1500 revolutions per minute.

In their paper, Vogelhuber and Parish (1968) indicated that 18 Werkspoor engines had been installed, seven on board ships for "no-loss storage of ethylene and methane," one at an LNG facility for modular upgrading of facility capacity, and 10 in other cryogenic applications. The unit was said to be capable of operating 2000 hours before replacement of the piston rod rolling seals was necessary (to prevent ingress of lubricant to the working

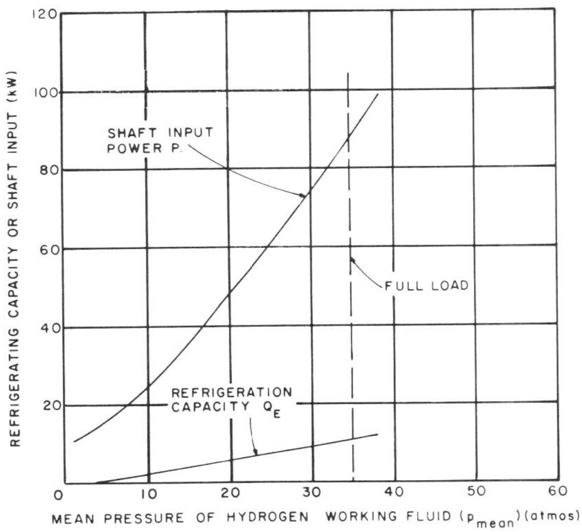

Fig. 3.53. Refrigeration capacity and power input of the Werkspoor cryocooler as a function of the mean pressure of the hydrogen working fluid.

space and contaminating the regenerator, and egress of working fluid from the working space). Replacement of the seals occupied four man-hours of effort.

Vogelhuber and Parish presented data on installed and operating costs of the Werkspoor engine liquefaction system and conventional Claude cycle and cascade cycle systems. They concluded the cost for the Stirling based system was generally favorable in liquefaction capacities up to 1.4 mmcfd. Such an installation would be a hefty liquefaction plant necessitating duplicate *four-cylinder* Werkspoor engine with two-stage Joule–Thompson expansion stages, precooling equipment, and all the associated purification and storage facilities.

Nothing further about the Werkspoor engine was published. It is not known if the engine is in production, if CVI still handles the engine in the United States, and how many of the units were built.

The rhombic drive for Stirling engines was invented by Rolf Meijer at Philips about 1952/53. To those familiar with Stirling engines, it is clear from Fig. 3.51 that the Werkspoor engine owed much to contemporary Philips engine prime movers. Werkspoor does much of the heavy machine work and fitting of the Dros hydraulic engine and probably was familiar with the earlier rhombic prime movers constructed by Philips.

Future Developments

There appear to be excellent future prospects for applications of intermediate capacity engines as reviewed earlier. It is unlikely that any of the machines discussed above will enter volume production. They are too complicated and expensive. There have to be better and easier ways to accomplish the task.

There is every prospect that double-acting Stirling engines will come to dominate the intermediate capacity field as inevitably as in the prime mover automotive field. Double-acting engines have half the moving parts of single-acting systems, so the economic motivation is overwhelming. Double-acting engines can use a variety of kinematic drives, wobble-plates, swash-plates, Scotch-yoke, crank and connecting rods. All these have been discussed in more detail elsewhere (Walker, 1980). Already, multistage four-cylinder wobble-plate helium compressors can be seen on prototype development trials at Cryogenic Technology Inc., Waltham, Massachusetts. With little imagination, one can foresee their adaptation to the Philips swash-plate engine configuration shown in Fig. 3.15. Such developments will relegate the Dros and Werkspoor engines to the status of historical curiosities.

It is more than likely that double-acting free-piston Stirling engines with hydraulic or linear electric power input will come to dominate the field. In large capacities, these could be made so compact, so reliable (elimination of seals, use of gas bearings, etc.), and so simple as to be virtually unassailable as a commercial product. Already the literature contains many references to free-piston Stirling engines as prime movers, and Philips have in production a miniature double-acting free-piston Stirling cooling engine (see Fig. 3.7). More significantly, a large U.S. manufacturer of large-scale cryogenic equipment is presently working hard on a prototype double-acting Stirling free-piston engine along the lines enumerated by Finkelstein (1978).

LARGE-CAPACITY STIRLING CRYOCOOLERS

No large-capacity Stirling cooling engines have been developed or seriously addressed as concepts in the literature. The breakpoint for economic superiority of one system compared with another is difficult to determine. Vogelhuber and Parish (1968) attempted to provide a comparison, but their data, now a decade old, are obsolete and not sufficiently comprehensive for general conclusions to be drawn.

Crystal ball gazing leads to the feeling that large Stirling engines could never be competitive with cascade cycles or Claude systems for base-load gas liquefaction. However, recalling with humility the times we have been wrong in previous forcasting, the opinion is ventured with much trepidation.

REFERENCES

Babcock, G. H. (1885). "Substitutes for Steam." *ASME Trans.* **7**, 680–741.

Beale, W. (1979). Private Communication. Sunpower Inc., Athens, Ohio.

Berchowitz, D. M., Rallis, C. J., and Urieli, I. (1977). "A New Mathematical Model for Stirling Cycle Machine." Proc. 12th I.E.C.E.C., pp. 1522–1527, Washington, D.C., August 28–September 2.

Cairelli, J. E., and Thieme, L. G. (1977). "Initial Test Results with a Single Cylinder Rhombic Drive Stirling Engine." Proc. ERDA, Highway Veh. Syst. Cont. Coord. Mtg., Dearborn, Michigan, October 4–6.

Chellis, F., and Stewart, R. W. (1972). "Multiple Cold Finger Refrigeration System for Cooling Infra-Red Detectors." AFFDL-TR-72-84, AD No. 903-939L (Cryogenic Technology Inc., Latham, Massachusetts).

Collins, S. C., and Cannaday, R. L. (1958). *Expansion Machines for Low Temperature Processes.* Oxford University Press, Oxford.

Daniels, A., and du Pre, F. K. (1971). "Triple-Expansion Stirling Cycle Refrigerator." *Adv. Cryog. Eng.* **16**, 178–184.

de Jonge, A. K. (1979). "A Small Free-Piston Stirling Refrigerator." Paper No. 799245, pp. 1136–1141, Proc. 14th I.E.C.E.C., Boston, Massachusetts.

Dros, A. A. (1965). "An Industrial Gas Refrigerating Machine with Hydraulic Piston Drive." *Philips Tech. Rev.* **26**(10), 297–306.

Finegold, J. G., and Vanderbrug, T. G. (1977). Stirling Engines for Undersea Vehicles." Final Report No. 5030-63, J.P.L., March.

Finkelstein, T. (1959). "Air Engines." *Engineer* **207**, 492–497, 522–527, 568–571, 720–723.

Finkelstein, T. (1959). "Development and Testing of a Stirling Cycle Machine with Characteristics Suitable for Domestic Refrigeration." English Electric Report W/M/3A.

Finkelstein, T. (1960). "Generalized Thermodynamic Analysis of Stirling Engines." S.A.E. Paper No. 118B, January.

Finkelstein, T., Walker, G., and Joshi, T. (1970). "Design Optimization of Stirling-Cycle Cryogenic Cooling Engines by Digital Simulation." Cryog. Eng. Conf., Paper K4, Boulder, Colorado, June.

Finkelstein, T. (1975). "Computer Analysis of Stirling Engines." *Adv. Cryog. Eng.* **20**, 269–282.

Finkelstein, T. (1978). "Modeling Design and Optimization of Stirling Engines." Proc. 13th I.E.C.E.C., San Diego, California, August 20–25.

Finkelstein, T. (1978). Pressure Compounding of Stirling Engines. Proc. 13th I.E.C.E.C., San Diego, California.

Gedeon, D. (1978). "Optimization of Stirling Cycle Machines." Proc. 13th I.E.C.E.C., San Diego, California.

Goldwater, B., and Morrow, R. B. (1977). "Demonstration of a Free Piston Stirling Linear Alternator Power Conversion System." Proc. 12th I.E.C.E.C., Paper No. 779249, pp. 1488–1495, Washington, D. C., August 28–September 2.

Haarhuiss, G. J. (1978). "The MC 80—A Magnetically Driven Stirling Refrigerator." Proc. 7th Int. Cryog. Eng. Conf., London, IPC Business Press, Guildford, U.K.

Higa, W. N. (1965). "A Practical Philips Cycle for Low Temperature Refrigeration." *Cryog. Technol.* **8**, 203–209, July-August.

Horn, S. B., Lumpkin, M. L., Walters, B. T., and Acord, T. T. (1973). "Miniature Cryogenic Cooler for TOW Night Sight." Proc. Closed Cycle Cryocooler Tech. and Appl., Vol. 1, pp. 55–72, AFFDL-TR-73-149, WPAFB, Ohio, AD No. 918234

Khan, M. I. (1962). "The Application of Computer Techniques to the General Analysis of the Stirling Cycle." M.Sc. thesis, Univ. of Calgary.

Kirk, A. (1874). "On the Mechanical Production of Cold." *Proc. Ins. Civil Eng.* **37**, 244–315, London.

Köhler, J. W. L. (1965). The Stirling Refrigeration Cycle. *Sci. Am.* **212**(4), 119–127.

Lee, K. (1976). "The Stirling Cycle with Adiabatic Compression and Expansion." M.Sc. thesis, Univ. of Calgary.

Lindale, E. (1978). "Stirling Cycle Refrigerators for Gamma Ray Detector." Report No. PL-42-Cr78-0713, Johns Hopkins Univ., Applied Physics Lab., Laurel, Maryland (Philips Laboratories).

Maki, E. R., and de Hart, A. O. (1971). "A New Look at Swash-Plate Drive Mechanisms." S.A.E. Trans. (Truck, Powerplant Fuels and Lubricants Mtg.), Vol. 80, Paper No. 710829, St. Louis, Missouri, October.

Martini, W. (1978). Design Manual for Stirling Engines. NASA CR 135382, DOE/NASA Contractor Report (NTIS, Springfield, Virginia). See also: Stirling Engine Design and Feasibility for Automotive Use. (1979). (ed. M. J. Collie), Noyes Data Corp., Park Ridge, New Jersey.

Prast, G. (1963). "A Philips Gas Refrigerating Machine for 20 K." *Cryogenics* **3**, 156–160, September.

Prast, G. (1965). "A Gas Refrigerating Machine for Temperatures down to 20 K and Lower." *Philips Tech. Rev.* **26**(1), 1–11, January.

Qvale, E. B. (1963). "An Analytical Model of Stirling-Type Engines." D.Sc. thesis, MIT.

Qvale, E. B., and Smith, J. L. Jr. (1969). "An Approximate Solution for the Thermal Performance of a Stirling Engine Regenerator." *Trans. A.S.M.E., J. Power* **91**, 109–112, April.

Rios, P. A., Qvale, E. B., and Smith, J. L., Jr. (1968). "An Analysis of the Stirling Cycle Refrigerator." *Adv. Cryog. Eng.* **13**, 332–342.

Rios, P. A., and Smith, J. L. Jr. (1969). An Analytical and Experimental Evaluation of the Pressure-Drop Losses in the Stirling Cycle. A.S.M.E. Paper No. 69-WA/Ener-8, Vol. 92, Ser. A., No. 2, pp. 182–188, April.

Schmidt, G. (1861). "Theorie der Geschlossenen Calorischen Maschine Von Laubroy und Schwartzkopff in Berlin." *Z. Ver. Oster. Ing.*, p. 79.

Schmidt, G. (1871). "Theorie der Lehmannschen Calorischen Maschine." *Z. Ver. dt. Ing.* **15**(1).

Schock, A. (1978). "Nodal Analysis of Stirling Cycle Devices." (a) Draft Report DOE; (b) Proc. 13th I.E.C.E.C., San Diego, California.

Stoddart, D. (1960). "Generalized Thermodynamic Analysis of Stirling Engines." B.Sc. thesis, Durham Univ.

Urieli, I. (1977). "A Computer Simulation of Stirling Cycle Machines." Ph.D. thesis, Univ. of Witwatersrand, Johannesburg, South Africa, February.

Urieli, I., Rallis, C. J., and Berchowitz, D. M. (1977). "Computer Simulation of Stirling Cycle Machines." Proc. 12th I.E.C.E.C., Paper No. 779252, Washington, D.C., pp. 1512–1521, August 28–September 2.

Urieli, I. (1979). A Review of Stirling Cycle Machine Analysis. Paper No. 799236, pp. 1086–1090, Proc. 14th I.E.C.E.C., Boston, Massachusetts.

Vogelhuber, W. W., and Parish, H. C. (1968). Compact LNG System Using Large Stirling Cycle Cold Gas Refrigerator. Paper No. 28, Session 5, Proc. First Int. L.N.G. Conf., Chicago, Illinois (Institute of Gas Technology, Chicago, Illinois).

Walker, G. (1962). An Optimization of the Principal Design Parameters of Stirling Cycle Machines. *J. Mech. Eng. Sci.* **4**(3).

Walker, G., And Khan, M. (1965). "The Theoretical Performance of Stirling Cycle Machines." S.A.E. Paper No. 949A, Int. Auto. Eng. Congress, Detroit, Michigan, January.

Walker, G. (1973). *Stirling Cycle Machines.* Oxford University Press, Oxford, England.

Walker, G., and Burn, K. (1976). "Exploratory Study of the Rainbow Variant Stirling Engine." Proc. 11th I.E.C.E.C., Lake Tahoe, New York, September.

Walker, G. (1978). "Split-Stirling Cooling Engine with Fluidic Control of Displacer Motion." Invention Disclosure, Calgary, Alberta, December.

Walker, G. (1980). *Stirling Engines.* Oxford University Press, Oxford, England.

West, C. (1971). "The Fluidyne Heat Engine." Report No. AERE-R-6776, Atomic Energy Research Establishment, Harwell, Berks, England.

Zimmerman, J. E., and Radebaugh, R. (1977). "Operation of a SQUID in a Very Low Power Cryocooler." Proc. of Conf. on Appl. of Closed Cycle Cryocoolers to Small Superconducting Devices, pp. 59–66, Boulder, Colorado, October (issued as NBS Spec. Pub. 508, April 1978—U.S. Govt. Printing Office).

Chapter 4

Vuilleumier Cryocoolers

INTRODUCTION

A Vuilleumier or VM cryocooler uses thermal energy at high temperatures to generate cyclic pressure fluctuations in the closed engine. The pressure fluctuations are used to produce refrigeration at low temperatures. The heat absorbed by the engine at high temperatures and at low temperatures is rejected at ambient temperatures. Figure 4.1 is a schematic diagram showing the essential elements of the engine.

The pressure fluctuations are produced by shuttling gas periodically from an ambient temperature region to a high-temperature region by the action of a reciprocating displacer in the hot cylinder. There is another displacer in the cold cylinder. The hot and cold displacers are coupled by some form of kinematic mechanism to maintain proper phasing and synchronous operation.

A little work is required to operate the displacers and overcome the fluid and mechanical friction of seals and bearings. In most cases the work input is supplied by an electric motor. Ingenious self-acting systems have been made which use some of the heat input to generate enough work to drive the displacers. Difficulties with seals limit the usefulness of this attractive feature.

Vuilleumier and Duplex Stirling Engines

A Vuilleumier engine can be thought of as a Stirling engine working as a refrigerator with a thermal compressor instead of a mechanical compressor. They should not be confused with the Stirling–Stirling duplex arrangement described in Chapter 3 in which one Stirling engine acting as a prime mover receiving heat at high temperature produces work to drive another Stirling engine operating as a refrigerator accepting heat at low temperature (the useful refrigeration effect). Both reject heat at ambient temperature.

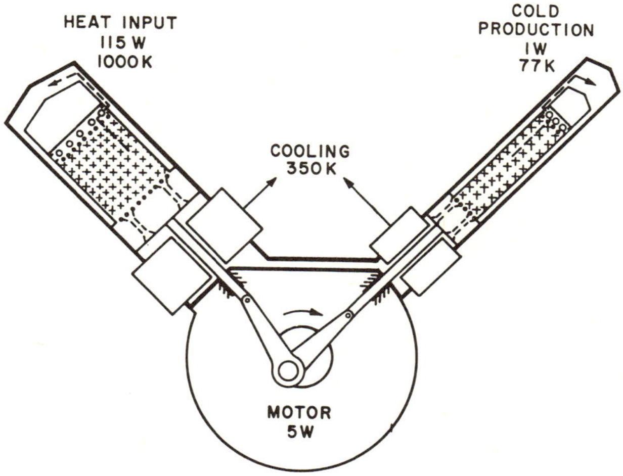

Fig. 4.1. Essential elements of a Vuilleumier cryocooler (after Pitcher and du Pre, 1970).

Advantages

The principal advantage claimed for Vuilleumier cryocoolers is the potential for long life made possible by the low mechanical forces acting on the drive mechanism, bearings, and seals. These low forces arise as a consequence of the low pressure ratios (1.1 to 1.2) in Vuilleumier engines compared with Stirling engines (2 to 2.5). The pressure ratios are low in Vuilleumier engines for they are generated solely as the result of changes in temperature, not both temperature and volume as in a Stirling engine.

Stirling cryocoolers are used where space and weight are at a premium. The design is constrained to a machine operating at high pressure and speed with maximum possible compression ratios to obtain the highest refrigerating effect per unit mass or volume. The price to be paid for this is a short operating life. It is often overlooked that it would be simple to relax the output requirements imposed on Stirling engines and obtain the same low speed, low compression ratio, low refrigeration output characteristic of Vuilleumier engines. This would endow Stirling engines with the same long life claimed as the principal attraction of Vuilleumier engines.

Disinterested comparative studies have shown the Stirling cryocooler as more efficient and lower in cost, volume, and weight than the equivalent Vuilleumier unit. The long-life characteristic of Vuilleumier engines has, in fact, proved elusive despite a decade of intensive development. One particular difficulty has been the high wear rates of the hot rider, the sliding guide ring, at the high-temperature end of the hot displacer.

The real advantage of the Vuilleumier cryocooler is the ability to use a thermal energy input to produce refrigeration. This can be supplied by concentrated solar energy, radioisotopes, combustion of fossil fuels or waste materials, and finally, waste thermal streams of industrial plants or power systems.

Solar and isotope energy made the Vuilleumier engine a strong candidate for cryogenic cooling in long-duration space exploration. This was recognized by NASA, and exploratory developments have been undertaken by the Goddard Spaceflight Center. Waste thermal streams or combustion of waste products can be combined with Vuilleumier engines to produce refrigeration at both cryogenic and higher temperatures for food processing and air conditioning. Terrestrial applications of Vuilleumier engines for air conditioning with solar energy are very attractive. In many locations, air conditioning refrigeration is required only when the sun is shining. This is a happier circumstance than the use of solar energy for power production, where the main requirement for power arises at night.

In all these potential applications for Vuilleumier engines, similar considerations apply to duplex Stirling systems, but so far as is known, no comparisons have been made. Self-starting and complete independence of an auxiliary electric power supply would weigh heavily in favor of duplex Stirling systems. There are good prospects that an ingenious designer could instill the same desirable characteristics in the Vuilleumier engine.

Thermodynamic Aspects

Most Vuilleumier cryocoolers have used electric resistance heating to provide thermal input to the hot displacer cylinder and a motor to operate the engine. Resistance heating for Vuilleumier engines is thermodynamically inefficient. The second law of thermodynamics limits the fraction of heat supplied that can be converted into work, the Carnot penalty. Therefore, a high price is paid to produce electric power. Frequently, only 10% to 30% of the energy supplied is transformed to electricity. The electric power can be transformed in electric motors to useful work at relatively high efficiency without payment of further thermodynamic penalties. The work produced eventually becomes converted to low grade heat by way of frictional dissipation.

The use of electricity in resistance heating produces no useful work. All the energy is converted (degraded) to heat directly. If now that heat is used in an engine, the Carnot penalty must be paid again. If the work is used to drive a refrigerator, the refrigerator itself sustains a further

Carnot penalty. The use of electric resistance heating in Vuilleumier cryocoolers requires payment of an extra Carnot penalty over and above the two exacted if electric motors are used to provide the work to drive a refrigerating machine.

Clearly, energy utilization in a resistance heated Vuilleumier cryocooler has to be less effective than the equivalent Stirling engine driven by an electric motor. The electric resistance-heated duplex Stirling engine suffers the same thermodynamic penalty. Electric resistance heating should never even be contemplated where efficiency is a prime factor.

Most of the resistance heated Vuilleumier cryocoolers constructed to date have been miniature engines for infrared night vision and heat-seeking guidance applications. In these sizes, efficiency is of little significance for the power levels are relatively low. Electric power can then be justified on grounds of convenience and instant availability in the military vehicles, aircraft, and missiles where such units are found.

Comparative Data

Chellis (1973, 1977) has given comparative data, reproduced in Tables 4.1 and 4.2 for a number of cryocoolers of different sizes and types. These clearly show the weight, size, and efficiency penalties of the Vuilleumier engines in comparison with equivalent Stirling machines.

HISTORICAL REVIEW

The earliest description of this type of machine was given by Rudolph Vuilleumier (1918) in a U.S. patent. Vuilleumier is described as a citizen of the United States, resident in New York. No other information is known about him or his work. A search of the literature* has failed to produce a single reference to a working model or any subsequent development.

The engine or a minor variant seems to have been reinvented independently 20 years later by Vannevar Bush (1938), then a professor at the Massachusetts Institute of Technology. Again, little is known of his work on the concept. Later on, Professor K. W. Taconis (1951) of the Cryogenic Laboratory at the University of Leiden in Holland described, in a U.S. Patent, yet another form of heat-powered refrigerator with regenerative heat exchangers. These systems and their potential for large-scale cryogenic

* I would be very glad to receive any information readers may have about the work, life, or machines of Rudolph Vuilleumier, or about similar machines invented earlier in other parts of the world.—G.W.

Table 4.1. Comparison of Closed-Cycle Cryocoolers[a]

Cryocooler type	Cooling capacity		Input power (Watts)	Compressor		Expander		MTBF (hr)	Application
	Watts	K		Size (in.)	Wt[b] (lb)	Size (in.)	Wt (lb)		
Gifford–McMahon	3	20	2,100	$28 \times 16 \times 16$	175	$18 \times 10 \times 6$	22	20,000	Ground based paramp
Gifford–McMahon	1	26	790	$10 \times 10 \times 10$	20	$11 \times 6 \times 2.6$	5	1,000	Airborne IR
Integral Stirling	1	26	475	$14 \times 10 \times 6$	26	—	—	250	Airborne IR
Integral Stirling	1	77	100	$11 \times 4.5 \times 3$	3.5	—	—	250	Airborne IR
Split Stirling	1	77	135	$10 \times 5.5 \times 2.5$	5	$3 \times 2.5 \times 2$	0.3	1,000	Airborne IR
Split Vuilleumier	1	77	350	$10 \times 4 \times 3$	10.5[c]	$3 \times 2.5 \times 2$	0.3	2,200	Airborne IR

[a] After Chellis (1977).
[b] Includes fan.
[c] Includes motor inverter and electronic temperature controller for hot end.

Table 4.2. Comparison of Small 77 K Refrigerators[a]

	Smaller units 1 to 2 watts							Larger units 3 to 5 watts		
Power source	400 Hz ac				28 V dc			400 Hz ac		
Cycle[b]	ST	ST	VM	VM	ST	VM	VM	ST	ST	GM
Capacity at 77 K (W)	0.9	1.5	2.0	1.0	1.0	1[c]	1.0	3.5	3	5.0
Power input (W)	90	120	340	125	42	120	90	180	530	575
System weight (lb)	3.25	4.0	14	6.3	7.7	7.0	6.2	6.5	12	15
Operating speed (RPM)	1750	1750	600	600	1620	—	450	1750	—	333
Watts in/watt at 77 K	100	80	170	125	42	120	90	51	176	115
Lb weight/watt at 77 K	3.6	2.6	7	6.3	7.7	7.0	6.2	1.85	4	3

[a] After Chellis (1973).
[b] ST, Stirling; VM, Vuilleumier; GM, Gifford–McMahon.
[c] 1 W at 85 K.

cooling were reviewed by Yendall (1960). He concluded they were not attractive economically for large-scale cryogenic applications.

The next significant event was the patent application by Hogan (1964) and a paper by Chellis and Hogan (1964) (of A. D. Little, Cambridge, Massachusetts), describing a low-temperature heat-powered Vuilleumier-type refrigerator, for cooling in the 10–20 K range. It absorbed heat at ambient temperature and heat was rejected from the system at 77 K to liquid nitrogen acting as the coolant.

Hughes Vuilleumier Cryocoolers

Hughes Aircraft Co., Culver City, California, commenced work on Vuilleumier cycle engines in the mid 1960s. A patent outlining modifications to the Vuilleumier engine to drive its own displacers and even produce useful work was granted to Ken Cowans (1968), then of Hughes Aircraft. Shortly thereafter, came a voluminous report (Magee and Doering, 1968) detailing Hughes' work on the theoretical development, historical aspects, and laboratory evaluation of "bread board" Vuilleumier cycle engines. It was this report that established the claim of the Vuilleumier engine for very high reliability. They declared the engine to have the potential for long operating life, and this reputation has been perpetuated thereafter despite variously reported failures of the engine to actually achieve long life (Harkless, 1973).

Other reports containing much important information about experimental performance and materials used in Vuilleumier cooling engines have been prepared by Hughes for various sponsors, particularly the U.S. Air Force. Because no papers about the Hughes Aircraft cryocooler developments are known to the author, it is necessary to resort to contractor reports prepared by Hughes. Enough work has been done for others that a good appreciation of the state of development can be gained.

Following the Magee–Doering report, Leo (1970) prepared an interesting summary of Vuilleumier cycle characteristics, and later Leo (1971) gave an expanded and updated version of the earlier Magee report. Doody (1971) described a double-expansion Vuilleumier cycle engine for low-temperature infrared applications. This is of interest as an early, if not the original, application of double-expansion techniques in Vuilleumier cryocoolers to achieve temperatures lower than were possible with single-expansion engines. The first use in Stirling engines of double-expansion techniques occurred a decade earlier (Prast, 1963). Globe *et al.* (1971) and Rhia (1971) presented informative material on Vuilleumier cryocoolers for infrared missile guidance systems. Later, Rhia (1971), in three consecutive reports, included design and test data for a long-life, high-capacity Vuilleumier engine for space applications.

About the same time Morgan (1971) reported on a comparative study of Stirling and Vuilleumier cryocoolers to liquefy 1 to 12 kg liquid air per hour for infrared and other sensor applications aboard bomber planes. This document is notable as a serious attempt to compare Stirling and Vuilleumier machines for the same cryogenic application. The Stirling configuration was an enlarged version of the two-piston arrangement favored by Hughes for their small Stirling cooling engines. This would not be the "optimum" Stirling engine if, say, Philips were to design it for this specific application. Nevertheless, in the comparison by Morgan, the Stirling cryocooler emerged the winner on grounds of cost, size, weight, and efficiency.

Despite this, work continued on Vuilleumier engines at Hughes. In 1973, Berry contributed to the U.S. Army Night Vision Laboratory study (Horn, Cowans, and Berry, 1973) of low-production-cost Vuilleumier cooling engines. Simultaneously, Berry (1973) described a new family of modular split-Stirling and split-Vuilleumier cryocoolers. At the same time, Renyer (1973) outlined progress on the high-capacity, long-life Vuilleumier refrigerator shown in Fig. 4.2 and, in cross-section, in Fig. 4.3. This is a three-stage unit capable of 12 W at 75 K, 10 W at 33 K, 3 W at 12 K with a total power input of 2700 W. The outline design for the engine was given by Rhia (1971) in the reports referenced above. Work on the engine over the five year period 1970/75 was summarized by Doody (1975) with extensive test data on the limited test accomplished.

At the other end of the scale, Berry (1974) described the design and development of an "ultraminiature" split-Vuilleumier cooling engine designed for a capacity of 0.75 W at 75 K of minimal size and weight in the cold region to permit mounting in a gimbaled sensor.

Work began on a large-capacity Vuilleumier cooling engine for satellite sensors and preliminary design aspects of this machine were reported by Morgan (1975). An unclassified, unrestricted report for NASA was prepared by Russo (1976) describing a Vuilleumier cryocooler for detector cooling on the limb-scanning infrared radiometer. This report is recommended for readers interested in Hughes work but unable to gain access to the restricted Air Force reports.

One Air Force report available without restriction is that of White (1976). It is of exceptional value to beginners. White has been involved in the Air Force Vuilleumier program virtually from the start and is, therefore, able to write meaningfully over the whole range of material. Substantial extracts of this report are contained in this chapter. Elsewhere, White (1973) has outlined a digital simulation program for the three-stage Hi-Cap Vuilleumier cooling engine. This was developed for predicting the off-

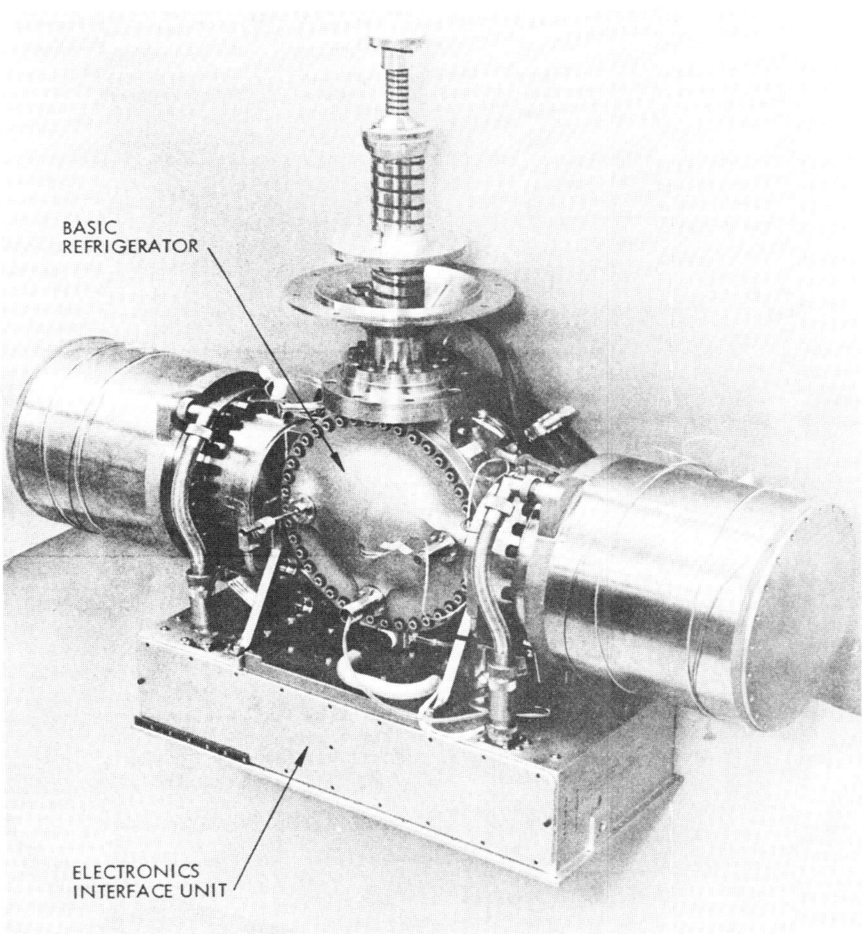

Fig. 4.2. Triple-expansion high-capacity Vuilleumier cryocooler (after Renyer, 1973).

design performance of the Hi-Cap cooler. White gave various results to show the effect of variation on the design and operating parameters of the engine. More recently, White (1976) gave full details of the computer program, including a program listing, but the report is, unfortunately, restricted in circulation. White (1976) has also reported on the preliminary tests carried out on the high-capacity Vuilleumier cooling engines developed by Hughes and Philips Laboratories.

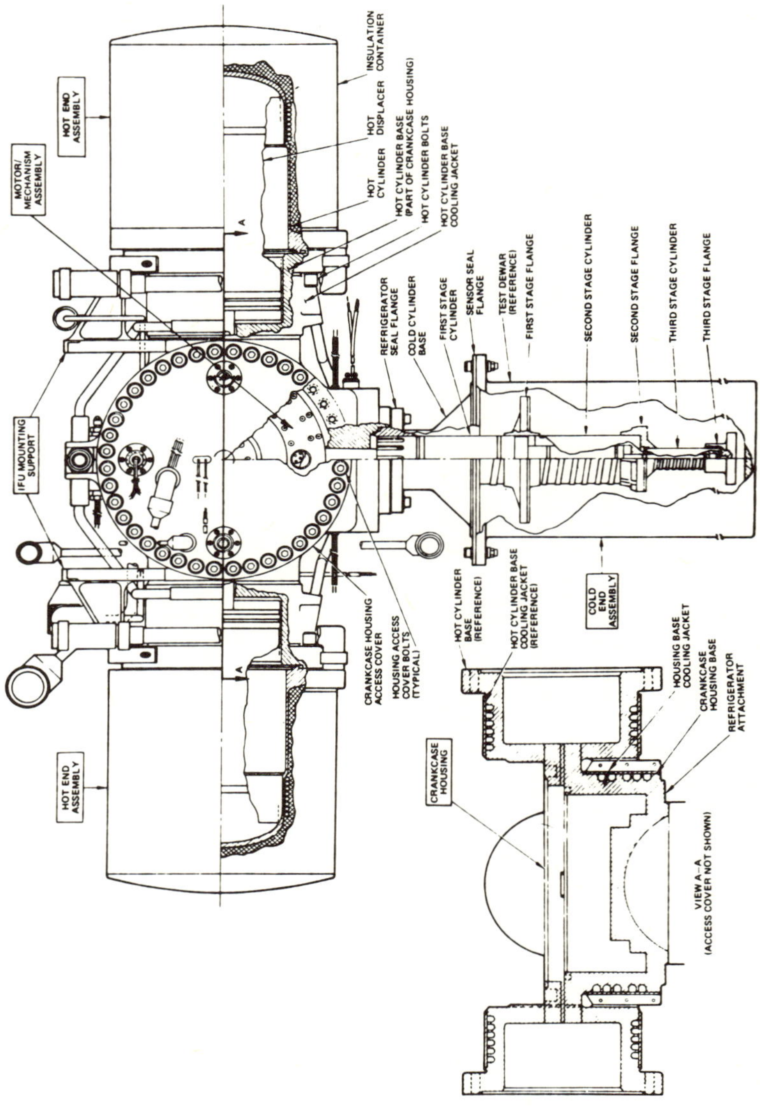

Fig. 4.3. Triple-expansion high capacity Vuilleumier cryocooler (after Renyer, 1973).

Philips Vuilleumier Cryocoolers

The development of a high-capacity Vuilleumier cooler was carried out by the Philips Laboratories, Briarcliff Manor, New York, in parallel with the Hi-Cap program of Hughes Aircraft Co. discussed above. The sponsor was again the U.S. Air Force Flight Dynamics Laboratory of the Wright–Patterson Air Force Base, Dayton, Ohio, under the "Advanced Surveillance Technology" program.

The Philips development was reported in detail by Pitcher (1975) with interim technical reports presented earlier (Pitcher, 1971, 1972). General review material was presented at the 1973 Cryocooler Conference by Lehrfeld and Pitcher (1973) and was supplemented by an excellent review paper (Pitcher, 1973) on life limiting components and systems for spacecraft cooling. Following construction and evaluation of a prototype high-capacity cooler, the development of a flight design unit was undertaken. This was reported by Pitcher (1977).

The Philips Vuilleumier engine was constructed to the same technical specification as the Hughes Aircraft Hi-Cap cooler. This included three stages of refrigeration, 0.3 W at 11.5 K, 10 W at 33 K, 12 W at 75 K with a maximum power input of 2700 W, a life of 20,000 hr unattended operation, a total weight of 205 lb (including electronics but excluding the heat rejection system) and with liquid cooling having inlet cooling temperatures of 10 to 45°C. The Philips Vuilleumier cryocooler conforming to this specification is shown in Fig. 4.4 and schematically in Fig. 4.5. The Philips engine was markedly different from that of the Hughes Aircraft Hi-Cap cooler shown in Figs. 4.2 and 4.3.

A notable feature of interest in the Philips engine was the use of the rhombic drive mechanism (see Chapter 3), a Philips invention associated with the development of Stirling engines. This permitted the opposed-piston configuration, shown in Fig. 4.4, to be perfectly balanced and eliminated the "side thrust" forces on cylinder walls found in other kinematic drives. Another Philips Stirling engine development, the roll-sock seal, was incorporated in the design to hermetically seal the working spaces from the crankcase. The hermetic seal allowed the use of oil lubrication in the crankcase and with it an assurance of the drive mechanism to endure 20,000 hr. The low frequency of operation (7 Hz) supported the expectation of long life from the seal and drive mechanism.

Problems of oil–helium management were foreseen in long-term operation in the zero gravity environment, for the flexible diaphragms of the rolling seals were slightly permeable to helium. Some prospect was foreseen of a gas bubble causing the lubrication oil pump to lose suction. To prevent this, the crankcase was entirely filled with oil, and to compensate for volume

Fig. 4.4. Philips Laboratories triple-expansion high capacity Vuilleumier cryocooler (after Pitcher, 1975).

changes due to the ingress and egress of displacer rods, a bellows device was incorporated in the crankcase system.

The rigidity of the rhombic drive mechanism permitted the use of close tolerance regenerator seals (operating at ambient temperatures) in both the hot and cold cylinders. Rubbing surfaces were entirely eliminated in the "dry" helium-filled spaces of the engine to further enhance the potential for extremely long lifetime.

On reading the Philips reports, one cannot help admiring the high level of engineering competence displayed therein. At the same time, they convey the impression they could have done a better job with a Stirling cryocooler. Replacement of the Vuilleumier thermal compressor (heated electrically) by a mechanical compressor (driven electrically) would have converted the system to a Stirling cryocooler. There is little doubt it would have been more efficient, lighter, smaller, and cheaper, with no sacrifice—in fact, an enhanced potential—of long life.

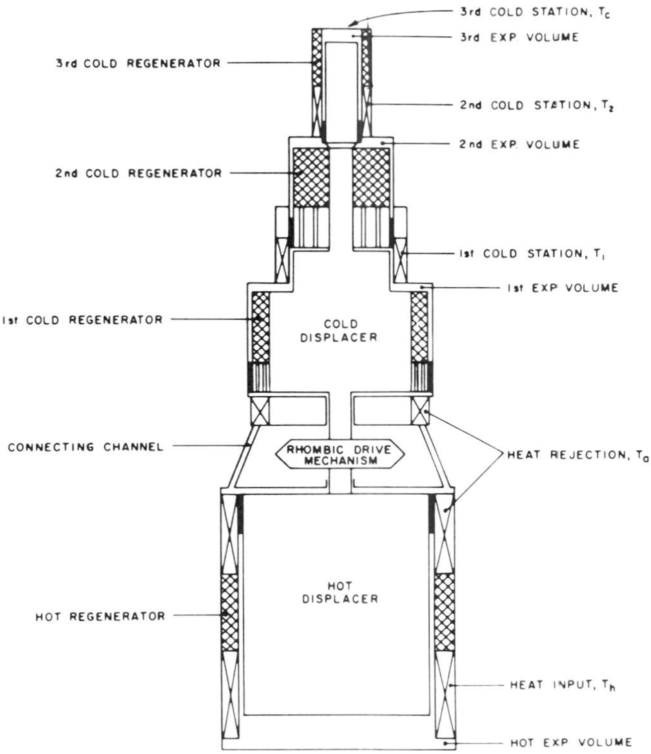

Fig. 4.5. Schematic arrangement of Philips Laboratories triple-expansion high-capacity Vuilleumier cryocooler (after Pitcher, 1975).

Support for this comes from the detailed failure analysis given by Pitcher (1975). He concluded that the potential for a mechanical failure rate was 6.72 per 10^6 hours of operation and an electrical/electronic failure rate was 26 per 10^6 hours of operation. Of the total electrical/electronic failures, 9.6 were attributed to the heater control. Elimination of the heater would, therefore, reduce the electrical/electronic failure to 16 per 10^6 hours of operation. The use of a large motor for the increased work input would reduce the prospect of motor controller failures (estimated at 9 per 10^6 hours of operation). At the same time, the mechanical failure rate might increase to a balanced electrical and mechanical failure rate of 10 per 10^6 hours of operation.

Evidence of the more effective use of electric power in a Stirling engine is contained in results presented by Pitcher (1975) as an appendix to the main report. The results are reproduced in Table 4.3. It contains estimates

Table 4.3. Estimate of Power Requirements for Multistage Stirling and Vuilleumier Cooling Engines with Different Temperature Levels and Cooling Capacities[a]

Entry no.	Temperature (K)			Cold production (W)			Power input	Inlet coolant temp.	Comments
	1	2	3	1	2	3	(W)	(°C)	
1	110	30	13	5.0	5.0	0.5	2000	25	VM
2	110	30	13	5.0	5.0	0.5	1200	25	Stirling
3	112	30	11.5	5.0	8.0	0.3	2200	25	VM
4	87	30	11.5	5.0	8.0	0.3	2350	25	VM
5	75	30	11.5	5.0	10.0	0.3	2500	25	VM
6	75	30	11.5	12.0	10.0	0.3	2700	25	VM
7	75	30	11.5	20.0	10.0	0.3	2900	25	VM
8	—	55	12	—	11.2	0.9	1606	25	Stirling
9	—	39	12	—	11.2	1.0	1500	25	Stirling
10	110	55	12	13.0	7.8	0.9	1900	25	Stirling
11	110	39	12	13.0	7.8	1.0	1800	25	Stirling

[a] After Pitcher (1975).

made by Philips of eleven cooling engines, of Stirling and Vuilleumier types, having varying cooling capacities at different temperature levels. Entry numbers 1 and 2 are directly comparable in terms of temperature and cold production. The power input to the Stirling engine (entry No. 2) is 1200 W. The power input to the Vuilleumier engine is 66% greater at 2000 W.

Entry No. 6 is the prototype Vuilleumier engine described by Pitcher in his reports. Other important data, unfortunately not included in Table 4.3, include the pressure level of operation and the speed, volume, and mass of the machines. So far as is known, these data have not been disclosed by Philips.

The high-capacity cooling engine described above was by no means the only Vuilleumier cryocooler made by Philips. Daniels and du Pre (1971) briefly discussed the miniature engine shown in Figs. 4.6 and 4.7. This was earlier discussed by Pitcher and du Pre (1970) and in more detail by Pitcher (1970), prepared for the Night Vision Laboratory, Fort Belvoir, Maryland. This report is unclassified, unrestricted, and highly recommended for anyone interested in small Vuilleumier cooling engines.

Another interesting task performed by Philips for the Night Vision Laboratory was reported by Cohen and Daniels (1973). This was the design and development of a split-Vuilleumier cooling engine of 2 W cooling capacity at 75 K in an ambient environment at a temperature of 320 K. The engine was intended to be the self-acting type capable of producing sufficient work to drive the displacers (patent by Cowans, 1968).

Fig. 4.6. Miniature prototype Vuilleumier cryocooler (after Pitcher, 1970).

Difficulty was experienced in achieving the mass and self-acting goals for the hot side of the engine. The development did not proceed beyond the "bread board" stage. The report is notable for the comprehensive discussion of both the analytical and experimental aspects, the inclusion of a computer program listing for the analytical work, the inclusion of test results, and, above all, a remarkably competent exposition of the problem of the cold-finger seal that besets all split-Stirling or Vuilleumier cryocoolers.

AiResearch Vuilleumier Cryocoolers

A Vuilleumier cryocooler was developed by the AiResearch Manufacturing Co., Torrance, California for the NASA-Goddard Space Flight Center (GSFC), Greenbelt, Maryland. The objective was to achieve a long

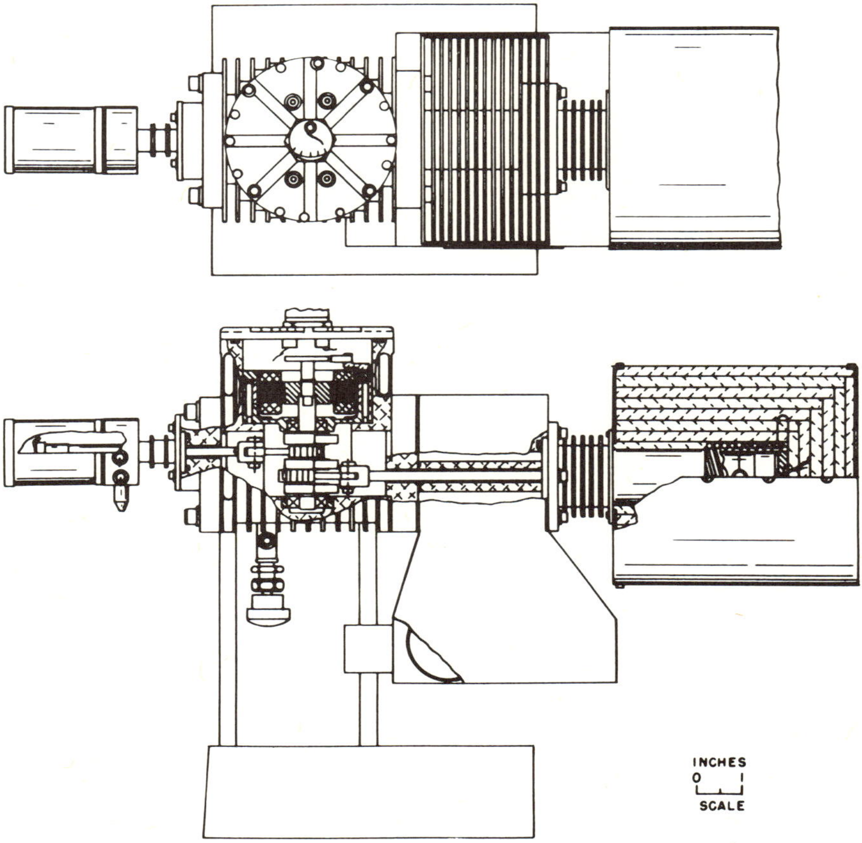

Fig. 4.7. Cross section of miniature Vuilleumier cryocooler (after Pitcher, 1970).

life (2 to 5 year reliability) for a small cryocooler having a capacity of 5 W at 75 K. Later, a miniature engine was required having a capacity of $\frac{1}{4}$ W at 65 K. The prototypes were heated electrically but the ultimate thermal source was to be isotope or concentrated solar energy.

The GSFC–AiResearch development was brilliantly executed by a competent research development team familiar with working at the frontiers of technology. They were without a large body of previous experience in the field and, as a consequence, were able to develop innovative approaches without the constraints imposed by past successes. Their work has been reported in the reports and papers referenced below.

Figure 4.8 is a cross section of one of the AiResearch engines. It is essentially the same general configuration as the Philips opposed piston machine discussed earlier. However, AiResearch chose to use a crank-connecting rod drive instead of the Philips rhombic drive. They used hard, dry rubbing, inorganic materials for the crankshaft bearings and the linear crosshead bearings. Noncontacting close tolerance seals were used on the displacers.

Much effort was devoted to experimental studies of dry-rubbing materials for bearings and seals. The bearing material selected was the proprietary Boeing compact 6-84-1 sliding against Inconel 718 flame sprayed with chrome carbide or tungsten carbide. Browning *et al.* (1973) advise that

> . . . selection of the Boeing material was based on wear rates, the predictability of this material over extended period of operation and, to a lesser extent, a low coefficient of friction. Flame-sprayed material was selected for the mating surface because of wear rates, coefficient of friction and fabrication considerations, primarily the ease of flame spraying the desired bearing mating material.
>
> Chrome carbide is used at the hot end of the refrigerator which may operate at temperatures as high as 1200°F. Tungsten carbide is used for the remainder of the bearings which operate at temperatures below 200°F. Tungsten carbide cannot be used at the elevated temperatures of the hot end due to the difference in thermal expansion between the tungsten carbide and the base material Inconel 718.

AiResearch paid close attention to the development of highly effective heat exchangers. These include the recuperative heater, cooler, and freezer heat exchangers and the two regenerative exchangers, one for the hot side, one for the cold side. Some of the heat exchanger configurations adopted or evaluated are shown in Figs. 4.9 and 4.10 for the freezer, Fig. 4.11 for the ambient temperature cooler, and Fig. 4.12 for the high-temperature heater. Results measured on a prototype 5-W AiResearch Vuilleumier cooling engine were given by Browning *et al.* (1973) and are reproduced in Fig. 4.13. The review paper by Browning *et al.* (1973) discusses the AiResearch–NASA program from the aspects of long life, thermal design, and cold end movement.

Earlier, Gasser, Yoshikawa, and Browning (1972) presented a similar general review of the NASA–AiResearch program. Detailed reports of various aspects of the work were presented by Yoshikawa (1970), Browning and Potter (1972), and Browning, Miller, and Potter (1972). The design of a miniature Vuilleumier engine for NASA by AiResearch was described by Miller and Potter (1973). Al Sherman (1972) of NASA GSFC discussed the Vuilleumier cooling engines from the aspect of selected performance characteristics, and earlier, Sherman (1971) presented a comprehensive mathematical analysis of a Vuilleumier refrigerator.

Sherman (1978) has presented a contemporary review of the NASA development program relating to spacecraft cryocoolers. A progress report of this current activity was given by Sherman *et al.* (1979). It is of interest to note that further work on Vuilleumier cooling engines has been abandoned in favor of Stirling engine developments.

RCA Vuilleumier Cryocoolers

NASA interest in Vuilleumier engines appears to have started in the late 1960s. The earliest known manifestation of this was the ICICLE (integrated cryogenic cooled isotope engine) study carried out for NASA by the R.C.A. Corp. (Anon.—RCA, 1970).

Later on, Crouthamel and Shelpuk (1972) reviewed research on a combustion-heated Vuilleumier refrigerator for Army field-portable infrared night vision systems. The machine was designed to be self sustaining

Fig. 4.8a. AiResearch Vuilleumier cryocooler.

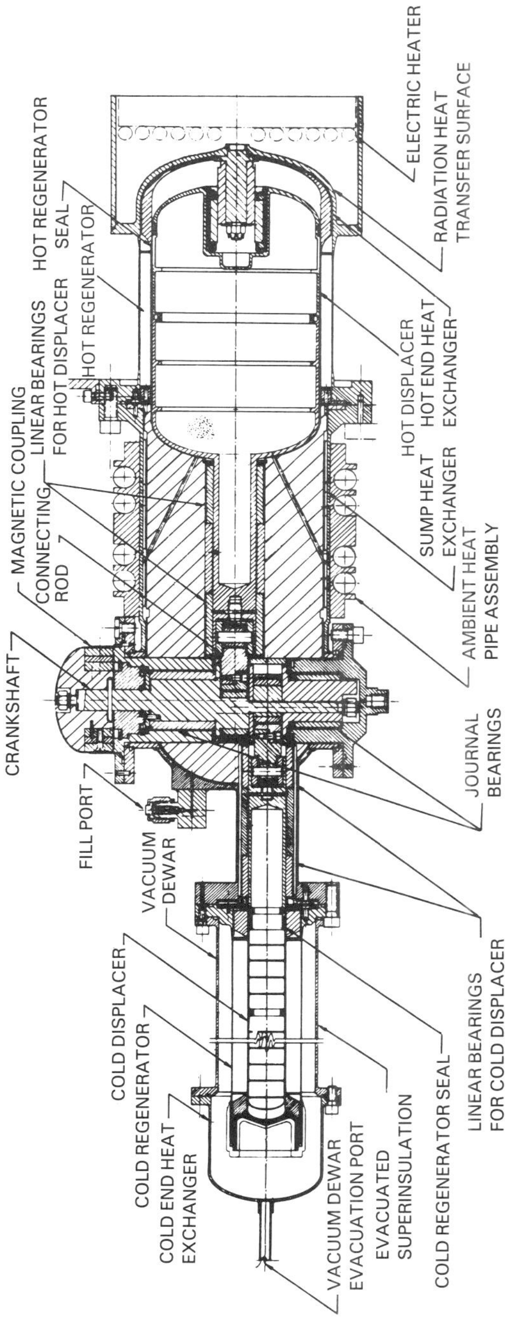

Fig. 4.8b. Cross section of AiResearch Vuilleumier cryocooler, 5 W capacity at 77 K, for NASA-Goddard Space Flight Center (after Browning *et al.*, 1973).

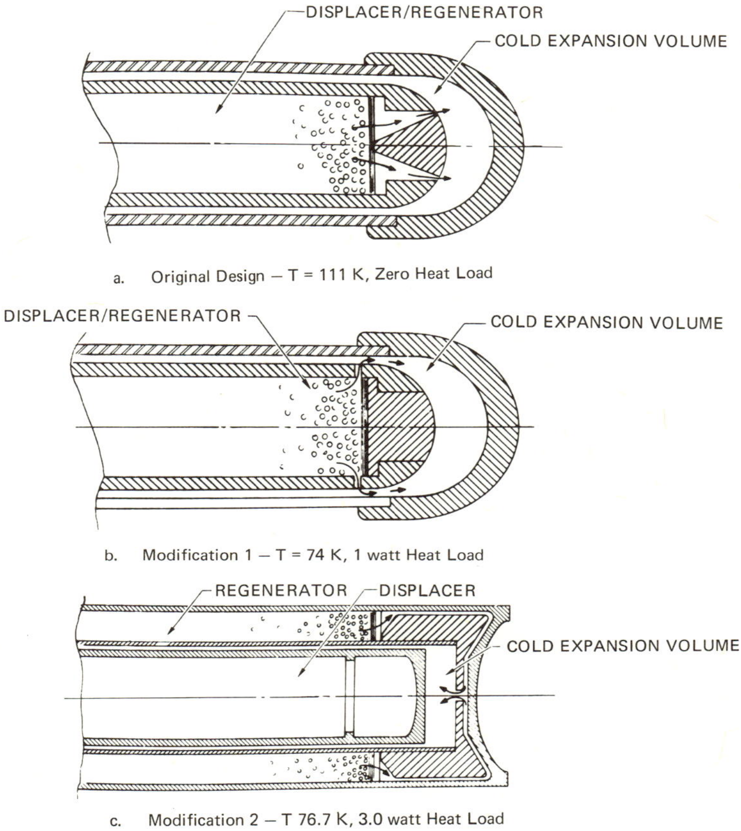

Fig. 4.9. Different freezer configurations for the AiResearch/NASA Vuilleumier cryocooler (after Browning *et al.*, 1973).

with no requirements for a drive motor. A feasibility model was constructed and tested. The thermocatalytic combustion of 40 g/hr of propane provided the sole energy input to produce

i. 1.6 W of refrigeration at 77 K,
ii. 16 W of internal power to operate the refrigerator,
iii. 2 W of external power to operate the external heat fan.

The combustion heated system had appreciable weight advantages compared with the electrical heated system (including storage battery for say a 12-hr mission). Total system weight for the electrical heated system was 3.1 kg for the engine plus 16.8 kg for storage batteries. The propane fuel and container for the same mission would weigh only 0.9 kg and the burner

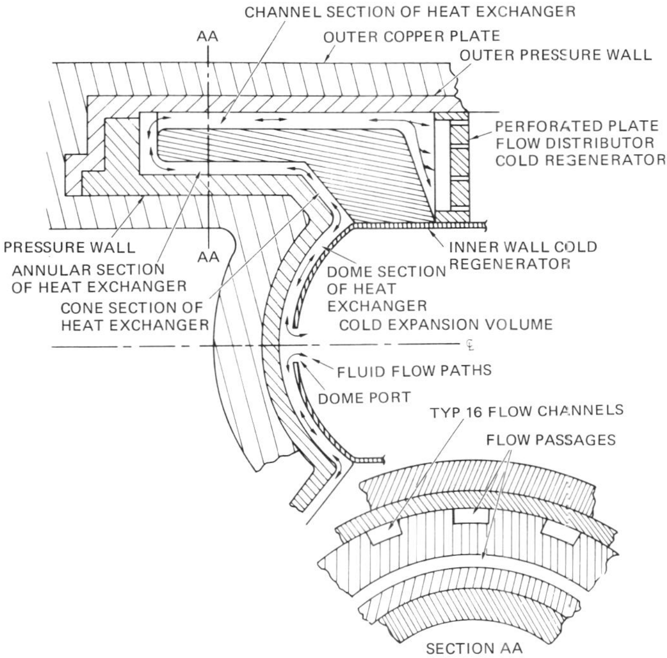

Fig. 4.10. Preferred freezer arrangement for the AiResearch/NASA Vuilleumier cryocooler (after Browning *et al.*, 1973).

system might increase the mass of the engine by 0.5 to 3.6 kg. The saving in mass would, therefore, be $(19.9-4.5)=15.4$ kg.

Ackermann (1971) made a theoretical analysis and parametric optimization of Vuilleumier cooling engines for the Night Vision Laboratory, Fort Belvoir, Virginia. An analytical paper for steady state operation of an idealized Vuilleumier cooling engine was contributed by Rule and Qvale (1969). Qvale, then an assistant professor at Purdue University and now at the Technical University of Copenhagen, had earlier worked with Professor Smith at the Massachusetts Institute of Technology in the development of a simulation program for Stirling engines.

EDITOR'S NOTE: *The following sections—Cycle of Operation, Inherent Thermodynamic and Heat Transfer Losses, VM Cooler Variations, and VM Accessories and Components—are reproduced, with minor editorial changes, from the unclassified United States Air Force report by Ronald White* (1976). *The report was prepared by White to satisfy*

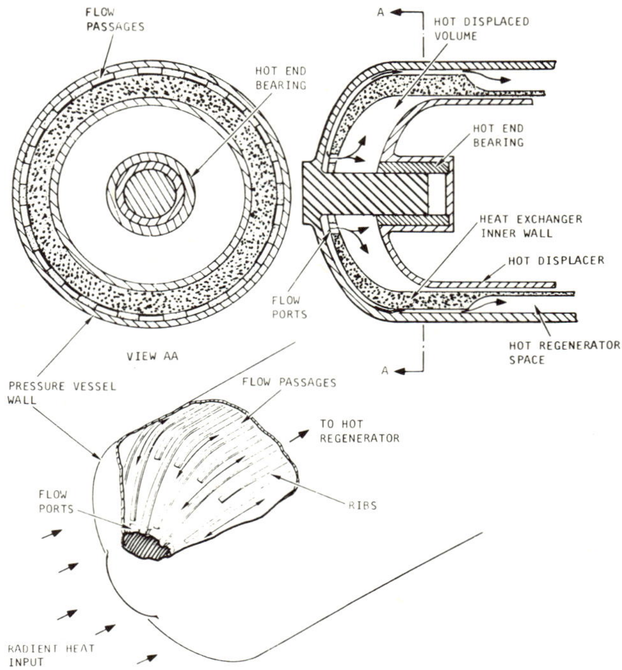

Fig. 4.11. Cooler arrangement for the AiResearch/NASA Vuilleumier cryocooler (after Browning *et al.*, 1973).

the requirements for a Master of Mechanical Engineering degree at the University of Dayton, Ohio.

I am most appreciative of the contribution to the public domain by Mr. White of the material in this report and for his permission for its reproduction here. For those wishing to study Vuilleumier engines beyond the level presented herein, this report is recommended as the best place to start. Considerations of space precluded reproduction of the entire report, and recourse to the original material is highly recommended.—G.W.

CYCLE OF OPERATION

The continuous production of refrigeration at low temperature requires the expenditure of a certain amount of energy. In most refrigeration cycles this is supplied as mechanical energy by an electric motor. In the Vuilleumier (VM) cycle the energy is supplied in the form of heat. Externally, the VM refrigerator is viewed as a machine which absorbs heat at both a high and low temperature and then rejects this heat at some intermediate temperature. The heat input at the hot end of the machine provides the energy required to produce the cooling effect at the cold end, while the heat from both the hot and cold ends is rejected at the ambient (intermediate) end.

The VM refrigerator, in its simplest form, is composed of two cylinders with displacers, two thermal regenerators, three heat exchange areas, connecting passages, and a mechanism to drive the displacers in the proper sequence.

The displacer is a long loose-fitting pistonlike device whose function is to move (displace) gas from one end of the cylinder to the other (see Fig. 4.14). The cylinder–displacer combination is arranged so that there is a gas passage connecting the two ends of the cylinder. With this connection, the effect of moving the displacer is to displace the gas from one end of the cylinder to the other. Note that the total volume of the gas remains constant (assuming an infinitely thin push rod). This is in contrast to a piston–cylinder combination that would change the gas volume. Since the

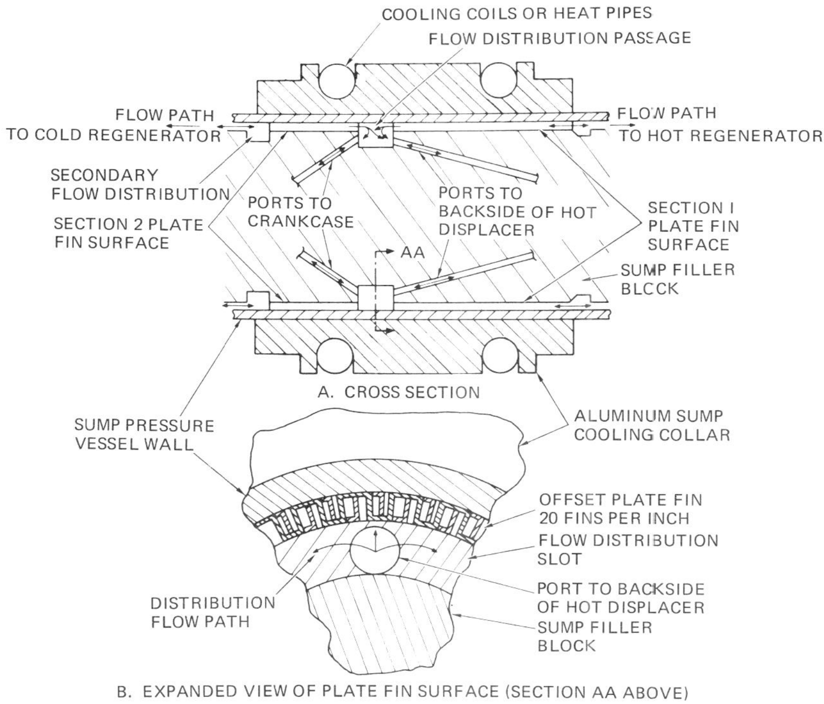

Fig. 4.12. Heater arrangement for the AiResearch/NASA Vuilleumier cryocooler (after Browning *et al.*, 1973).

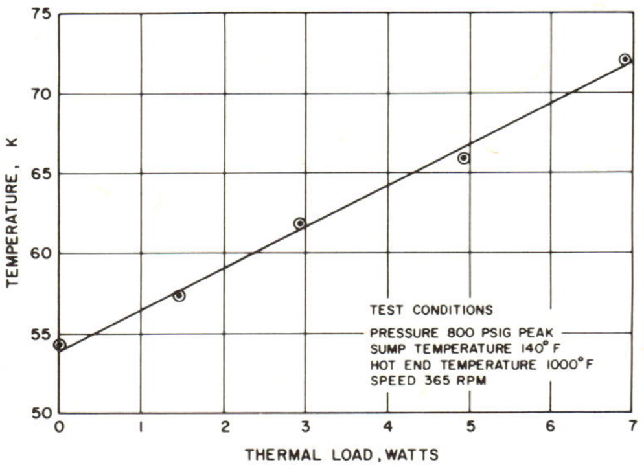

Fig. 4.13. Performance characteristics of the prototype AiResearch/NASA Vuilleumier cryocooler (after Browning *et al.*, 1973).

only VM that utilizes a piston is the special-case self driven VM, the term "*displacer*" will be used throughout this report as a reminder that the total gas volume of the system is not affected by the movement of the displacers. The pressure on each end of the displacer is theoretically the same (in real refrigerators it differs by the pressure drop through the regenerator and connecting passages). Therefore, the energy needed to move the displacer is very low, especially when compared to a refrigerator using pistons. Since the two ends of the cylinder will be at significantly different temperatures the displacer is long with thin walls and packed with insulation to reduce the end-to-end heat conduction.

Next, a thermal regenerator is arranged to connect the two ends of the cylinder so that gas passing from one end of the cylinder to the other must pass through the regenerator. The thermal regenerator is an energy storage device whose heat capacity greatly exceeds that of the gas. In operation, gas passing through a regenerator is heated or cooled depending on flow direction. For example, gas flowing from the ambient-temperature end to the high-temperature end of a cylinder is heated by the hot regenerator. The energy added to the gas was stored in the regenerator packing,

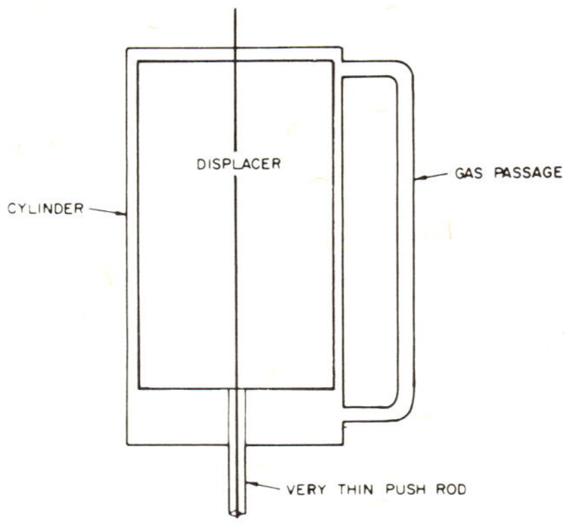

Fig. 4.14. Displacer and cylinder (after White, 1976).

or matrix, by gas flow in the reverse direction during a previous part of the cycle. The regenerator matrix consists of a porous material with a high ratio of surface area to volume, a high heat capacity, and low heat conductivity in the direction in which the gas flows through it. Examples of regenerator matrix materials used in VM refrigerators are balls, stacked screens, and "wool" made of (depending on temperature) stainless steel, monel, bronze, copper, or lead. At each point in the regenerator there is a very high heat transfer coefficient between the matrix and the gas. The regenerator connects two regions of different temperatures, and therefore has a temperature gradient along its matrix. At one end, the matrix has the temperature of one region and at the other end the temperature of the second region. Because of the presence of the regenerator, the gas can flow back and forth between the two volumes without serious loss or gain of heat. This thermal isolation (although not strictly true) tends to maintain each of the volumes at nearly constant, but different, temperature levels. The energy storage capability of the regenerator is generally great enough that its temperature profile is only slightly altered by providing energy to the gas.

To better understand how these components function together to produce refrigeration, the power input section will be discussed first, followed by the refrigeration section, and then the total refrigerator.

The power input section, the first cylinder–displacer, is composed of the hot cylinder, hot displacer, hot regenerator, hot heat exchanger, and part of the ambient temperature heat exchanger. Theoretically if it were separated from the remainder of the refrigerator it would look like Fig. 4.15. For this theoretical discussion, the regenerator is assumed to be 100%

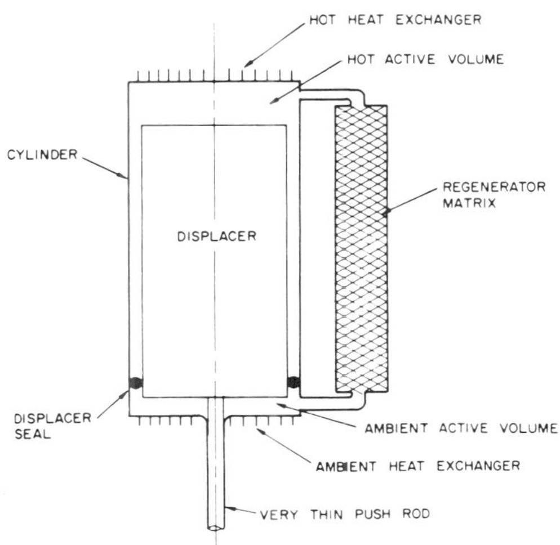

Fig. 4.15. Power input section of Vuilleumier cryocooler (after White, 1976).

efficient with no significant gas void spaces. Therefore the gas in the system is distributed between the two (hot and ambient) active volumes. The term "active volume" is used to describe the volume swept by the displacer. When the displacer is in the top position the hot active volume is nearly zero and all of the gas is in the ambient active volume. The ambient volume is held at an ambient temperature (significantly below the hot temperature) by the ambient heat exchanger. Using the perfect gas law, the gas is at low temperature (volume is constant), therefore it is at a low pressure. When the displacer is in the bottom position the ambient active volume is nearly zero and all of the gas is in the hot active volume. Since this volume is maintained at a high temperature (usually about 1200°F, 920 K) by addition of heat through the hot end heat exchanger, the gas is hot and therefore at high pressure. Since there are no valves in the system, all of the gas throughout the system including any at the other end of the cylinder (or in any dead volumes to be discussed later) is at the same pressure. Therefore, it can be seen that by moving the hot displacer, the system pressure can be varied from a maximum to a minimum in a cyclic way and that heat is supplied to the hot end and rejected at the ambient end. The hot section acts as a thermal compressor (whose effect on pressure is similar to a piston in a dead ended cylinder). A similar thermal compressor was patented by V. Bush (1939).

The manner in which the cold section, the second cylinder–displacer, of the VM refrigerator uses the changing pressure to produce a cooling effect is similar to that of the cold section of the Stirling cycle refrigerator, since the method by which the cyclic pressure is produced (thermal compressor or mechanical compressor) has little effect on the cold section.

Referring to Fig. 4.16, if the displacer is in the top position and pressure from the compressor is increasing, gas will flow into the ambient active volume (the cold active volume is zero) and increase in pressure. This compression causes the temperature of the ambient volume gas to increase, but this heat is rejected by the ambient heat exchanger so that at the end of the compression process the ambient active volume is filled with high-pressure ambient temperature gas. The cold displacer is then moved to its lower position and the gas is displaced from the ambient end, through the cold regenerator to the cold end. As the gas passes through the cold regenerator, it deposits its heat in the regenerator matrix and emerges at the cold end as cold high-pressure gas (any decrease in volume of the gas is made up by the compressor). After all the gas had been displaced to the cold active volume, the compressor displacer is moved so that the pressure decreases to the minimum. This decreases the pressure throughout the refrigerator including the cold section. The gas in the cold active volume undergoes an expansion (some of it goes back through the regenerator

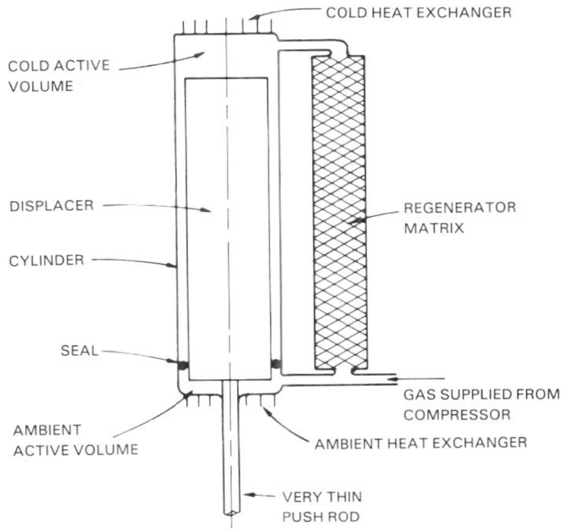

Fig. 4.16. Cold section of Vuilleumier cryocooler (after White, 1976).

toward the compressor) becoming colder low-pressure gas. This colder gas then has the ability to absorb heat from the device to be cooled (refrigeration). After the gas has absorbed the heat, the cold displacer is moved to the upper position, displacing the gas from the cold volume through the regenerator where it picks up the heat that was deposited on the previous half cycle and emerges at the ambient end where it will reject the heat picked up from the refrigeration load. After this displacement is completed, the compressor starts to increase the pressure and the cycle begins again.

From this description it can be seen that by combining two cylinder–displacer–regenerator assemblies and phasing their motion properly, useful refrigeration can be obtained.

To this point, nothing has been said about what sort of mechanism drives the displacer push rods. In the previous discussion a stop–start square wave motion was implied to simplify the discussion of the events in each section. In almost all of the real VM refrigerators built to date, the displacers have been driven harmonically. Harmonic drives (such as a crankshaft with connecting rods) are easy to fabricate and avoid the problem of high acceleration at the end of the stroke. However, harmonic drives do complicate the discussion of the VM cycle. A schematic of a harmonically (sinusoidal) driven VM refrigerator is shown in Fig. 4.17.

The cycle operates through the use of displacers moving the gas from one section to another without the requirement to compress the gas in a closed volume. Therefore, the pressures throughout the system are nearly equal at any moment. The seals shown in the schematic are to force all of the displaced gas through the regenerators. Since the pressure drop across the regenerators in real refrigerators is only a few pounds per square inch, the loading on the seals and displacer drive bearings is minimal, which

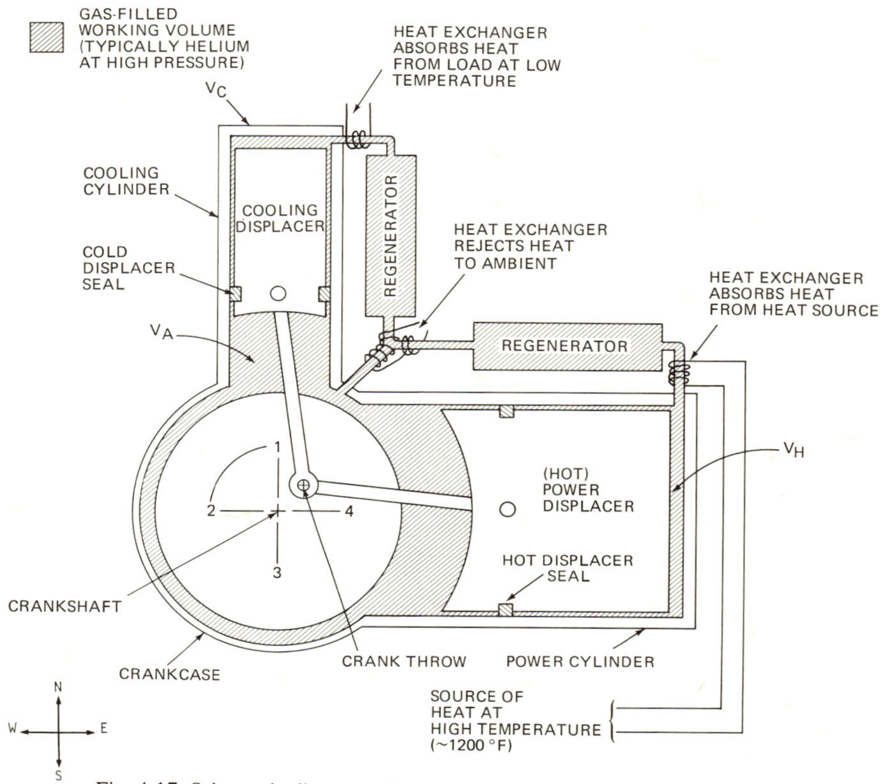

Fig. 4.17. Schematic diagram of Vuilleumier cryocooler (after White, 1976).

contributes significantly to the long life of VM refrigerators. Only a small (few watt) timing motor is needed to drive the mechanism, since the forces are so small.

The equations to be presented in this report assume isothermal operation in the refrigerator expansion and compression volumes since such an assumption makes it possible to derive a set of thermodynamic equations that are relatively simple and that are fairly representative of the thermodynamics of the VM refrigerator. Isothermal operation is not achieved in an actual refrigerator, but can be approached by careful design.

As shown in Fig. 4.17, the idealized VM refrigerator has two cylinders fitted with displacers which separate the refrigerator into three volumes V_H, V_A, and V_C. The subscript H is the hot temperature level, A is the ambient temperature level, and C is the cold temperature level. V_A actually consists of the summation of the two active volumes at the ambient end of the two cylinders. The crankcase is fitted with filler blocks so that the volume in the crankcase region is only the ambient active volume. It is

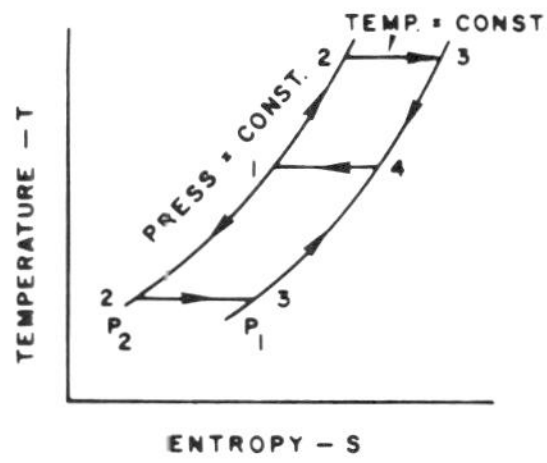

Fig. 4.18. Theoretical temperature–entropy diagram for Vuilleumier cryocooler (after White, 1976).

assumed that there are no pressure drops within the refrigerator in the idealized model. The thermal regenerators are assumed to be perfect; i.e., no temperature difference is required between gas flows in each direction in order to transfer heat, and therefore no heat flows through the regenerators over a complete cycle. The regenerators are assumed to have an infinite heat capacity and therefore the temperature of the regenerator is invariant with time. In the idealized refrigerator it is assumed that there are no dead volumes, that is, all the volume inside the refrigerator is active volume (V_H, V_A, and V_C). This means the gas volume in the regenerators, heat exchangers, around the sides of the displacers, around crankcase parts, at the ends of cylinders, etc. is assumed negligible (this will be modified later). The heat conduction axially through the regenerator matrix, and along the walls of the displacers and cylinders, is assumed negligible.

Attempts have been made to describe the cycle using a temperature–entropy (T–S) diagram such as Fig. 4.18. It is apparent that the diagram describes more than one unit mass of gas and the diagram is somewhat artificial since a great many simultaneous operations occur. For an example, let us look at three masses of gas that cycle between the ambient end and the hot end. The mass of gas in the hot end nearest the hot regenerator entrance will be the last mass in and (if no mixing occurs) the first mass out. Its T–S diagram would look like Fig. 4.19a. The second mass (between the first and last) would look like Fig. 4.19b. The third mass of gas is the first one in the hot end and the last one out. It would have a T–S diagram that looks like Fig. 4.19c. It can be seen that in order to use the T–S diagram to describe this cycle a large number of individual T–S diagrams for both ends of the refrigerator would need to be summed (mixing could further complicate the analysis).

However, the pressure–volume (P–V) diagram for each volume provides a clearer and more accurate picture of the operation than does the T–S diagram.

The P–V or indicator diagrams for the three sections of a well-designed V–M refrigerator ($P_1 \simeq P_2 = P_{max}$ and $P_3 \simeq P_4 = P_{min}$) are shown in Figs. 4.20a–4.20c. Note that the ambient volume is the sum of the ambient end volumes of the hot and cold cylinder.

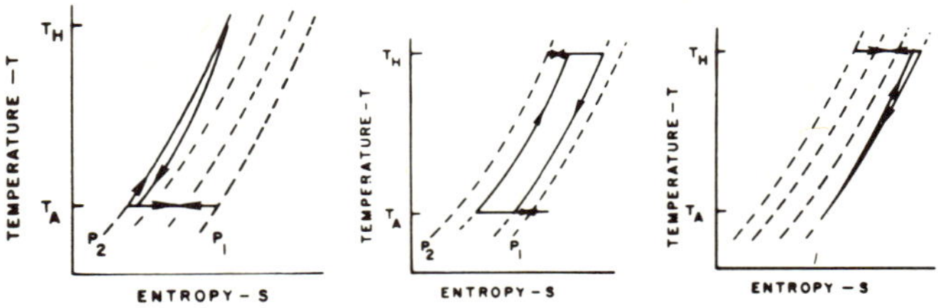

Fig. 4.19. Temperature–entropy diagram for the mass of gas in the hot space that is (a) last to enter, first to leave the hot space; (b) arbitrarily intermediate the first and last elements of gas to enter or leave the hot space; and (c) first to enter, last to leave the hot space (after White, 1976).

At each volume for a complete cycle, the first law reduced to

$$Q = (h_2 - h_1) + W$$

where Q and W represent energy per unit mass of gas flowing into or out of the active volume. The temperature of the gas crossing the boundaries does not vary with time (since ideal regenerators were assumed), so $h_2 = h_1$ and $Q = W$ or

$$Q = W = \oint P\,dV$$

The heat flow is equal to the work at that volume. There is, of course, no net work resulting from the idealized refrigerator because no pressure differences exist within the machine at any given time. Since the pressure at each end of each displacer is equal and by elementary geometry the change of volume at one end of a cylinder is exactly equal to the negative of the volume change at the other end, we have

$$\oint P_A\,dV_A = \oint P_H\,dV_H + \oint P_C\,dV_C$$

Since the displacer motion has been assumed harmonic, the pressure changes are assumed to occur isothermally, and if the phase separation angle is assumed to be 90°, the volume at each of the ends can be described

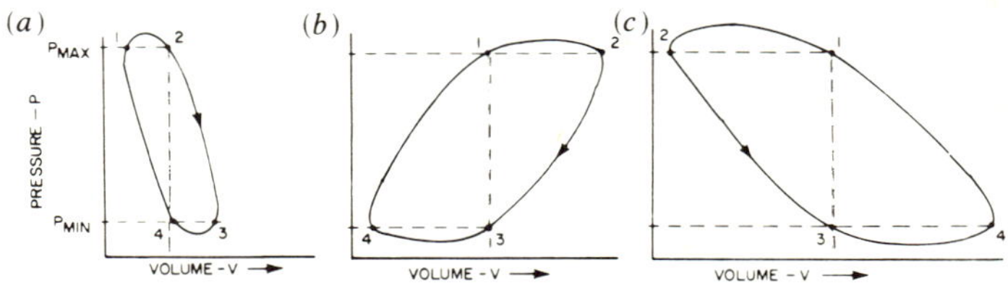

Fig. 4.20. Pressure–volume (indicator) diagram for the (a) cold volume; (b) hot volume, and (c) ambient volume of a Vuilleumier cryocooler (after White, 1976).

by the maximum active hot and cold volumes:

$$V_H = \tfrac{1}{2}V_{HM}(1-\cos\theta) \tag{4.1}$$

$$V_C = \tfrac{1}{2}V_{CM}(1-\sin\theta) \tag{4.2}$$

$$V_A = \tfrac{1}{2}V_{HM}(1+\cos\theta)+\tfrac{1}{2}V_{CM}(1+\sin\theta) \tag{4.3}$$

Here V_{HM} is the maximum swept volume of the hot end and V_{CM} is the maximum swept volume of the cold end.

If the gas is assumed to be ideal, $n = pV/RT$ and

$$n = \frac{p}{R}\left\{\frac{V_{HM}(1-\cos\theta)}{2T_H}+\frac{V_{CM}(1-\sin\theta)}{2T_C}+\frac{V_{HM}(1+\cos\theta)}{2T_A}\right.$$
$$\left.+\frac{V_{CM}(1+\sin\theta)}{2T_A}\right\} \tag{4.4}$$

rearranging, the equation becomes

$$n = \frac{p}{R}\left[V_{HM}\left(\frac{1}{2T_H}+\frac{1}{2T_A}\right)+V_{CM}\left(\frac{1}{2T_C}+\frac{1}{2T_A}\right)\right.$$
$$\left.+V_{HM}\left(\frac{1}{2T_A}-\frac{1}{2T_H}\right)\cos\theta+V_{CM}\left(\frac{1}{2T_A}-\frac{1}{2T_C}\right)\sin\theta\right]$$

multiplying numerator and denominator by $2T_A$ the equation becomes

$$n = \frac{P}{2T_AR}\left[V_{HM}\left(\frac{T_A}{T_H}+1\right)+V_{CM}\left(\frac{T_A}{T_C}+1\right)\right.$$
$$\left.+V_{HM}\left(1-\frac{T_A}{T_H}\right)\cos\theta+V_{CM}\left(1-\frac{T_A}{T_C}\right)\sin\theta\right] \tag{4.5}$$

This equation can be simplified by grouping the parameters, using the symbols a, b, and c defined by

$$a = V_{HM}\left(1+\frac{T_A}{T_H}\right)+V_{CM}\left(1+\frac{T_A}{T_C}\right) \tag{4.6}$$

$$b = V_{HM}\left(1-\frac{T_A}{T_H}\right) \tag{4.7}$$

$$c = V_{CM}\left(1-\frac{T_A}{T_C}\right) \tag{4.8}$$

Where a reflects the charge of gas in the refrigerator, b reflects the effect of the movement of the hot end displacer on the total pressure, and c reflects the effect of the cold-end displacer on the pressure.

The pressure, P, at any instant and at every point in the working medium is

$$P = \frac{2nRT_A}{a + b\cos\theta + c\sin\theta} \tag{4.9}$$

To obtain the angle θ at which P is maximum or minimum, differentiate the equation and set $dP/d\theta = 0$, and then $\tan\theta = c/b$. Using Eq. (4.9) to calculate the area inside the P–V diagrams:

$$Q_H = \oint P\,dV_H = nRT_A V_{HM} \int_0^{2\pi} \frac{\sin\theta}{a + b\cos\theta + c\sin\theta}\,d\theta \tag{4.10}$$

$$Q_C = \oint P\,dV_c = nRT_A V_{CM} \int_0^{2\pi} \frac{\cos\theta}{a + b\cos\theta + c\sin\theta}\,d\theta \tag{4.11}$$

where Q_H is the heat input to the hot end and Q_C is the heat input to the cold end.

These equations can be integrated using formula 2.558-2. from *Table of Integrals Series and Products* by I. S. Gradshteyn and I. M. Ryshik. Note that $a^2 > (b^2 + c^2)$ and that $\int d\theta/(a + b\cos\theta + c\sin\theta)$ must be integrated from 0 to π and from π to 2π with $-\pi/2 \leq \arctan\theta \leq \pi/2$. The result is

$$Q_H = \frac{2\pi cnR}{b^2 + c^2} V_{HM} T_A \left[1 - \frac{a}{(a^2 - b^2 - c^2)^{1/2}}\right] \tag{4.12}$$

$$Q_C = -\frac{2\pi bnR}{b^2 + c^2} V_{CM} T_A \left[1 - \frac{a}{(a^2 - b^2 - c^2)^{1/2}}\right] \tag{4.13}$$

These are the equations for the heat input to the hot end and the heat absorbed by the cold end (per cycle) for an ideal VM refrigerator.

The relationship between the maximum hot active volume and the maximum cold active volume is

$$\frac{V_{HM}}{V_{CM}} = \frac{T_H(T_A - T_C)}{T_C(T_H - T_A)} \tag{4.14}$$

This equation is derived from the ideal work and the ideal refrigeration that takes place in the machine during a cycle, when the pressure at crank position 1 is assumed equal to the pressure at crank position 2 as indicated in Fig. 4.17.

The coefficient of performance of this ideal refrigerator is

$$\begin{aligned} \mathrm{COP} &= \frac{Q_C}{Q_H} = \left(-\frac{V_{CM}}{V_{HM}}\right)\left(\frac{b}{c}\right) = \left(-\frac{V_{CM}}{V_{HM}}\right)\left\{\frac{V_{HM}[(1 - T_A)/T_H]}{V_{CM}[(1 - T_A)/T_C]}\right\} \\ \mathrm{COP} &= \left(\frac{T_H - T_A}{T_H}\right)\left(\frac{T_C}{T_A - T_C}\right) \end{aligned} \tag{4.15}$$

which is the same as a Carnot engine, driving a Carnot refrigerator, therefore the figure of merit (FOM) is unity.

The maximum pressure ratio is obtained by differentiating the pressure equation:

$$\frac{P_{\max}}{P_{\min}} = \frac{a + (b^2 + c^2)^{1/2}}{a - (b^2 + c^2)^{1/2}} \tag{4.16}$$

In order to better understand this ideal refrigerator an example was calculated and plotted in Fig. 4.21. The assumptions were as follows:

Temperature of hot end = 812 K
Temperature of ambient end = 366 K
Temperature of cold end = 77 K
V_{HM} is 87.23% of the total volume
V_{CM} is 12.77% of the total volume
Phase angle is 90°
$\theta = 0$ at hot displacer top dead center

The volumes V_H, V_A, and V_C were plotted to show how they vary with crank position, θ. The percentage of the total number of molecules N_H, N_A, N_C, in each of the volumes vs. crank position was also plotted, as was the resulting pressure (in arbitrary units).

From Fig. 4.21, it can be seen that although the maximum cold volume is only 12.77% of the total, at one point of the cycle 52% of the molecules of the system are in the cold end. Also the maximum percentage of molecules ever in the hot end is 55%. The maximum pressure occurs 45° before the maximum hot volume occurs and the minimum pressure occurs 45° before the minimum hot volume occurs. Noting the shape of the pressure curve in the region of 180° to 270°, it can be seen that the pressure curve is not quite sinusoidal.

All of the above equations and discussion assumed an ideal refrigerator; now the various nonideal factors will be added to these equations until a real refrigerator is discussed.

There are a number of void volumes in a real refrigerator. These include clearances around the displacers to prevent scraping, clearances at the ends of the cylinders to prevent hitting the end and to allow for thermal expansion, gas flow spaces in the regenerators, heat exchangers, gas transfer passages, and clearances in and around the mechanism. The void volumes must undergo the pressure variations of the cycle but do not contribute to the cooling effect. The void volumes decrease the maximum pressure ratio produced and, therefore, the amount of cooling produced.

To illustrate the magnitude of some of these void volumes, the cold regenerator flow passage void volume is usually 1.5 to 3.5 times the active

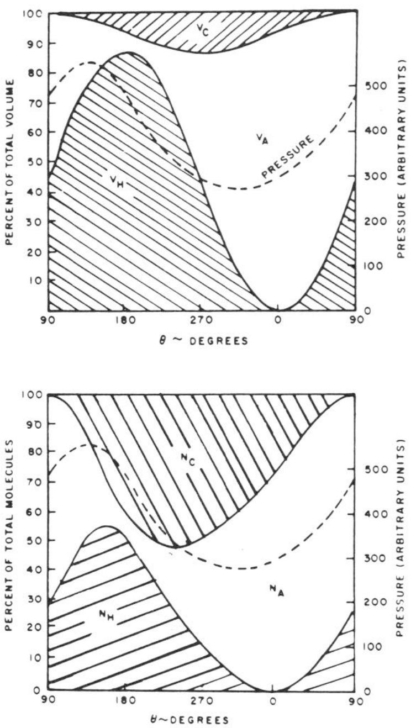

Fig. 4.21. Pressure–volume interactions in a Vuilleumier cryocooler (after White, 1976).

cold volume, while the displacer side clearance void volume is usually about 0.17 times the active cold volume. The effect of void volumes is related to the temperature of the void volume. The void volumes at the coldest temperatures have the greatest effect on the refrigerator performance while the void volumes at the hot end have the least effect on refrigerator performance. The effect of void volumes is more pronounced in miniature VM refrigerators, due in part to ordinary manufacturing tolerances. In miniature VM refrigerators, decreasing the void volume is both essential and expensive. Reduction of void volume requires close tolerances and unusually shaped parts (especially in the crankcase). Void volume reduction techniques used in current refrigerators even include the use of epoxy to fill the screw slots in the internal screws.

The effect of void volume can be accounted for by adding terms to Eq. (4.4) of the form:

$$\frac{V_{v1}}{T_{v1}}+\frac{V_{v2}}{T_{v2}}+\cdots\frac{V_{vn}}{T_{vn}}$$

where V_{v1} is a void volume and T_{v1} is the temperature of that void volume, or V_v/T_v where V_v is the sum of void volumes and

$$T_v = \frac{\int T_v \, dV_v}{\int dV_v}$$

The addition of void volumes will change Eq. (4.5) to

$$a = V_{HM}\left\{1+\frac{T_A}{T_H}\right\} + V_{CM}\left\{1+\frac{T_A}{T_C}\right\} + 2V_v\left\{\frac{T_A}{T_v}\right\} \tag{4.17}$$

The other equations (4.9)–(4.16) remain the same except the numerical value of a has changed. Typical values of the reduced void volume ratio (reduced void volume to cold end volume) for small VM refrigerators built to date are in the range of 1.5 to 3.7, where reduced void volume is defined as

$$V_{v\text{ reduced}} = V_{v\text{ actual}} \times \frac{T_C}{T_{v\text{ actual}}} \tag{4.18}$$

Another correction needed for the basic VM equations is the compressibility factor for the working fluid, which is helium. Helium has been used as the working fluid in all VM refrigerators built to date since it behaves as a nearly perfect gas and has the lowest temperature capability. The effect of compressibility of helium is slight until temperatures below 70 K are reached. Adding the compressibility factor (Z) to Eq. (4.4) gives

$$n = \frac{P}{2R}\left[\frac{V_{HM}(1-\cos\theta)}{T_H Z_H} + \frac{V_{CM}(1-\sin\theta)}{T_C Z_C} + \frac{V_{HM}(1+\cos\theta)}{T_A Z_A} + \frac{V_{CM}(1+\sin\theta)}{T_A Z_A} + \frac{2V_v}{T_v Z_v}\right] \tag{4.19}$$

which changes Eqs. (4.6) or (4.17), (4.7), (4.8), and (4.14) to

$$a = \frac{V_{HM}}{Z_H}\left(\frac{Z_H}{Z_A} + \frac{T_A}{T_H}\right) + \frac{V_{CM}}{Z_C}\left(\frac{Z_C}{Z_A} + \frac{T_A}{T_C}\right) + \frac{2V_v T_A}{Z_v T_v} \tag{4.20}$$

$$b = \frac{V_{HM}}{Z_H}\left(\frac{Z_H}{Z_A} - \frac{T_A}{T_H}\right) \tag{4.21}$$

$$c = \frac{V_{CM}}{Z_C}\left(\frac{Z_C}{Z_A} - \frac{T_A}{T_C}\right) \tag{4.22}$$

$$\frac{V_{HM}}{V_{CM}} = \frac{T_H Z_H}{T_H Z_H - T_A Z_A}\left(\frac{T_A Z_A - T_C Z_C}{T_C Z_C}\right) \tag{4.23}$$

$$\text{COP} = \left(\frac{T_H Z_H - T_A Z_A}{T_H Z_H}\right)\left(\frac{T_C Z_C}{T_A Z_A - T_C Z_C}\right) \tag{4.24}$$

Equations (4.9)–(4.13) and (4.16) are still valid, however, the values of *a*, *b*, and *c* within these equations have changed slightly.

To summarize the theoretical VM section, if compressibility effects and void volumes are to be accounted for, the equations to use are (4.9)–(4.13), (4.16), and (4.18)–(4.24). Equation (4.13), Q_C, is the gross refrigeration produced by the refrigerator. To obtain the net refrigeration produced by the refrigerator, all of the various cold end losses must be subtracted from the gross refrigeration. Equation (4.12), Q_H, is the P–V heat input to the hot cylinder. To obtain the actual heat (power) input needed, the various hot end losses must be added to the P–V input.

INHERENT THERMODYNAMIC AND HEAT TRANSFER LOSSES

Since it is impossible to build perfect regenerators and to eliminate all undesirable heat transfer processes in the refrigerator, these losses must be subtracted from the gross refrigeration to determine the net refrigeration available at the cold end. Similar losses must be added to the heat input to the gas to determine the required heat input to the hot end.

Shuttle Loss

Shuttle loss is caused by the mismatch of thermal gradients between the displacer and the cylinder. The cold cylinder wall is at ambient temperature at one end and at cryogenic temperature at the other. It has a fixed length and a gradient from warm to cold that is approximately linear. The displacer is shorter than the cylinder wall by the length of the stroke, however, it has the same temperature extremes: warm at one end and cold at the other. When the displacer is at one extreme of its travel the temperature gradients are somewhat mismatched. As it passes through its stroke and reaches the other extreme the gradients are again mismatched, but now they are mismatched in the opposite direction so that the displacer picks up heat from the cylinder when it is at the warm end and it gives off heat to the cylinder when it is at the cold end of its stroke. Hence, there is a picking up of heat at the warm end, a shuttling of the displacer to the cold end, where it drops off the energy to the cylinder. It is thus termed a shuttle loss.

If the motion of the displacer is approximately harmonic and if the thermal time lag of the cylinder and displacer materials is small compared with the reciprocating time, the shuttle heat transfer can be computed from the following equation (Leo, 1971):

$$Q_{SH} = 0.186\, Y^2 C \frac{k_g}{S} \frac{(T_w - T_c)}{L_{cy}} \tag{4.25}$$

where Q_{SH} is the shuttle heat loss, Y is the stroke, C is the wetted perimeter ($\pi \times$diameter), S is the radial clearance between displacer and cylinder, T_w is the temperature of the warm end of the cylinder, T_c is the temperature of cool end of cylinder, L_{cy} is the length of cylinder, k_g is the thermal conductivity of the gas (helium). The terms "warm" and "cool" are used since this equation is used to compute the shuttle losses of both the hot and the cold ends of VM refrigerators.

Pumping Loss

The pumping loss is due to the fixed clearance volume that must exist between the displacer and the cylinder so that the displacer can move without rubbing. This volume is bounded on one end (usually the ambient end) by a displacer seal and is open on the other end. Because this volume around the displacer allowed for running clearance is fixed, the mass of gas in this volume at any one time is proportional to the (cycling) pressure. At the minimum pressure point there is a minimum mass of gas in this volume. As the pressure increases toward the maximum pressure point, gas flows into this volume proportional to the increasing pressure. Therefore, using the cold cylinder as an example, cold gas flows into this volume and down into the warm areas where it picks up energy. Then as the gas pressure falls from the maximum back to the minimum pressure, some warm gas flows out of this volume into the cold regions. This causes a loss of refrigeration called the pumping loss. This loss occurs on both the cold end and the hot end of VM refrigerators when displacer seals are used to force gas flow through the regenerators. The following equation for pumping loss was derived by Leo (1971):

$$Q_{\mathrm{PU}} = \frac{2(\pi D_c)^{0.6} L_{cy} (P_{\max} - P_{\min})^{1.6} N^{1.6} C_p^{1.6} (T_w - T_c) S^{2.6}}{1.5 Z R^{1.6} k_g^{0.6} [(T_w + T_c)/2]^{1.6}} \tag{4.26}$$

where Q_{PU} is the pumping loss, D_c is the diameter of the cylinder, L_{cy} is the length of the cylinder, $P_{\max}$ is the maximum pressure, $P_{\min}$ is the minimum pressure, N is the cycle speed, C_p is the specific heat of the gas, T_w is the temperature at the warm end, T_c is the temperature at the cool end, S is the radial clearance between displacer and cylinder, Z is the compressibility factor for gas, R is the gas constant, and k_g is the conductivity of the gas.

By examining Eqs. (4.25) and (4.26) it can be seen that there is an optimum gap (clearance) for best performance (lowest losses). If the gap is too small the shuttle loss will be high and if the gap is too large the pumping loss will be high. Hence, for a fixed diameter and stroke, there is an optimum gap for least losses and there is much to be gained by maintaining the gap accurately. To accomplish this the materials of the cylinder and displacer are selected so that their coefficients of thermal expansion are matched.

Heat Transfer through Displacer

This loss is the heat transfer through the displacer due to the difference in the temperatures of the ends of the displacer. It is expressed as

$$Q_D = \frac{k_D A_D}{L_D}(T_w - T_c) \tag{4.27}$$

or if the displacer is hollow

$$Q_D = \frac{\pi k_D (D_{\mathrm{OD}}^2 - D_{\mathrm{ID}}^2)(T_w - T_c)}{4L_D} \tag{4.28}$$

where Q_D is the conduction heat loss through the displacer, k_D is the conductivity of the displacer material, A_D is the area of the end of the displacer, L_D is the length of the displacer, T_w is the temperature of the warm end, T_c is the temperature of the cool end, D_{OD} is the outside diameter of the displacer, and D_{ID} is the inside diameter of the displacer.

Heat Transfer through Cylinder Wall

This loss is the heat transfer due to the difference in the temperatures of the ends of the cylinder. If the inside and outside diameters of the cylinder are known Eq. (4.28), with the appropriate conductivity and dimensions, can be used. If the refrigerator is being designed and the thickness of the cylinder wall is not known the following equations can be used:

$$Q_{\mathrm{CY}} = \frac{\pi k_{\mathrm{cy}} P_{\max} D_{\mathrm{cy}}^2 (T_w - T_c)}{2\sigma L_{\mathrm{cy}}[(1 - P_{\max})/2\sigma]} \tag{4.29}$$

where Q_{CY} is the conduction heat loss through the cylinder, k_{cy} is the conductivity of the cylinder material, $P_{\max}$ is the maximum pressure, D_{CY} is the inside diameter of the cylinder, L_{CY} is the length of the cylinder, and σ is the allowable stress.

Heat Generated by Friction between Displacer and Cylinder

This loss is due to friction between the cylinder wall and the displacer riders and seals:

$$Q_f = \oint F_f \, dY$$

where F_f is the friction force in the direction of travel and Y is the stroke. It is rather difficult to evaluate this loss since much must be known about the seal forces, rider loads, and the locations at which this loss is generated.

Regenerator Losses

The regenerators of the VM refrigerator must be very efficient if the refrigerator is to be a practical device for producing refrigeration at temperatures below 100 K. In fact, the overall performance of the machine depends directly on the efficiency of the regenerators. The regenerator losses can be divided into two types:

Heat Load Due to Friction in the Regenerator. This refrigeration loss is caused by aerodynamic heating when the working medium flows through the regenerator:

$$Q_{RF} = \frac{\Delta P_{\text{reg}}(P_{\text{max}} + P_{\text{min}}) V_{CM} N}{\rho J Z R T_c} \tag{4.30}$$

where:

$$\Delta P_{\text{reg}} = \frac{(P_{\text{max}} + P_{\text{min}})^2 V_{CM}^2 N^2 f L_R}{2 Z^2 R^2 T_C^2 g_c A_{\text{reg}}^2 D_e \rho} \tag{4.31}$$

or

$$Q_{RF} = \frac{(P_{\text{max}} + P_{\text{min}})^3 V_{CM}^3 N^3 f L_R}{2 Z^3 R^3 T_C^3 g_c A_{\text{reg}}^2 \rho^2 D_e J} \tag{4.32}$$

where Q_{RF} is the loss due to aerodynamic friction in the regenerator, P_{max} is the maximum system pressure, P_{min} is the minimum system pressure, V_{CM} is the maximum displaced cold volume (if a hot regenerator is being calculated this would be maximum displaced hot volume), T_C is the temperature of the displaced volume (hot or cold), N is the cycle speed, f is the friction factor, L_R is the length of the regenerator, Z is the compressibility factor for the working fluid, R is the gas constant for the working fluid, A_{reg} is the cross-sectional area of the regenerator, ρ is the density of the working fluid, D_e is the equivalent hydraulic diameter, g_c is the gravitational conversion factor, and J is the conversion factor.

Heat Load Due to Limiting Value of Film Coefficient in Regenerator. The limiting value of the film coefficient in the regenerator prevents sufficient cooling of the working fluid as it flows from the hot to the cold end of the regenerator. Leo (1971) discussed in general the derivation of the following equations that evaluate the amount of energy that remains in the working fluid because of this limiting value of film coefficient:

$$Q_{RH} = (1 - N_R) C_p (T_w - T_c) \frac{(P_{\max} + P_{\min}) V_{CM} N}{2 R T_C Z} \tag{4.33}$$

where

$$N_R = \frac{h' h'' L_R^2}{h' h'' L_R^2 + h'' L_R C_p G r_h + h' L_R C_p G r_h} \tag{4.34}$$

Where Q_{RH} is the loss due to the limiting value of the film coefficient in the regenerator, N_R is the regenerator efficiency, C_p is the specific heat of the working fluid, $(T_w - T_c)$ is the temperature difference across the regenerator, V_{CM} is the maximum displaced cold volume (if a hot regenerator is being calculated this would be the maximum displaced hot volume), T_C is the temperature of the displaced volume (hot or cold), N is the cycle speed, R is the gas constant of the working fluid, Z is the compressibility factor, h' is the heat transfer coefficient—flow out, h'' is the heat transfer coefficient—flow in, L_R is the length of the regenerator, G is the mass velocity (mass flow rate per unit cross-sectional area), and r_h is the hydraulic radius. The value of N_R for cold regenerators is usually in the range from 0.995 to almost 1.0 and for hot regenerators is in the range of 0.9 to 1.0.

These equations for the regenerator losses are only approximate since they are based on average mass flow rates rather than instantaneous mass flow rates. Another way to compute the performance of the VM refrigerator was partially outlined by Globe *et al.* (1971). It involves cutting the VM refrigerator into a large number of control volumes and determining the mass flow rate into and out of each of these control volumes as a function of crank position. The losses, especially the regenerator and pumping losses, can be calculated and their effect on the pressure can be iterated with the mass flow equations so that a more accurate description of the refrigeration at every crank position is obtained. This can be done for an existing refrigerator design with the aid of a computer but is extremely difficult to do when optimizing a new design since the control volumes themselves are being changed during the optimization process. Therefore, the equations presented above or similar equations are usually used for optimizing new designs.

Net Refrigeration

To obtain the net refrigeration, the cold end losses for each stage are summed and subtracted from the gross refrigeration for that stage. In multistage refrigerators some of the losses from colder stages appear as increased refrigeration at warmer refrigeration stages. These include conduction losses, regeneration heat loss, shuttle loss, and pumping loss. These losses from a colder stage are called interstage heat flow and should be added to the gross refrigeration of the warmer stage. Other cold end losses that were not discussed here but might be applicable depending on refrigerator design are regenerator conduction, insulation, seal leakage losses, and heat leaks down instrumentation leads.

Total Heater Power Input

To calculate the heater power required at the hot end, the hot end losses are summed and added to the heat input to the gas. Losses not discussed above that should be included depending on the actual design are the hot end insulation loss, conduction losses down heater and instrumentation leads, and heat leaks through the insulation inside the hollow hot displacer.

VM COOLER VARIATIONS

Multistage VM Coolers

In many applications there is a need to produce useful refrigeration at more than one temperature level at the same time. An example of this is cooling an electronic device to a very low temperature while cooling a Dewar heat shield surrounding the device to an intermediate temperature to intercept the Dewar heat leaks. This saves considerable power since the majority of the heat is removed at the higher temperature for much less input power (and refrigerator size) than if all the heat had been removed at the lower temperature.

Two mechanisms have been used to add additional cold stages to VM refrigerators. The parallel cylinder system (Fig. 4.22) adds another smaller diameter cold cylinder parallel to the first cold cylinder. In this configuration the displacer seals are at the ambient temperature region of the displacers (an advantage) but fabrication of the cylinder assembly is more difficult. Keeping the two cylinders straight and parallel from brazing temperature down through cryogenic temperature is difficult and expensive.

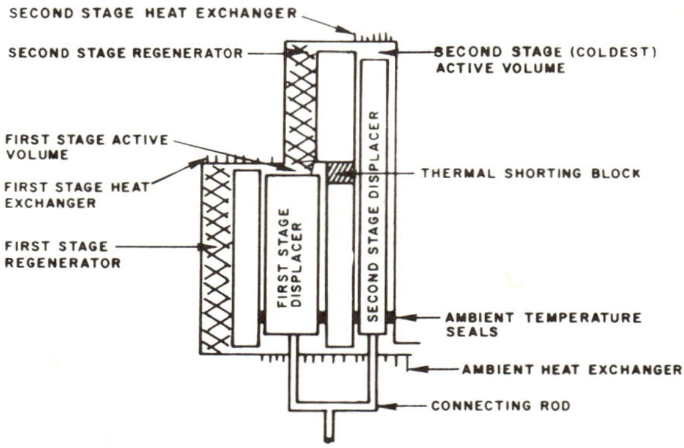

Fig. 4.22. Multiple stage Vuilleumier cryocooler arrangement with parallel cylinders (after White, 1976).

The other configuration is the series cold cylinder configuration (Fig. 4.23). This configuration is easier to fabricate but requires a displacer seal at the base of the second stage displacer that seals properly at the first stage (cryogenic) temperature.

Adding additional cold stages affects the previously developed equations in the following ways. Additional cold stage factors (second term) are added to Eq. (4.20). Additional terms are added to Eq. (4.22) and additional equations similar to Eq. (4.13) are written to describe the gross refrigeration of the additional stages. The ratio between the cold active

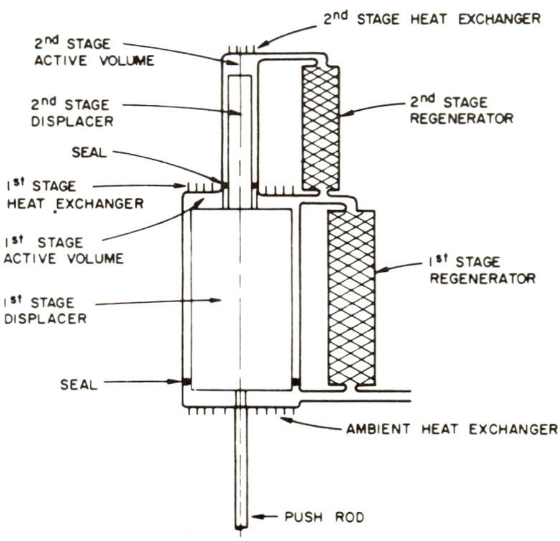

Fig. 4.23. Multiple-stage Vuilleumier cryocooler arrangement with tandem cylinders (after White, 1976).

volumes of a two-stage refrigerator is obtained by taking the ratio of the gross refrigeration equations [similar to Eq. (4.13)]. The result is

$$\frac{Q_{C_1}}{Q_{C_2}} = \frac{V_{CM_1}}{V_{CM_2}} \tag{4.35}$$

Phase Angle

Phase angles other than 90° were investigated by E. B. Qvale and T. T. Rule (1968) and by B. Leo (1971). Phase angles other than 90° complicate the Eqs. (4.1)–(4.13) considerably and make the fabrication of parts more difficult. The optimum phase angle is a function of the active swept volumes of the hot and cold cylinders and of the thermal boundary conditions. The investigations showed that for refrigerators with heat rejection temperatures near room temperature, the optimum phase angle would be in the range from 90° to 102° depending on cold end temperature (the lower the temperature the greater the angle). Multistage refrigerators further complicate these relationships and reduce the range of optimum phase angles. Since the gross refrigeration is changing very slowly with respect to phase angle near the optimum, most manufacturers are using a 90° phase angle.

Similar Cycles

There are other heat powered refrigeration cycles similar to the VM cycle. One by Bush (1938) is quite similar to the VM except that the two ambient volumes (one at the ambient end of the cold cylinder and one at the ambient end of the hot cylinder) are separated by a thermal regenerator and reject heat to heat sinks at different temperatures. Another heat-powered refrigerator was patented by Taconis (1951). It differs from the VM in the timing of the movements (three instead of four motions) of the displacers. Another heat-powered refrigerator was patented by Hogan (1964). It produced cooling in the 10–20 K range, while the hot end absorbed heat at room temperature and the heat rejection was at 77 K (the heat was rejected to liquid nitrogen). A patent by Cowans (1968) describes a modification to the VM refrigerator that allows it to drive its own displacers and produce useful shaft power. This is done by increasing the cross-sectional area of either the hot displacer connecting rod or both connecting rods, so that with the addition of connecting rod seals, and by lowering the crankcase pressure below the minimum pressure in the VM cycle, a net force can be created to drive the refrigerator. This has the advantage that the small timing motor used on most VM refrigerators is not needed (but something must give it a shove to get it started). However, this adds the life-limiting problem of dynamic connecting rod seals that must be able to seal against the full cycle pressure (several hundred pounds per square inch). This type of sealing problem is avoided by most VM refrigerators since in a "pure" VM cycle the only dynamic seals in the system are the displacer seals. Displacer seals usually experience very small pressure differences of 5 to 15 psi which contributes to their very long life.

VM ACCESSORIES AND COMPONENTS

Hot End Temperature Controller

One important accessory required by electrically heated VM refrigerators is the hot end temperature controller. By examining the VM theoretical equations, it can be seen that the higher the temperature of the hot end, the higher the efficiency of the refrigerator. VM refrigerators are usually designed to operate at the highest temperature possible, consistent with metallurgical limits. The most popular hot end material is Inconel 718. The strength of this material falls off quite rapidly above 1250°F, so VM refrigerators are usually designed to operate at about 1200°F. However, there are several problems with trying to operate at 1200°F. A change of input voltage to the heater can change the heater power and the hot end temperature. Aircraft power supply voltages can vary as much as $\pm\frac{1}{6}$ of the mean voltage. In addition, the ambient heat rejection temperature aboard an aircraft can vary as much as 200°F, which will affect the power requirement and therefore the hot end temperature. Cold end load changes also have an effect on the power required and the hot end temperature. Refrigerator malfunctions such as loss of working fluid or a stalled motor prevent the working fluid from absorbing sufficient heat and cause the hot end to overheat. The heater is usually sized to supply the correct power at the minimum voltage and maximum ambient temperature. To prevent hot end overheating problems, a hot end temperature controller is used.

Proportional controllers are usually used since maintaining the heater at a nearly uniform temperature reduces the heater stresses and improves the life of this component. Frequently a simple on–off controller, set at a higher temperature, backs up the primary controller as an additional safety measure. A variety of controller concepts have been used. One of the most popular is pulse width modulation, due to its high efficiency. However, this type controller requires considerable filtering and shielding to prevent the electromagnetic interference it creates from affecting nearby equipment. Other concepts include linear proportional control, zero voltage (ac) switching, slow on–off switching, and mechanical devices such as Curie point switches and vapor bulb thermometers. On large VM refrigerators the controller problem is reduced by calculating both the minimum power and the maximum power required. The minimum power is then supplied by a large heater with a simple on–off controller for malfunction protection only. The difference between the minimum and maximum power is supplied by a smaller heater with a proportional controller and necessary shielding. This arrangement is more efficient and reduces the size and weight of the controller and electronic filters.

Heaters

Two types of electric hot end heaters have been used in VM refrigerators. The furnace type is a ribbon of heater wire wrapped on a ceramic mandrel and held in place with cement. The ceramic furnace surrounds the hot end of the hot cylinder and transfers heat by radiation. The Calrod-type sheathed heater has been the most popular. Heat is transferred either by radiation (Yoshikawa, 1970) or by brazing the heater sheath directly to the hot cylinder. Since the watt density of the heaters required by most VM designs is very high (for this type of heater) the heater must be properly heat sunk to the hot cylinder or burnouts will occur. Care also must be taken to be sure that the active (heat-producing) portion of the heater terminates while still thermally connected to the hot cylinder. The larger-diameter low-resistance lead-in wire that runs between the heater wire and the terminal (inside the sheath) must be of a material that does not embrittle or corrode when exposed to insulation or atmospheric contaminants. Single-ended straight wire heaters have caused numerous failures and have been largely abandoned in favor of two-ended helically wound single wire heaters. Straight wire heaters are available in smaller sheath diameters but must use smaller-diameter heater wire since the total length of the heater wire is less. This, coupled with the possible nonuniform reduction of heater wire diameter during the swaging of the heater sheath and higher stresses during heater cycling have contributed to numerous straight wire heater failures. The helical single-wire heaters have larger heater sheaths (less convenient for the refrigerator designer) but have a larger-diameter longer heater wire that does not change cross-section (the helix angle changes) during the swaging of the heater sheath. The heater wire is closer to the sheath (less temperature drop) and is less sensitive to thermal cycling. These heaters are much more reliable than straight wire heaters. In critical applications additional redundant heaters are added to avoid scrapping an expensive hot cylinder assembly due to a burned out heater.

Motors

As mentioned earlier, ac induction motors with the rotor inside the helium space and the stator with its windings outside the helium space have been successfully used in a large number of applications. These motors are either two or three phase and are purchased with a matching inverter. Total efficiency for the motor and inverter is about 25%. In small VM refrigerators, the motor power is a small fraction of the total power, therefore cleanliness and reliability are more important than efficiency. In a few applications an inverter could not be used (due to space or ambient temperature problems) and since life was less critical a dc brush-type motor was used. Special brush materials were used along with special commutator coatings. The motor windings were potted to reduce the generation of contaminates. Very few data have been gathered on this motor, so its limitations are still unknown. A brushless dc motor is being used on one refrigerator. The motor efficiency is expected to be at least 55%. It will be in the oil-filled crankcase of the refrigerator and should present no contamination problem.

Regenerators

The cold regenerators have been previously discussed and are discussed in considerable detail in all of the references. The term "internal regenerator" is used for regenerators inside the displacer and the term "external regenerator" is used for regenerators attached to or a part of the cylinder. Cold regenerator matrix materials are usually screens of 100 mesh to 500 mesh in copper alloys or stainless steel and balls of monel or lead in sizes down to 0.002 in. Lead balls are usually used for temperatures below 50 K since lead is one of the few materials with appreciable specific heat at these temperatures.

The hot regenerators are described as internal or external also, however, the forms of this matrix are more varied. Stacked screens, balls, tubes, and the annulus have been used for the hot regenerator matrix. The internal annular regenerator is composed of the walls of the displacer and the cylinder. In this configuration the gas flows between the displacer and the cylinder. A displacer seal is not used. This eliminates one wearing part (the seal) and eliminates the pumping loss, but makes the radial location of the displacer within the cylinder very critical, which in turn makes hot rider ring wear extremely critical. As an example, if the regenerator is designed with a 0.007-in. radial gap between the displacer and cylinder, and if rider wear allows the displacer to be out of concentricity by 0.002 in., 15% of the gas flow is on the narrow side of the regenerator while 85%

of the flow is on the wide side. This causes the regenerator loss to be doubled (Leo, 1971). Since lubrication and wear of the hot rider is a serious problem the internal annular regenerator is rarely used anymore. The external annular regenerator is composed of the walls of the cylinder and one or more liner sleeves. The liner sleeves must be very thin and concentric with the cylinder. A displacer seal riding on the sleeve assures gas flow through the regenerator. Concentricity is a problem with this regenerator also.

The tubular regenerator is composed of small-diameter thin wall tubes constrained (usually in a single layer) between the cylinder (or displacer) and a cylindrical liner with the tube axis in the direction of the cylinder axis. The gas flows in the axial direction either through the tubes or in the triangular-shaped spaces between the tubes and the liner (or cylinder). This type of hot regenerator is easy to fabricate and is being used in several VM refrigerators. The screen and ball regenerators are similar to the cold regenerators except matrix elements are larger and of materials such as monel and stainless steel.

Heat Rejection

Heat rejection has been accomplished in several ways. These include rejection to forced ambient air (Globe *et al.*, 1971), rejection to a pumped liquid which in turn rejects the heat to air or a radiator (Doody, 1971), and rejection by heat pipes (Yoshikawa, 1970; Browning and Potter, 1972; AiResearch Mfg. Co., 1972, 1974; Browning *et al.*, 1972; Miller and Potter, 1973, 1974).

SPLIT-VUILLEUMIER CRYOCOOLER

In the above discussion of Vuilleumier cryocoolers, White considered only integral units, particularly high-capacity units intended for spacecraft. The omission of split-Vuilleumier cryocoolers was surprising for they are an important class of miniature cooler much used in military applications.

Figure 4.24 shows a split-Vuilleumier cryocooler manufactured by Cryogenic Technology Inc. Table 4.4 shows the specifications of the same unit designated Model No. VM-1 and described as having the following advantages:

i. The U.S. Air Force standard cryogenic cooler;
ii. Highest MTBF (mean time before failure) of existing coolers of similar size and capacity;

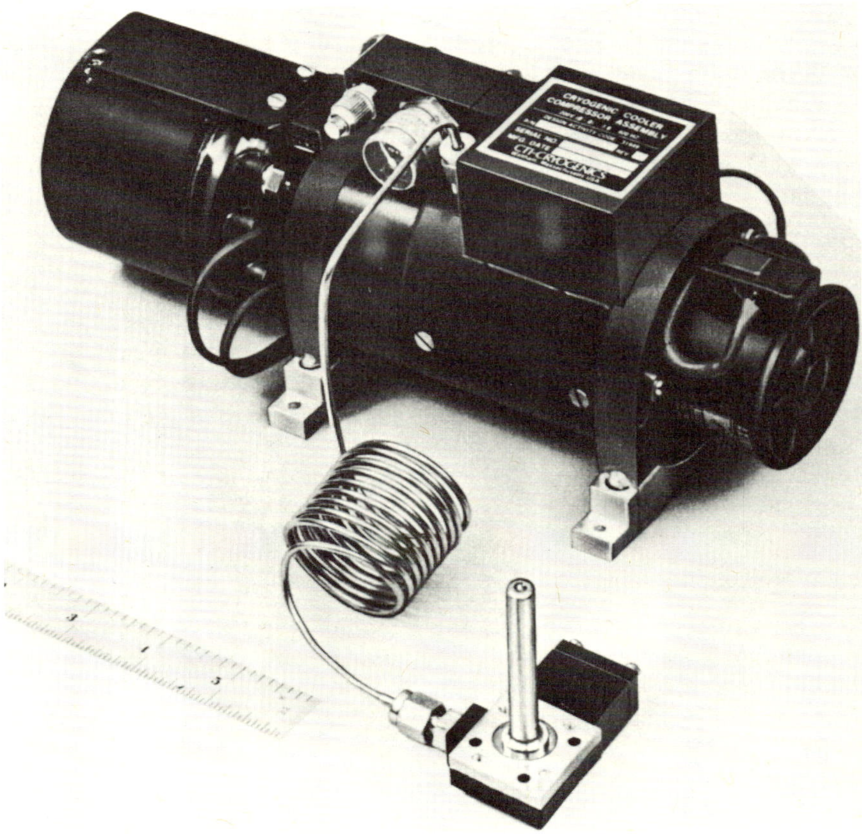

Fig. 4.24. Photograph of split-Vuilleumier cryocooler, Model No. VM-1. Capacity 1 W at 77 K (courtesy Cryogenic Technology, Inc.).

iii. Shock and vibration resistance ratings higher than any other existing coolers;
iv. Split configuration for remote cooling;
v. Low operating cost over a long in-service life;
vi. Simplified maintenance and service requirements.

Split coolers are attractive for applications where minimum vibration and mass are important as in gimbaled, swiveling infrared sensors, or missile sights. The compressor and expander units are coupled by a thin flexible tube permitting independent motion of the expander cylinder. The void volume of the connecting tube increases the dead volume and thereby

Table 4.4. Specifications for CTI Model No. VM-1 Split-Vuilleumier Cooling Engines[a]

Refrigeration types	Split-Vuilleumier cycle
Refrigeration capacity	1.0 W @ 77 K (25°C ambient) 0.7 W @ 77 K (55°C ambient)
Cooldown time to 85 K	<10 min (55°C ambient)
Weight	10.0 lb (4.54 kg)
Power requirements	115 Vac, 3ϕ, 400 Hz, 370 W
Ambient temperature limits	Operating: −54°C to 55°C Storage: −54°C to 71°C
Scheduled maintenance	2200 operating hours
Military specifications	Meets applicable U.S. Military specifications

[a] From CTI Technical Specification Sheet.

decreases the amplitude of pressure fluctuations with deleterious effects on the cooling capacity.

The regenerative displacer of the expander may be activated by an electric motor or solenoid or by fluidic control. Electric drive increases the size and weight of the expander unit but is a reliable system for maintaining proper synchronization of the compressor and expander displacers. Fluidic control is attractive to eliminate the complexities of the additional electronics and reduce size and weight. However, the same problems of maintaining displacer synchronization exist in Vuilleumier cryocoolers as were discussed earlier (Chapter 3) in connection with split-Stirling systems. Split-Vuilleumier cryocoolers were described by Berry (1974), Cohen and Daniels (1973), and Horn *et al.* (1973).

REFERENCES

Ackermann, R. (1971). "Vuilleumier Refrigerator Analysis, Low Temperature Refrigerator Optimization." Contract No. DAAK02-70-C-0291, USAECOM, Fort Belvoir, Virginia, AD 724770 (Cryomech Inc., Jamesville, New York).

AiResearch Manufacturing Co. (1972). "Vuilleumier Program Engineering Notebook." Contract NAS 5-21096, Goddard Space Flight Center, Greenbelt, Maryland, August.

AiResearch Manufacturing Co. (1974). "*Fractional Watt Vuilleumier Cryogenic Refrigerator Program Engineering Notebook*," Vol. 1, *Thermal Analysis*. Contract NAS 5-21715, Goddard Space Flight Center, Greenbelt, Maryland, May.

Anon.—RCA (1970). "ICICLE Feasibility Study Final Report." NASA CR 109880, AD-N70-34012 (Radio Corp. of America, Camden, New Jersey).

Berry, R. L. (1973). "Modular Cryogenic Refrigerators." *Closed Cycle Cryogenic Cooler Technology and Applications*, pp. 119–136, AFFDL-TR-73-149, Vol. 1, WPAFB, Dayton, Ohio.

Berry, R. L. (1974). "Ultraminiature Vuilleumier Refrigeration System." AFFDL-TR-74-16, WPAFB Dayton, Ohio (Hughes Aircraft Co., Culver City, California).

Block, R. F. (1973). "The Design, Development and Test of a Heat Pipe Interface for the Hot Cylinder of a V-M Refrigerator." *Closed Cycle Cryogenic Cooler Technology and Applications*, pp. 271–288, AFFDL-TR-73-149, Vol. 1, WPAFB, Dayton, Ohio.

Browning, C. W., Miller, W. S., and Potter, V. L. (1972). "75 K Vuilleumier Cryogenic Refrigerator Task IV, Model Fabrication." Report No. 72-8687, NASA GSFC, Greenbelt, Maryland (AiResearch Manufacturing Co., Torrance, California).

Browning, C. W., and Potter, V. L. (1972). "75 K Vuilleumier Cryogenic Refrigerator. Task II, Analytical and Test Program," Report No. 72-8497, NASA GSFC Greenbelt, Maryland (AiResearch Manufacturing Co., Torrance, California).

Browning, C. W., and Potter, V. L. (1972). "75 K Vuilleumier Cryogenic Refrigerator, Final Report for Task II—Analytical and Test Program." Contract NAS 5-21096, Goddard Space Flight Center, Greenbelt, Maryland, August.

Browning, C. W., Miller, W. S., and Potter, V. L. (1972). "75 K Vuilleumier Cryogenic Refrigerator—Final Report for Task IV Model Fabrication." Contract NAS 5-21096, Goddard Space Flight Center, Greenbelt, Maryland, November.

Browning, C. W., Potter, V. L., Miller, W. S., and Gasser, M. G. (1973). "Developments Toward Achievements of Long-Life Cryogenic Vuilleumier Refrigeration Systems." *Closed Cycle Cryogenic Cooler Technology and Applications*, pp. 225–244, AFFDL-TR-73-149, Vol. 1, WPAFB, Dayton, Ohio.

Bush, V. (1938). "Apparatus for Heat Transfer." U.S. Patent No. 2, 127,286, August 16.

Bush, V. (1939). "Apparatus for Compressing Gases." U.S. Patent No. 2,157,229, May 9.

Chellis, F., and Hogan, W. H. (1964). "A Liquid-Nitrogen-Operated Refrigerator for Temperatures Below 77 K." *Adv. Cryog. Eng.* **9**, 545–551.

Chellis, F. (1973). "The Future of the Stirling Cycle Refrigerator in Airborne I-R Applications." *Closed Cycle Cryogenic Cooler Technology and Applications*, pp. 11–118, AFFDL-TR-73-149, Vol. 1, WPAFB, Dayton, Ohio.

Chellis, F. (1977). "Design Compromises in the Selection of Closed Cycle Cryocoolers." *Proc. Appl. of Closed Cycle Cryocoolers to Small Superconducting Devices*, Boulder, Colorado, pp. 109–122, October (NBS Spec. Pub. 508).

Cohen, B., and Daniels, A. (1973). "Design and Development of a Free Displacer/Thermal Compressor Cryogenic Refrigeration System." Final Report, Contract No. DAAK 02-71-C-0421, N.V.L. USAECOM Fort Belvoir, Virginia, AD 920028 (Philips Laboratories).

Cowans, K. K. (1968). "Heat Powered Engine." U.S. Patent No. 3,379,026, April 23.

Crouthamel, M. S., and Shelpuk, B. (1973). "A Combustion-Heated Thermally Activated Vuilleumier Refrigerator." *Adv. Cryog. Eng.* **18**, 339–351.

Daniels, A., and du Pre, F. K. (1971). "Miniature Refrigerators for Electronic Devices." *Philips Tech. Rev.* **32**(2), 49–56.

Doody, R. D. (1971). "Two-Stage Vuilleumier Cycle Cryogenic Refrigerator System for Advanced Forward Looking Infrared (AFLIR) Applications." AFFDL-TR-71-17, DDC Number AD886822, August.

Doody, R. D. (1975). "Long Life High Capacity Vuilleumier Refrigerator for Space Applications." AFFDL-TR-75-108, WPAFB, Dayton, Ohio (Hughes Aircraft Co., Culver City, California).

Gasser, M. G., Yoshikawa, D. K., and Browning, C. W. (1972). "An Approach to Long-Life VM Cryogenic Refrigerators for Space Applications." *Appl. Cryog. Technol.* **4**, 416–434.

Globe, J. B., Rhia, F. J. III, Gainey, R. T., Aske, H. D., and Doody, R. D. (1971). "Vuilleumier Cycle Cryogenic Refrigeration System for Infrared Scanner Applications." AFFDL-TR-71-18, WPAFB, Dayton, Ohio (Hughes Aircraft Co.).

Harkless, L. B. (1973). "Reliability Test Results on V–M Coolers." *Closed Cycle Cryogenic Cooler Technology and Applications*," pp. 93–110, AFFDL-TR-73-149, Vol. 1, WPAFB, Dayton, Ohio.

Hogan, W. H. (1964). "Closed-Cycle Cryogenic Refrigerator and Apparatus Embodying Same." U.S. Patent No. 3,151,446, Oct. 6.

Horn, S. B., Cowan, K. C., and Berry, R. L. (1973). "Low Production Cost VM Coolers." *Closed Cycle Cryogenic Cooler Technology and Applications*, pp. 73–82, AFFDL-TR-73-149, Vol. 1, WPAFB, Dayton, Ohio.

Horn, S. B., Lumpkin, M. L., Walters, B. T., and Acord, T. T. (1973). "Miniature Cryogenic Cooler for TOW Night Sight." *Closed Cycle Cryogenic Cooler Technology and Applications*, pp. 52–70, AFFDL-TR-73-149, Vol. 1.

Lehrfeld, D., and Pitcher, G. K. (1973). "An Oil-Lubricated Triple Expansion VM Cooler for Long Duration Space Missions." *Closed Cycle Cryogenic Cooler Technology and Applications*, pp. 189–210, AFFDL-TR-73-149, Vol. 1, WPAFB, Dayton, Ohio.

Leo, B. (1970). "Designer's Handbook for Spaceborne Two-Stage Vuilleumier Cryogenic Refrigerators." AFFDL-TR-70-54, Dayton, Ohio (Hughes Aircraft Co., Culver City, California).

Leo, B. (1971). "Vuilleumier Cycle Cryogenic Refrigeration System Technology Report." AFFDL-TR-71-85, WPAFB, Dayton, Ohio (Hughes Aircraft Co., Culver City, California).

Magee, F. N., and Doering, R. D. (1968). "Vuilleumier Cycle Cryogenic Refrigerator Development." AFFDL-TR-68-2, WPAFB (Hughes Aircraft Co., Ltd.).

Miller, W. S., and Potter, V. L. (1973). "Fractional Watt Vuilleumier Cryogenic Refrigerator—Final Report for Task I Preliminary Design." Report No. 72-8846, NASA GSFC, Greenbelt, Maryland (AiResearch Manufacturing Co., Torrance, California).

Miller, W. S., and Potter, V. L. (1974). "Fractional Watt Vuilleumier Cryogenic Refrigerator—Final Report for Task II Analytical and Test Programs." Contract NAS 5-21715, Goddard Space Flight Center, Greenbelt, Maryland, March.

Morgan, N. E. (1971). "Analysis and Preliminary Design of Airborne Air Liquefiers." AFFDL-TR-71-171, WPAFB, Dayton, Ohio, AD 892500 (Hughes Aircraft Co., Culver City, California).

Morgan, N. E. (1975). "Preliminary Design of a Cryocooler for Satellite Sensors (CCSS)." AFFDL-TR-75-154, Vol. 1, WPAFB, Dayton, Ohio (Hughes Aircraft Co., Culver City, California).

Pitcher, G. K. (1970). "Design and Development of Laboratory Model Vuilleumier Cycle Cryogenic Refrigerator." Report No. 243, DA Prof. No. 663704P05201, Contract No. DAAK02-69-C-0364, N.V.L. USAECOM, Fort Belvoir, Virginia (Philips Laboratories).

Pitcher, G. K., and du Pre, F. K. (1970). "Miniature Vuilleumier Cycle Refrigerator." *Adv. Cryog. Eng.* **15**, 447–451.

Pitcher, G. K. (1971). "Development of Spacecraft Vuilleumier Cryogenic Refrigerators—Part I—Thermodynamic Design and Study of Life Limiting Components." AFFDL-TR-71-147, Part 1, WPAFB (Philips Laboratories).

Pitcher, G. K. (1972). "Development of Spacecraft Vuilleumier Cryogenic Refrigerators Part II—Development and Fabrication." AFFDL-TR-71-147, Part II, WPAFB (Philips Laboratories).

Pitcher, G. K. (1973). "Mechanical Life of Space Cryocoolers." *Closed Cycle Cryogenic Cooler Technology and Applications*," pp. 211–224, AFFDL-TR-73-149, Vol. 1, WPAFB, Dayton, Ohio.

Pitcher, G. K. (1975). "Spacecraft Vuilleumier Cryogenic Refrigerator Development—Final Report." AFFDL-TR-75-114, WPAFB (Philips Laboratories).

Pitcher, G. K. (1977). "Development of a Flight Design for an Oil Lubricated VM Cryocooler." AFFDL-TR-77-95, WPAFB (Philips Laboratories).

Prast, G. (1963). "A Philips Gas Refrigerating Machine for 20 K." *Cryogenics* **3**, 156–160, September.

Renyer, B. L. (1973). "High Capacity Long-Life Vuilleumier Cycle Refrigerator." *Closed Cycle Cryogenic Cooler Technology and Applications*, pp. 147–162, AFFDL-TR-73-149, Vol. 1, WPAFB, Dayton, Ohio.

Rhia, F. J., III. (1971). "Development of Long Life High Capacity Vuilleumier Refrigeration System for Space Applications." AFFDL-TR-71-92: Part I—AD 887399L. Part II—"Description of Preliminary Refrigerator Design and Results of Thermal Analyses and Life Studies," AD 890601L. Part III—"Refrigerator Design and Thermal Analyses," AD 901235L. Part IV—"Design and Mechanical Analyses of Refrigerator and Interface Unit," AD 920 255.

Richter, R., and Mahefkey, E. T. (1973). "The Applicability of a Solar Collector Thermal Power System to a Vuilleumier Cooler." *Closed Cycle Cryogenic Cooler Technology and Applications*, pp. 245–270, AFFDL-TR-73-149, Vol. 1, WPAFB, Dayton, Ohio.

Rule, T. T., and Qvale, E. B. (1969). "Steady State Operation of the Idealized Vuilleumier Refrigerator." *Adv. Cryog. Eng.* **14**, 343–352.

Russo, S. C. (1976). "Study of a Vuilleumier Cycle Cryogenic Refrigerator for Detector Cooling in the Limb Scanning Infrared Radiometer." NASA CR 145078 AD/N77-1211 (Hughes Aircraft Co., Culver City, California).

Sherman, A. (1971). "Mathematical Analysis of a Vuilleumier Refrigerator." NASA TM-X-65534, Acc. No. N-71-25812.

Sherman, A. (1972). "Selected Vuilleumier Refrigerator Performance Characteristics." *Adv. Cryog. Eng.* **18**, 352.

Sherman, A. (1978). "Cryogenic Cooling for Spacecraft Sensors, Instruments, and Experiments." *Astronaut. Aeronaut.*, November.

Sherman, A., Gasser, M., Goldowsky, M., Benson, G., and McCormick, J. (1979). "Progress on the Development of a 3–5 year Lifetime Stirling Cycle Refrigerator for Space." Cryogenic Eng. Conf., Madison, Wisconsin, August.

Taconis, K. W. (1951). "Process of and Apparatus for Heat Pumping." U.S. Patent No. 2,567,454, September 11.

Vuilleumier, Rudolph. (1918). "Method and Apparatus for Inducing Heat Changes." U.S. Patent No. 1,275,507, August 13.

White, R. (1973). "Program for Predicting V–M Cooler Off-Design Performance. Closed Cycle Cryogenic Cooler Technology and Applications," pp. 163–188, AFFDL-TR-73-149, Vol. 1, WPAFB, Dayton, Ohio.

White, R. (1976). "Computer Program for Optimizing Three-Stage Vuilleumier Cycle Cryogenic Refrigerators." AFFDL-TR-76-27, WPAFB, Dayton, Ohio.

White, R. (1976). "Vuilleumier Cycle Cryogenic Refrigeration." AFFDL-TR-76-17, WPAFB, Dayton, Ohio, AD/A-027-055.

Yendall, E. F. (1960). "A Novel Refrigerating Machine." *Adv. Cryog. Eng.* **2**, 188–196.

Yoshikawa, D. K. (1970). "75 K Miniature Vuilleumier Cryogenic Engine—Task 1." Report No. 70-6854, NASA GSFC, Greenbelt, Maryland (AiResearch Manufacturing Co., Torrance, California).

Chapter 5

Gifford–McMahon, Solvay, and Postle Cryocoolers

INTRODUCTION

This chapter is concerned with cryocoolers having regenerative heat exchangers and valves to control the flow of working fluid, systems defined earlier as Ericsson engines.

Within the generic family, there are two main branches: (a) Solvay and (b) Gifford–McMahon or Postle engines. The essential difference between them is that Solvay engines incorporate a *piston* as the reciprocating element in the expansion space. Work is extracted from the gas during expansion. The Gifford–McMahon or Postle engines incorporate a displacer as the reciprocating element in the expansion space. No work is extracted from the gas during expansion.

The Solvay engine was invented in Germany about 1887 (Solvay, 1887), but according to Collins and Cannaday (1958), the prototype was not properly executed and never attained cryogenic temperatures. It was subsequently neglected until reinvented in the late 1950s by William Gifford and Howard McMahon of A. D. Little, Inc., Cambridge, Massachusetts (1960).

The Postle engine invented by Davy Postle of Stanthorpe, Queensland, Australia was described in a British patent of 1873 (Thomas, 1873). Postle was apparently not eligible to apply for the patent, as it was issued to William Henry Thomas of New Cross, England.

The Postle engine was never intended to be a cryocooler, but rather a marine refrigerator for the emerging meat trade between England and Australia, following the opening of the Suez Canal. It was a very advanced machine using hydrogen as the working fluid, regenerative free-piston displacers, and fluidic control of the inlet/outlet valves and displacer motion. It is not known if actual engines were constructed and put into service.*

The reinvention of the Solvay and Postle engines appears to have occurred simultaneously. Some development of Solvay engines occurred at A. D. Little Inc. but effort was concentrated principally on the Postle displacer concept. Successful prototypes were developed and put into production. They are widely known as Gifford–McMahon cryocoolers.

No Solvay engines are known to be in production anywhere. Cooling engines made by Air Products and Chemicals Inc. of Allentown, Pennsylvania are described as "modified Solvay" engines but have displacers rather than expansion engines and are, therefore, Gifford–McMahon engines. The lack of development effort given to Solvay engines is hard to understand. It presumably arises from the difficult problem of a dynamic seal on the piston to contain the working fluid in the system. The same problem exists with Stirling and Claude cryocoolers and the seal technology developed there could be applied to Solvay machines. Possibilities include the Philips rolling seal, welded metal bellows, or the hydrodynamic lubricated, close tolerance pistons used in Kaptiza, Collins, and Doll-Eder Claude cycle expanders. There are many variants of Solvay machines. It is likely these will receive development attention with the increasing need for intermediate-capacity cooling systems as superconducting applications materialize.

The Gifford–McMahon cryocooler was conceived at A. D. Little in the mid-1950s and rapidly became a commercial product. The engine is characterized as a large, heavy, slow-running system requiring minimal maintenance. Higa *et al.* (1978) recently described an installation of 30 Gifford–McMahon cryocoolers with a cumulative total of one million hours of operation over the preceding decade.

The durability of the Gifford–McMahon cryocooler makes it the system of choice where size, weight, and efficiency are not of prime importance. The coolers come in a variety of capacities and configurations. Table 5.1 summarizes the technical specification of Gifford–McMahon cryocoolers available from Cryomech Inc., Jamesville, New York.

Initial production versions of Gifford–McMahon cryocoolers were introduced by A. D. Little Inc. in the early 1960s and have remained in production there and at the successor company Cryogenic Technology, Inc., Waltham, Massachusetts. Figure 5.1 is a general view of a CTI Model No. 0120 Gifford–McMahon cryocooler. This is a two-stage machine capable of generating 1 W of refrigeration at 26 K.

The competing range of commercial cryocoolers was introduced by Cryomech, Inc. of Jamesville, New York, a company founded by William

* The author would appreciate any information readers could supply on this point.

Table 5.1. Technical Specifications for Gifford–McMahon Cryocoolers[a]

	Type of refrigerator					
	AL01	AL02	D101	GB02	GB05	GB12
Output						
Cooling capacity (W)	1 @ 25 K 3 @ 27 K 20 @ 80 K	40 @ 40 K 80 @ 80 K	0.5 @ 25 K 1 @ 27 K	1 @ 9.5 K 2 @ 11 K 4 @ 14.5 K	1 @ 15 K 8 @ 20 K —	4 @ 13 K — —
Lowest temperature	23 K	23 K	23 K	7.5 K	—	9 K
Power required (kW)	0.9	3.0	0.6	3.0	0.9	3.0
Refrigeration temperature	23 K–80 K	23 K–80 K	23 K–80 K	7.5 K–20 K	12 K–20 K	9 K–25 K
Expander	Concentric single stage	Dual single stage	Concentric single stage	Dual 2 stage	Concentric 2 stage	Dual 2 stage
Applications	Solid state detector cooling Air liquefaction Cryopumping Thermal shielding	Air liquefaction Cryopumping	Solid state detector cooling	Superconducting Nb to Nb_3Sn 1. Receiver (RF) 2. Detectors (sub mil-infrared) 3. Magnetometers	High-temperature superconductors	Target hydrogen recondensing
Weight (lb)						
Cold head	5	25	4	25	6	25
Compressor	125	175	35	175	75	175

[a] All require no cooling, are Gifford–McMahon cycles with heat exchangers, operate at 144 rpm, and are rated for 3000 hr endurance. Courtesy Cryomech, Inc., Jamesville, New York.

Fig. 5.1. Gifford–McMahon cryocooler CTI Model No. 0120 (courtesy Cryogenic Technology Inc.).

Gifford on leaving A. D. Little. As a professor at Syracuse University, Gifford pursued extensive studies of cryocoolers (see the extensive listing for Gifford in the Bibliography). He accomplished this with the aid of a number of able graduate students. One of the most capable, Dr. Ralph Longsworth, led Air Products and Chemicals, Inc. of Allentown, Pennsylvania, into the manufacture of small regenerative cryocoolers of both the Gifford–McMahon and Stirling varieties.

THE SOLVAY CYCLE

Figure 5.2 is a schematic diagram for a Solvay expansion engine. It consists of a cylinder with a piston connected to a crank–connecting rod assembly, a regenerative heat exchanger, inlet and outlet control valves, and a gas compressor. The engine operates in the inverted position with the crankshaft above the top end of the cylinder. The cold part is at the bottom end. The piston and cylinder are made long and thin and of low conductivity material. The bottom end operates at cryogenic temperatures whereas the top end is at the ambient atmospheric temperature. A dynamic

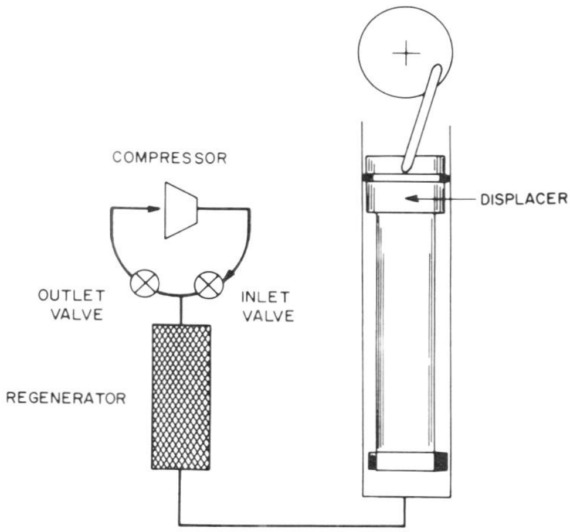

Fig. 5.2. Schematic diagram of Solvay cryocooler.

(sliding) gas seal is provided on the piston to stop leakage of gas from the cylinder. The seal is located near the top of the piston and operates at ambient temperatures. Location of the seal at the bottom of the piston would reduce the available refrigeration generated because of thermal degradation of the frictional work of the seal.

The regenerative heat exchanger consists of a porous matrix of finely divided material, usually metallic, contained in a cylinder of low-conductivity material. The function of the regenerator is to act as a thermal sponge abstracting heat from the gas when it is passing downwards through the matrix and releasing heat to the gas when it is passing upwards through the matrix.

Inlet and outlet valves control the flow of gas into and out of the expansion engine system. The valves are coupled to the crankshaft so their operation is synchronized with the piston position. The inlet and outlet valves are connected through a loop containing a gas compressor and adequate low-pressure and high-pressure reservoirs. A water cooler is included in the high-pressure side delivery duct of the compressor to remove the heat of compression.

The operation of the engine is as follows:

(1) Pressure Buildup. With the piston at the bottom of its stroke, the inlet valve opens and high-pressure fluid enters, compressing the fluid already in the engine from pressure p_1 to pressure p_2.

(2) Intake Stroke. With inlet valve open, the piston is raised from the bottom dead center position. Gas is admitted through the inlet valve and regenerator into the bottom of the cylinder. Passing through the regenerator, the gas cools by heat transfer to the matrix material of the

regenerator. The reduction in gas temperature causes a reduction in the gas pressure so that further charge is drawn into the cylinder through the inlet valve.

(3) Expansion Stroke. The inlet valve now closes and the piston continues to ascend in the cylinder expanding the gas which experiences a reduction in temperature. Some gas flows from the matrix to the cylinder and the pressure decreases. Expansion continues until the piston reaches top dead center position. The gas has expanded only partly and the pressure is still appreciably above the minimum of the cycle.

(4) Exhaust Stroke. When the piston reaches the top of its stroke the exhaust valve opens and fluid flows to exhaust until the pressure falls to the minimum value. The exhaust valve is kept open as the piston descends to the bottom of the stroke and gas is forced from the cylinder through the regenerator and exhaust valve to the gas compressor. In passing through the regenerator, the gas is heated by heat transfer from the matrix. This cooling of the matrix prepares it to cool the gas passing through it *en route* to the cylinder in the next cycle. When the piston reaches the bottom of the cylinder, most of the low-pressure gas has been exhausted from the engine. The exhaust valve closes, the inlet valve opens, and high-pressure gas enters the engine to repeat the cycle.

A work diagram for the Solvay engine is shown in Fig. 5.3. The area of the P–V diagram represents the net decrease in enthalpy which, for an ideal gas, is equivalent to the available refrigeration.

The void volume of the regenerator and connecting ducts are shown on the P–V diagram as the "clearance volume," V_r. Admission of high-pressure charge occurs until the volume of gas in the cylinder is V_1, at which point the inlet valve is closed and the gas begins to expand. Normally, in a closed piston and cylinder, expansion of gas follows a curved path on the P–V plane, similar to that shown by the broken line in Fig. 5.3, and may be an isothermal, adiabatic, or a polytropic process. In the Solvay engine, the process is markedly different because as the expansion proceeds

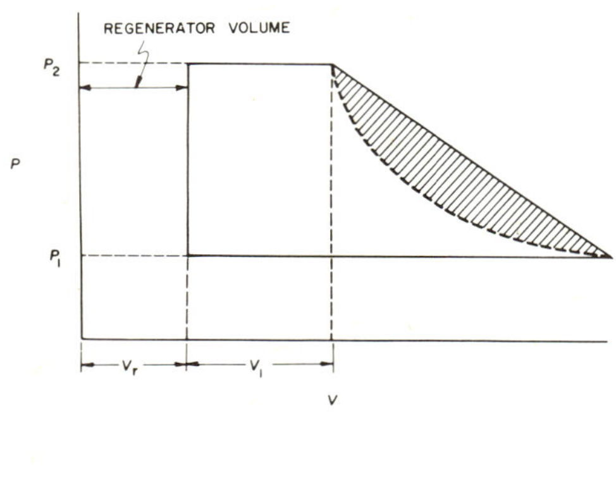

Fig. 5.3. Work diagram for Solvay cryocooler.

the mass of gas in the cylinder increases as fluid flows to the cylinder from the regenerator. In Fig. 5.3, the expansion process is arbitrarily represented by a straight line. The shaded area represents additional refrigeration attributable to expansion in the cylinder of gas in the regenerator. This partly compensates for the gas used to fill the regenerator void space which appears to expand in the system to no useful purpose.

The gas required to fill the regenerator is not so great as might be expected from a simple consideration of the void volume. The temperature of the regenerator varies from the low refrigeration temperature at the bottom end to ambient temperature at the top. The effective *average* value of temperature may, therefore, be twice that of the expansion cylinder so the mass required to fill the matrix is only *half* that required if it were all at the expansion space temperature.

For best performance, the piston should have a discontinuous motion with brief pauses at the end of each stroke. However, Gifford has indicated that it is possible to use harmonic piston motion for simplicity without significant loss of performance. Other constructional features are that the piston may be quite loose fitting in the cylinder with up to 0.5 mm (0.02 in.) clearance except for the gas tight seal at the warm end. To minimize heat conduction, the cylinder must be made of thin-wall stainless steel tube and the piston of low-conductivity, plastic material. The regenerator might be a stack of dense mesh wire screen of 0.03-mm (0.001-in.) diameter copper or phosphor bronze wire having 80 to 160 strands per cm (200 to 400 strands per in.) contained in a thin-wall stainless steel tube.

A miniature Solvay cooling engine described by Gifford *et al.* (1960) is shown in Fig. 5.4. The engine, intended for infrared detector applications, was capable of attaining 55 K in a single stage. The cylinder was a stainless steel tube 5 mm (0.2 in.) in diameter with a wall thickness of 0.13 mm (0.005 in.). The piston, made of linear base laminated phenolic, was designed to incorporate the regenerator consisting of fine wire mesh screen. The inlet and outlet pressure of the helium working fluid was 1725 and 345 kPa (250 and 50 psi) with a consumption of 6 to 9×10^{-4} m^3/sec (0.2 = 0.3 ft^3/min).

The unit was equipped with a subassembly at ambient temperature containing the inlet and exhaust valves, the piston seal, and cam shafts for the valve and piston actuation. A 400-Hz, 115-V motor was used to drive the crankshaft through a reduction gear at 500 rpm. The overall weight of the cooling unit was 0.14 kg (10 oz) including the timing motor but excluding the external gas compressor or compressed gas bottles.

The cooling characteristic for the device is shown in Fig. 5.5. It was possible to cool to operating temperatures within 2–3 min and to maintain temperatures in the 50–80 K range, indefinitely.

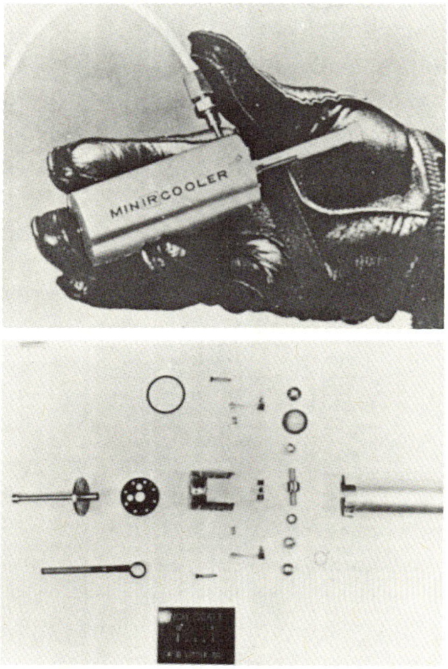

Fig. 5.4. Experimental miniature Solvay cryocooler (after Gifford *et al.*, 1960).

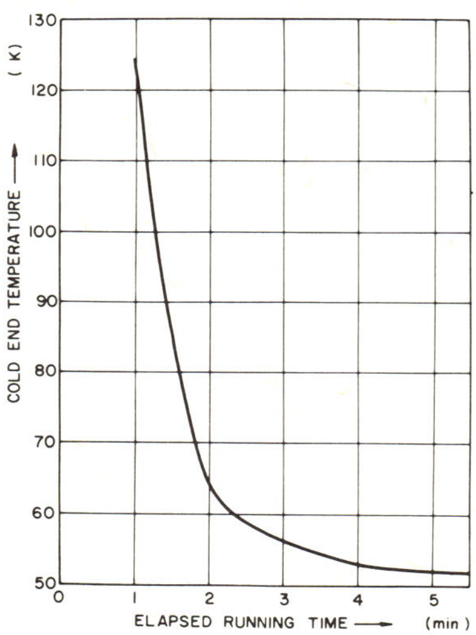

Fig. 5.5. Cooldown characteristic of the experimental miniature Solvay cryocooler (after Gifford *et al.*, 1960).

THE GIFFORD–McMAHON CYCLE

The Gifford–McMahon cycle, superficially similar to the Solvay cycle, is shown schematically in Fig. 5.6. It consists of a cylinder, closed at both ends, and containing a displacer having a length about three quarters of the cylinder. The displacer is connected to a light drive mechanism so that it can be moved up and down in the cylinder. The two volumes, one above and one below the displacer, can be varied from zero to maximum but the total volume remains constant.

The two spaces are connected through a thermal regenerator and to a gas supply. The gas supply system consists of inlet and outlet valves, a gas compressor, and high- and low-pressure reservoirs. The valves are coupled to the drive mechanism so their operation is synchronized with the position of the displacer. A heat exchanger is included downstream of the gas compressor to cool the gas to ambient temperature after compression.

The pressure above and below the displacer is the same except for small pressure drops across the regenerator when gas is flowing through it. Virtually no work is required to move the displacer in the cylinder. No work is done on the gas and the gas does no work on the displacer. Pressure in the system is increased or decreased by operation of the inlet or outlet valves.

The displacer is a loose fit in the cylinder except at the top and where it is equipped with a dynamic (sliding) seal to prevent leakage past the displacer from one space to the other. The pressure is virtually the same in both spaces so the seal problem is not severe and, furthermore, some leakage can be tolerated between the two spaces.

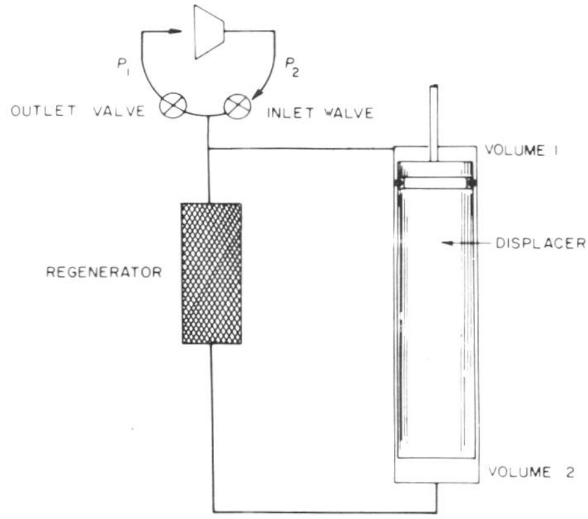

Fig. 5.6. Schematic diagram of Gifford–McMahon cryocooler.

The regenerator consists of finely divided metallic material which cools the gas when passing downwards to the cold space and heats the gas when passing upwards from the cold space.

The engine operates as follows:

(1) Pressure Buildup. With the displacer at the bottom of the cylinder and exhaust valve closed the inlet valve is opened, increasing the pressure level in the system. Fluid enters through the inlet valve to fill the regenerator and the space above the displacer, volume 1.

(2) Intake Stroke. With the inlet valve open, the displacer is moved from the bottom of the cylinder to the top. This displaces high-pressure fluid from the space above the displacer, volume 1, through the regenerator, to the space below the displacer, volume 2. In passing through the regenerator, the gas cools, causing the pressure to decrease and further gas to enter the system to maintain the maximum cycle pressure.

(3) Pressure Release and Expansion. With the displacer at the top of the cylinder, the inlet valve closes and exhaust opens. Fluid escapes and the pressure falls. The drop in pressure causes a reduction in the gas temperature level. The decrease in temperature of gas in the bottom cylinder space, volume 2, is the useful refrigeration process of the cycle.

(4) Exhaust Stroke. With exhaust valve open, the displacer moves from the top to bottom of the cylinder, displacing fluid from volume 2, below, to volume 1, above the displacer. As it flows through the regenerator, the fluid is heated by the matrix to near ambient temperatures. This process cools the matrix ready for the gas entering in the succeeding cycle.

During pressure buildup and expansion, the regenerator is most effective if gas can be made to flow through it at a constant rate. This will be achieved by progressive opening of the inlet and exhaust valves with the displacer stationary at the bottom and top of the cylinder, respectively. The inlet valve opening should be extended over about 75° and the exhaust valve about 110° of crank rotation.

Figure 5.7 shows the work diagrams for volumes 1 and 2 and Fig. 5.8 shows the temperature/time characteristic for gas in the engine. At the beginning of a cycle, volume 1 is maximum and volume 2 is minimum. The pressure is low, represented on Figs. 5.7 and 5.8 by state (a). The inlet valve opens and fluid enters volume 1 and the regenerator. The fluid contained therein is compressed, and its temperature is, consequently, increased. It mixes with incoming fluid to a final temperature of mixed fluid shown by state (b) on Fig. 5.7. Now the displacer moves from the bottom to top of the cylinder and fluid is displaced from volume 1 to volume 2 through the regenerator. In passing through the regenerator, the fluid cools to temperature (c) and more fluid enters through the inlet valve to maintain the pressure constant.

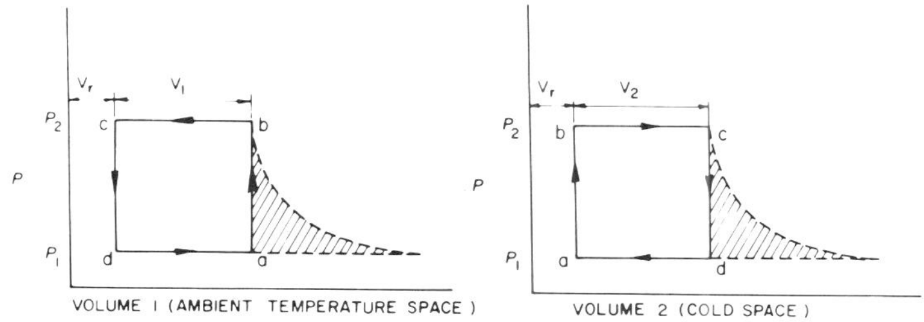

Fig. 5.7. Work diagrams for the Gifford–McMahon cryocooler.

At (c) the inlet valve closes quickly and the exhaust valve opens slowly and progressively. The pressure falls as fluid leaves the engine through the exhaust valve. Fluid in volume 2 expands as the pressure decreases and the temperature of gas in volume 2 decreases. Heat transfer from the cylinder walls increases the gas temperature to (d) in Fig. 5.7. Finally, the displacer moves from the top to bottom of the cylinder causing fluid to return back through the regenerator to volume 1 or exhaust. In passing through the regenerator, the temperature of the gas increases to a value slightly less than the *mixture*, following admission of high pressure gas. The gas leaving the exhaust is, therefore, warmer than the intake feed gas. Heat removed in this way is equal to the total refrigeration achieved. The net result of investing the compressor work is that heat is pumped from a low temperature to above the ambient level. The useful product is the refrigeration effect generated during expansion.

The work done by the gas *on* the displacer is shown by the work diagram for volume 2. The work done by the displacer *on* the gas is shown by the work diagram for volume 1. The pressures are the same in both spaces so the areas of the work diagrams are identical (neglecting the displacer rod area). For an ideal gas and adiabatic conditions, the area of the diagram for volume 2 is equal to the change in enthalpy of the gas in volume 2. It represents the net refrigeration available. Similarly, the area of the diagram for volume 1 represents the change in enthalpy of a perfect gas under adiabatic conditions.

Multiple-Expansion Gifford-McMahon Cycle

Additional expansion spaces to operate at different temperature levels can be added with very little increase in overall unit complexity. Figure 5.9 shows a schematic diagram for a multistage heat pump with three expansion spaces giving refrigeration at three different levels, 80, 35, and 14 K. Figure 5.10 shows a temperature/time history for the triple stage

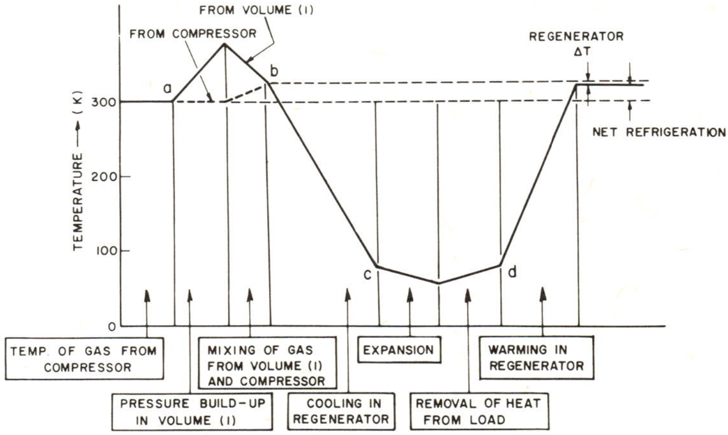

Fig. 5.8. Temperature/time history of gas in the Gifford–McMahon cryocooler.

device. The unit requires additional displacers, cylinders, and regenerators, but the same valves operating at room temperature may be used and the three displacers are all operated by the same actuator.

The system is equivalent to three expansion engines operating at different temperatures. The only low-temperature parts are closed circuits

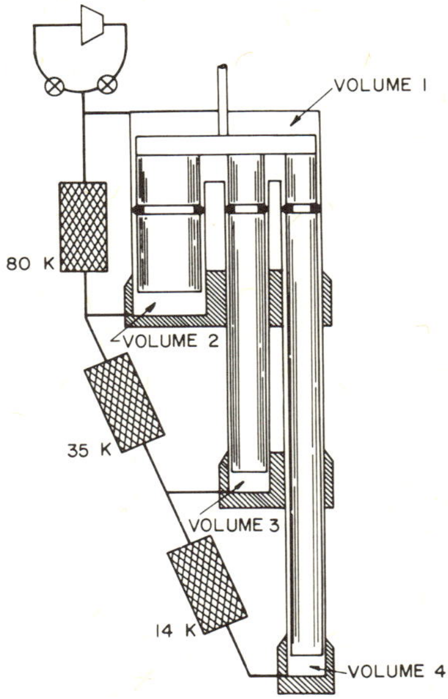

Fig. 5.9. Schematic diagram of multiple stage Gifford–McMahon cryocooler (after Gifford and Hoffman, 1961).

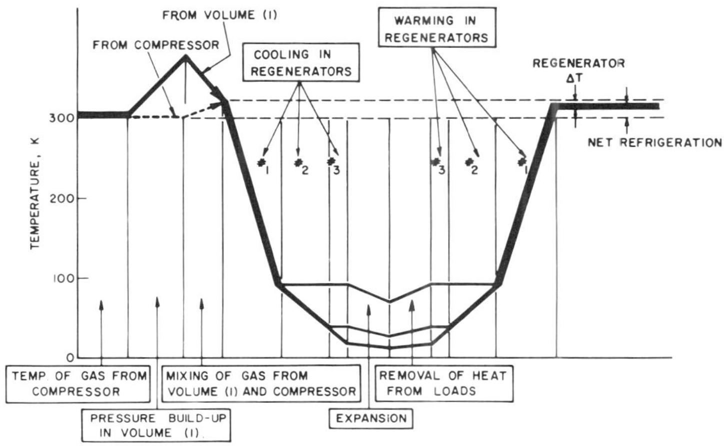

Fig. 5.10. Temperature/time history of gas in a multiple-stage Gifford–McMahon cryocooler.

and loosely fitting displacer/cylinder sets with little mechanical stress. The problems of construction are minimal. The additional stages operate exactly like the single-stage device with gas supplied at the temperature of the stage above. The expansion spaces describe P–V diagrams representing the useful refrigerating output of the stage and the regenerative heat exchangers prevent the loss of refrigeration to the stage above.

Multistaging is attractive because it provides a thermodynamically efficient process to attain cryogenic refrigeration temperatures. For a given refrigeration requirement, there is a decreased power requirement. Expansion over given pressure and volume ratios produces the same refrigeration regardless of temperature. However, the mass of gas involved is inversely proportional to the temperature. A unit of refrigeration at 20 K requires four times the gas and consequently four times the power, as at 80 K. Most cryocooler applications require the removal of heat at different temperature levels, such as precooling before liquefaction, or the use of cooled radiation shields to reduce heat leaks to the system at the lowest temperatures. It is, therefore, thermodynamically advantageous for refrigeration to be available at different temperature levels.

Furthermore, multistaging relieves the operating regimen of regenerators and heat exchangers. Consider the "refrigeration loss" Q_r:

$$Q_r = MC_p \, \Delta T_r \tag{5.1}$$

where M is the mass flow of gas, C_p is the heat capacity of the gas, and ΔT_r is the temperature difference between the gas entering and leaving the regenerator at the cold end. In expanding, the gas changes in temperature, ΔT_e. Now ΔT_e must be greater than ΔT_r, otherwise no refrigeration would be achieved. ΔT_e is proportional to the temperature T at which the expansion occurs, typically equal to $0.3T$. Thus, with expansion at 10 K, $\Delta T_e = 3$ K. For there to be some useful refrigeration achieved, ΔT_r would need to be less than 3 K, say 1–1.5 K. To achieve this in a single stage between 10 and 300 K requires an impractically high heat exchanger effectiveness. In a three-stage system with an intermediate expansion at 40 K, a ΔT_r of 1 K is possible for the regenerator operates only in the 10 to 40 K range.

Combination Gifford–McMahon Cryocooler and Joule–Thomson Expander

The application of a multistage Gifford–McMahon cryocooler to helium liquefaction was described by Gifford and Hoffman (1961). A schematic diagram of the unit is shown in Fig. 5.11. It consisted of the three-stage heat pump described above with additional recuperative heat exchangers below the three regenerators. The recuperative heat exchangers,

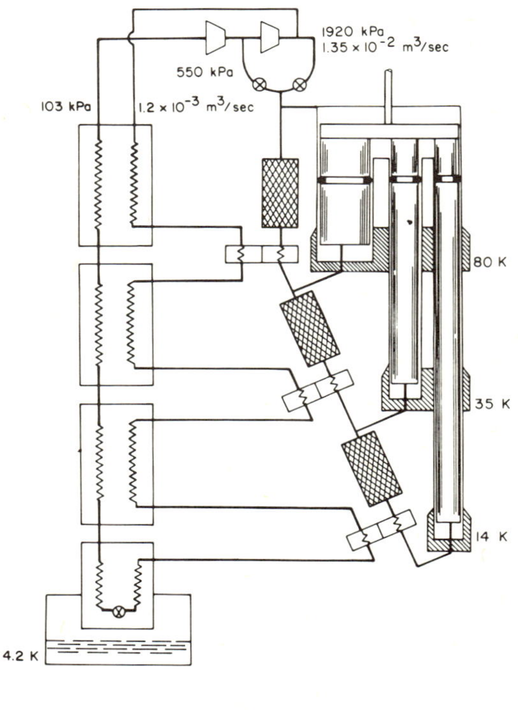

Fig. 5.11. Schematic diagram of multiple stage Gifford–McMahon cryocooler combined with a Joule–Thomson expander for helium liquefaction (after Gifford and Hoffman, 1961).

at progressively decreasing temperatures, cooled a stream of high-pressure helium gas below the inversion temperature for expansion in a Joule–Thomson (JT) valve (see Chapter 6). Additional counterflow heat exchangers were included in the JT circuit to maximize thermal performance between the high- and low-pressure gas streams.

Two compressors were incorporated to compress helium from the low-pressure JT circuit at 103 kPa (15 psi) to 550 kPa (80 psi) and compress helium from 550 to 1920 kPa (80 to 280 psi). The pressure range for operation of the multistage heat pump was 550 to 1920 kPa (80 to 280 psi) with a flow rate of $1.35 \times 10^{-2}\ m^3/s$ ($45\ ft^3/min$) and expansion at temperatures of 80, 35, and 14 K. The JT circuit operated over a pressure range of 1920 to 103 kPa (280 to 15 psi) with a flow rate of $1.2 \times 10^{-3}\ m^3/s$ ($4\ ft^3/min$). The net useful refrigeration available was 7000 mW at 4.2 K. The cool-down time was about five hours. The refrigerator unit was $0.38 \times 0.3 \times 0.71$ m ($15 \times 12 \times 28$ in.) and weighed about 41 kg (90 lb) (exclusive of vacuum pumps) and required about 100 W of power. The compressor unit was $0.55 \times 0.6 \times 1.9$ m ($22 \times 24 \times 75$ in.), weighed about 318 kg (700 lb), and required 3 kW of power and free access of air for cooling.

Fluidic-Driven Displacer

A refinement to the original Gifford–McMahon cryocooler occurred with application of a technique known as "cryomatic gas balancing." The displacer mechanical drive was eliminated and a free displacer used instead. Movement of the displacer was caused by differential fluidic (gas pressure) forces acting on the displacer arising from the inlet and exhaust of gas through a simple rotary cored valve.

A schematic diagram of a single-stage Gifford–McMahon cryocooler with fluidic-driven free displacer is shown in Fig. 5.12. It consists of two connected coaxial cylinders of different length and diameter containing the free displacer. The displacer is about three quarters the length of the larger cylinder and has a coaxial extension to fit the smaller cylinder. This creates three chambers of variable volume depending on the position of the displacer. Chamber 1 is the volume between the top of the displacer extension and the small cylinder head. Chamber 2 is the volume between the top of the displacer and the large cylinder. Chamber 3 is the volume between the bottom of the displacer and the large cylinder. Gas seals are provided on both the displacer and displacer extension to prevent the leakage of gas past the displacer from one chamber to another. Chambers 2 and 3 are interconnected by a thermal regenerator of finely divided metallic material. The pressure in both is, therefore, the same, apart from small differences due to the pressure drop in the regenerator matrix when

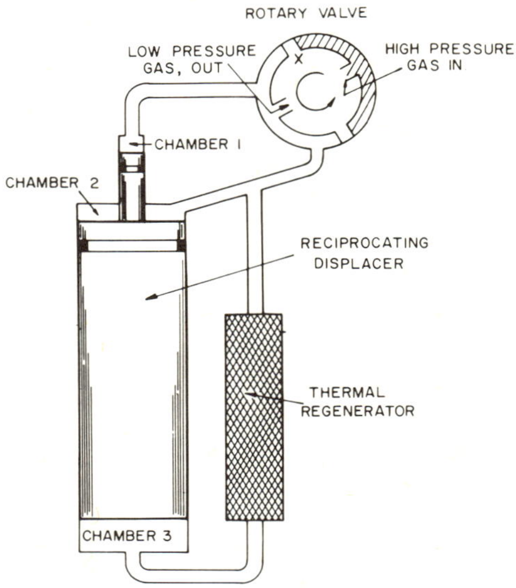

Fig. 5.12. Schematic diagram of a Gifford–McMahon cryocooler with a free, fluidic-driven displacer.

gas is flowing through. Gas is admitted and exhausted from chamber 1 and the regenerator conduit through a rotary cored valve. The same valve regulates both the inlet of high-pressure gas from the compressor and the exhaust of low-pressure gas.

The sequence of operations illustrated in Fig. 5.13 is as follows:

(1) Pressure Buildup. At the beginning of the cycle the displacer is at the bottom of the cylinder so the volume of chamber 3 is zero and the volumes of chambers 1 and 2 are at their maximum value. The rotary valve is in such a position that chamber 1 is full of gas at a high pressure and chambers 2 and 3 are connected to exhaust and at low pressure. This is represented in Fig. 5.13a. When the valve has turned to connect chambers 2 and 3 to the inlet port the pressure increases as represented in diagram (b). Fluid enters through the inlet port to fill chamber 2, the regenerator and the clearance space in chamber 3.

(2) Intake Stroke. Further rotation of the valve opens chamber 1 to exhaust and the pressure decreases. The upward force on the displacer due to pressure in chamber 3 now exceeds the downward force due to pressure in chambers 1 and 2, by an amount equal to Δp times the area of chamber 1. As a result, the displacer moves to the top of the cylinder (Fig. 5.12c), displacing fluid from chamber 2 through the regenerator to chamber 3. In passing through the regenerator, the fluid cools and additional fluid enters to fill chamber 3 and the regenerator.

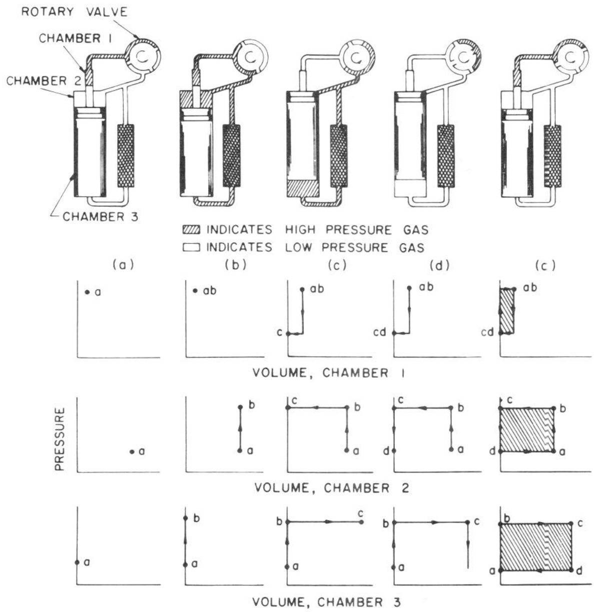

Fig. 5.13. Cycle of operation of the Gifford–McMahon cryocooler with free, fluidic-driven displacer.

(3) Pressure Release or Expansion. Further rotation of the valve opens chambers 2 and 3 to exhaust and the pressure decreases. Fluid flows out of chamber 3 through the regenerator during the expansion and useful refrigerating effect is developed in chamber 3. The situation is represented in Fig. 5.13d.

(4) Exhaust Stroke. The same low pressure now exists in all three chambers so the displacer will fall under gravity to the bottom of the cylinder. Should seal friction prevent this, the valve soon opens chamber 1 to the high-pressure gas supply. The pressure force acting downwards then moves the displacer to the bottom of the cylinder, displacing fluid from chamber 3 to chamber 2. In passing through the regenerator, the fluid is heated by the matrix back to near ambient temperature. This cools the matrix in preparation for the gas entering in the next cycle. The increase in temperature decreases the density of the gas and much of the fluid leaving chamber 3 passes to exhaust rather than remaining in chamber 2.

The work diagrams for the unit are similar to those discussed earlier for the engine with a mechanically driven displacer. The small diagram for chamber 1 represents the pneumatic work done in driving the displacer up and down the cylinder. The diagram for chamber 3 represents the change in enthalpy and, for adiabatic conditions with an ideal gas, the refrigeration effect of the process per cycle. The diagram for chamber 2 is virtually identical but reversed in direction to that for chamber 3. A temperature–time history of gas in the engine would be the same as that presented in Fig. 5.8.

In a description of the fluidic-drive displacer engine, Gifford (1966) does not discuss valve operation. It is virtually certain that the same careful attention to programmed valve opening would be required as with a mechanically driven displacer. This gains improved effectiveness of the regenerator resulting from steady flow during the pressure buildup and expansion processes. A good deal of expertise is concealed by Gifford's disarming comment:

> . . . timing between pressurizing and depressurizing the chambers may be set by proper selection of the components connecting the chambers with the valve . . .

Fluidic-driven displacers have been used in multistage units, where refrigeration is generated at different temperature levels. This requires an additional displacer/cylinder unit and regenerator per stage. A schematic digram of a two-stage cryomatic gas balancing refrigerator is shown in Fig. 5.14. An experimental unit of this type mentioned by Gifford (1966) had

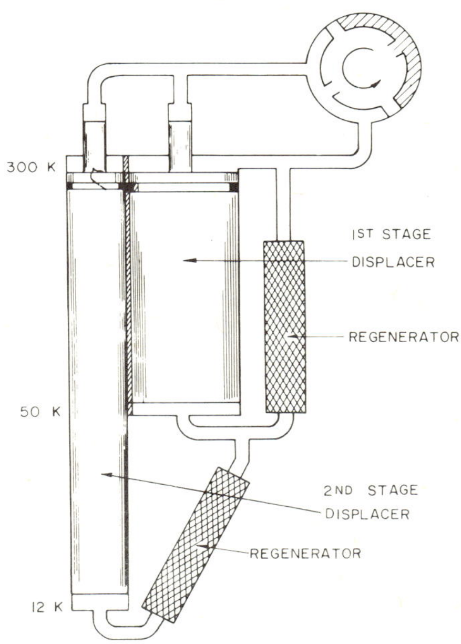

Fig. 5.14. Schematic diagram of two-stage Gifford–McMahon cryocooler with free, fluidic-driven displacer.

refrigeration capacities of 5 W at 15 K and 18 W at 50 K and was incorporated into a liquid helium 4.2 K refrigerator. He indicated that temperatures as low as 23 K had been reached with a single-stage machine having a 32-mm (1.25-in.) diameter displacer operating in the speed range 1 to 3 Hz (50–150 cycles per min).

Machine Design and System Optimization

Some of the interesting heat transfer, fluid friction and mechanical design problems of Gifford–McMahon refrigerating engines have been discussed by Gifford and Hoffman (1961). An important factor is the valve action and motion of the displacer. For best performance of the regenerator, the gas flow should be constant one way for half the cycle and constant the other way for the remainder of the cycle. Slight variations are permissible but very large surges of fluid through the matrix seriously affect regenerator effectiveness.

The requirements are not compatible with the flow patterns necessary to maximize the refrigeration effect in the expansion space for a displacer moving with continuous harmonic motion. Here, sudden changes in pressure from one extreme to the other are required at the time the displacer is stationary at the ends of its stroke. To maximize the area of the work diagram for the expansion space, a rapid increase in pressure is required when the displacer is at the bottom of the cylinder and a sudden decrease when the displacer is at the top of the cylinder. Efforts to achieve this would result in rapid surges of gas through the system and lead to a serious decline in regenerator performance.

An effective solution to the problem was achieved by (1) causing the displacer to move in a discontinuous manner so that there are lengthy pauses at each end of the stroke, and (2) increasing and decreasing the pressure through valves which open slowly to bleed gas in and out at an approximately constant rate. A composite diagram of the cyclic displacer motion, valve action, and mass flow for a typical case is shown in Fig. 5.15.

Another important problem is the design and construction of the regenerator. The regenerator should have the smallest possible void volume and pressure drop yet have high thermal effectiveness. Regenerators made from 100 to 200 mesh wire screen punchings contained in a thin wall cylinder have been found satisfactory. Several hundred screens may be required to obtain a regenerator a few inches long. Effective regenerator matrices have also been made from small spheres or balls simply poured into a small cylinder.

The matrix material must be selected with the operating temperature in mind. The heat capacity of all solids decreases at low temperatures

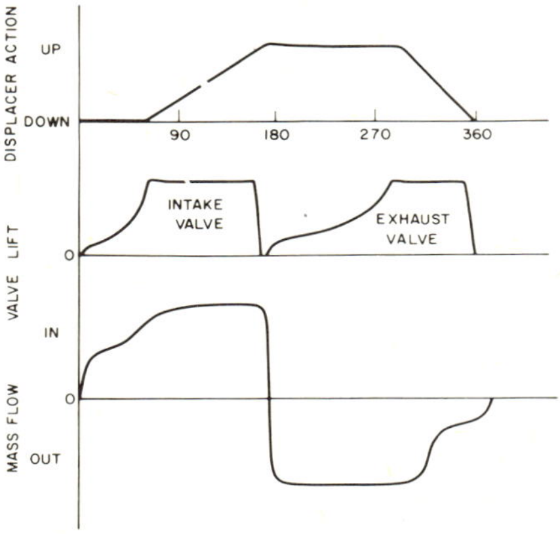

Fig. 5.15. Cyclic history of displacer motion, valve motion, and mass flow to and from the cylinder of a Gifford–McMahon cryocooler (after McMahon *et al.*, 1960).

whereas the heat capacity of helium increases. The operation of a regenerator depends on the matrix material having a high heat capacity compared with the gas flowing through. The problem of decreased matrix heat capacity becomes severe at temperatures of 50 K or less with the normal matrix materials of iron, steel, brass, and copper. For lower temperatures, lead is the preferred matrix material because of its relatively high heat capacity. The lead must be prepared as fine shot, powder, or some other finely divided form.

To optimize design, one must establish the best compromise of a number of factors:

(1) Refrigeration. The cold capacity of the machine is equal to the area of the *P–V* diagram for the expansion space. Pressure drops in the regenerator decrease the pressure excursion in the expansion space and so reduce the refrigeration.

(2) Regenerator Pressure Drop. This may be minimized by utilizing a *short, fat matrix* loosely filled with large diameter filling.

(3) Regenerator ΔT Loss. Due to imperfect operation of the regenerator, there is a difference in the temperature at which gas enters and leaves the regenerator which reduces the refrigeration available. This may be minimized by utilizing a *large matrix of finely divided densely packed solids* to give maximum area for heat transfer and a high matrix heat capacity.

(4) Regenerator Volume Loss. For each cycle, the regenerator void volume must be filled with high-pressure gas. In the Solvay engine, a fraction of the potential refrigeration in the void volume gas is gained, but not in the Gifford–McMahon engine. The gas in the void volume has to be compressed so its presence in the regenerator represents a direct loss

of potential refrigeration. The loss may be minimized by a *small densely packed matrix*.

(5) System Heat Leaks. Thermal leaks to the cold regions of the machine may occur by conduction, convection, and radiation. Care must be taken to minimize: (a) the convective losses by the use of vacuum enclosures or insulation, (b) the radiation losses by radiation shields or spectrally selective surfaces, and (c) conduction losses by the use of long thin wall enclosures of low-conductivity material.

Advantages and Disadvantages of Gifford–McMahon Cryocoolers

Gifford–McMahon cryocoolers have an important advantage over conventional expansion engines in Claude cycle systems. All the valves and sliding seals operate at room temperature. The operation of rapidly opening and closing valves for long periods without failure and with adequate lubrication at low temperatures is a difficult problem. Furthermore, the valves and seals must prevent gas leakage, otherwise there will be a decrease in refrigeration effectiveness. At low temperatures, the problem of leakage is aggravated by the decrease in gas viscosity. Operation at ambient temperature permits the use of the elastomeric materials for valves and seals and the use of standard lubricants to reduce wear.

Preventing gas leakage during the operation of a piston or displacer at low temperatures is difficult without lubrication. Clearances must be reduced to very small values and special dry sealing rings must be used. If the seals operate at room temperature, a wide choice of materials and lubricants is available and piston clearances below the seal may be relaxed, virtually without limit.

Another advantage of Gifford–McMahon cryocoolers is their use of regenerative heat exchangers. Regenerators are cheap, simple, easy to build, and highly effective in small sizes. They will operate with appreciable amounts of condensable contaminants in the working fluid with no serious impairment of cycle efficiency or blocking of the flow passages. Condensables deposited during the inlet flow are reevaporated and swept out during the exhaust flow. In systems with recuperative counterflow heat exchangers, condensable constituents, i.e., water, oil vapor, etc., must be eliminated from the working fluid; otherwise they will accumulate in the low-temperature regions and block the flow passages. Reversing heat exchangers are used to periodically exchange intake and exhaust channels to free the flow passages from accumulated condensed deposits. Regenerators automatically achieve the effect of reversing heat exchangers.

Solvay engines are inherently more efficient than Gifford–McMahon engines because the fluid expands more completely. However, the piston seal of the Solvay engine has to withstand the full pressure difference between the cylinder and the "crankcase." In the Gifford–McMahon engine, the displacer seal has to sustain only the difference due to pressure drop in the regenerator. Moreover, the difference reverses direction and tends to cancel any leakage. A moderate leakage can be tolerated. The displacer drive system is not required to sustain the large forces of a piston drive and may be structurally simple and lightweight with fewer vibration problems.

The principal disadvantage of the Gifford–McMahon cycle is the high work input per unit of refrigeration (W/Q) characteristic. This arises from the incomplete expansion and the mass of compressed gas consumed in the regenerator void volume with little or no return of refrigeration.

The ideal work per unit of refrigeration for various cycles is shown in Fig. 5.16. This was prepared by Gifford (1966) for refrigeration at 80 K with helium as the working fluid, heat rejection at 300 K and a range of pressure ratios p_2/p_1 up to 5. The W/Q characteristic is drawn for the Gifford–McMahon cycle, the Joule–Brayton cycle, and the ideal Stirling and Carnot cycles. In discussing the figure, Gifford suggests that because of the speed of operation, the reference cycle for Stirling cryocoolers should be the Brayton cycle, the ideal reference cycle for small reciprocating and turbine expanders. While allowing that the Gifford–McMahon system does require more work for a given level of refrigeration, he suggests the difference is not great at the small compression ratios of 1.5 to 2.5 used in practice. Furthermore, the practical advantages outweigh the additional work. The principal advantage is high reliability because the units are simple, slow moving, and unstressed. The gas seal on a typical machine

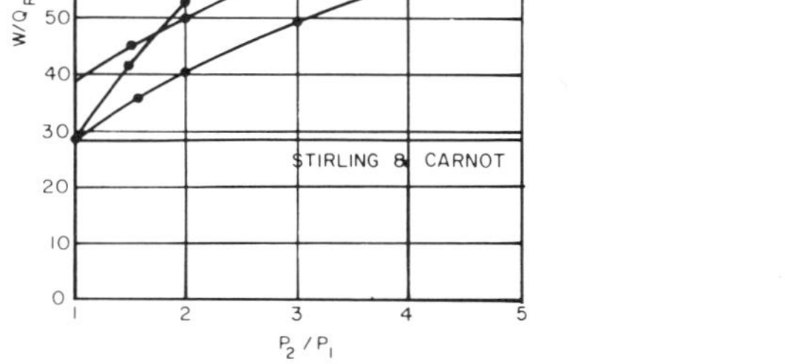

Fig. 5.16. Comparison of the work input to heat lifted ratio for various ideal thermodynamic cycles (after Gifford, 1966).

may move at the rate of 0.05 to 0.15 m/s (2 to 5 in./s) compared with the 0.75 to 1.25 m/s (30 to 50 in./s) of small Stirling cryocoolers. Continuous operating times of 5 to 10,000 hr are not uncommon and Gifford's cryocoolers are noted for their high reliability and extended maintenance-free operation.

Heat-Balance Analysis

Ackerman and Gifford (1971) presented data on the experimental performance of a Gifford–McMahon engine with a fluidic-drive free displacer. The experimental refrigerator is shown in Fig. 5.17. The machine was a single stage expansion engine with a displacer 19 mm (0.75 in.) in diameter, 0.1 m (4 in.) long and with a stroke of 12 mm (0.5 in.). With helium as the working fluid and pressures ranging from 689 to 1960 kPa (100 to 285 psi) it was found that the best operating speed was 3 Hz (190 cycles/min).

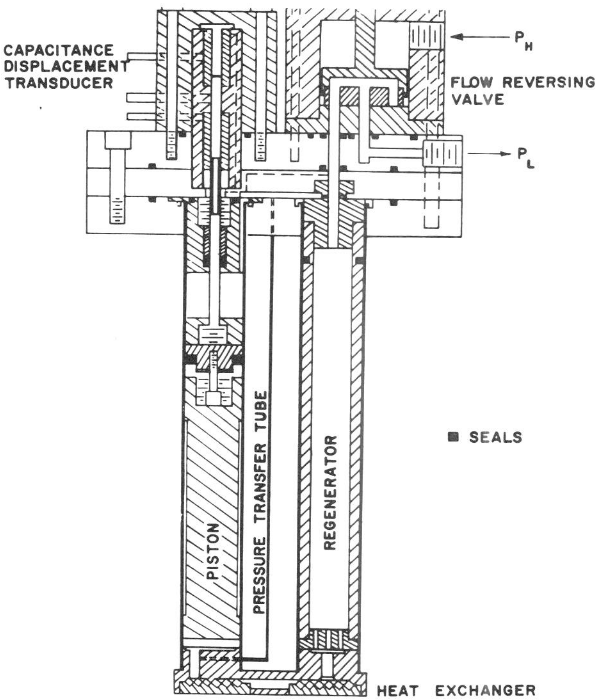

Fig. 5.17. Cross section of experimental Gifford–McMahon cryocooler (after Ackermann *et al.*, 1971).

The ideal refrigeration of 14.5 W, was calculated from

$$Q = (p_2 - p_1)V_D \times N \tag{5.2}$$

where V_D is the displacer swept volume and N is the number of cycles per minute. The actual pressure–volume diagram was quite different from the ideal case. Specimen pressure displacement diagrams for expansion space temperatures of 50 and 300 K are shown in Fig. 5.18. At 77 K the indicated diagram refrigeration was 6.5 W, and the actual refrigeration was measured as 3.9 W, this being the heat supplied electrically to the expansion space heat exchanger to maintain constant temperature.

The difference between the ideal and actual diagram refrigeration, i.e., 14.5 and 6.5 W, respectively, was attributed to losses due to pressure drop, void volume, and valve losses which degraded the pressure–displacement diagram. The difference between the indicated diagram and actual refrigeration was attributed to thermal losses including motional heat leak (shuttle heat transfer or the bucket brigade loss), thermal conduction, and imperfect regeneration. Estimates of the thermal conduction losses gave a total of 1.58 W. Loss due to imperfect regeneration was estimated as 0.45 W and to motional heat leak as 1.07 W. The total thermal loss estimate was 3.11 W compared with 2.6 W determined experimentally. The seal blow-by

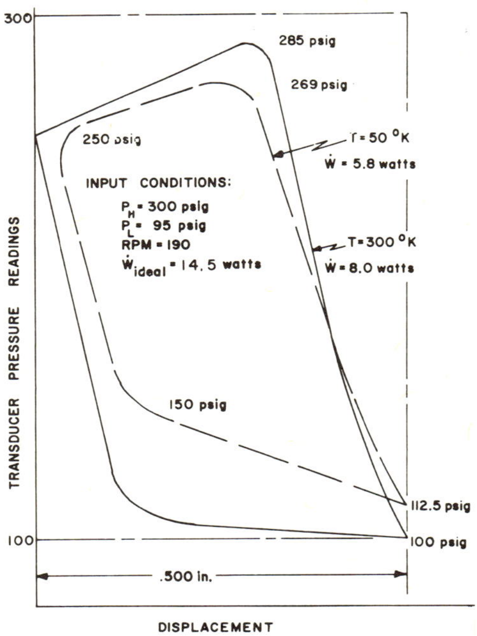

Fig. 5.18. Comparison of ideal and actual work diagrams for the experimental Gifford–McMahon cryocooler (after Ackermann *et al.*, 1971).

losses were found to be small and were ignored. No account was taken of radiation losses.

Much of the large difference, 8 W, between the ideal and indicated diagram refrigeration was attributed to the large void volume in the cold region necessary to accommodate the instrumentation. In commercial machines, the regenerative heat exchangers are frequently included in the displacer to reduce the cold end voids and conduction heat loss because the regenerator tube is eliminated. Motional heat leak then remains the prime thermal loss.

THE POSTLE ENGINE

The refrigerator invented by Davy Postle in 1873 was not intended as a cryocooler, indeed it is not definitely known that the concept was ever reduced to practice. A short description is included here to stimulate future development and because the patent is not readily accessible (Thomas, 1873).

A diagram of the engine is shown in Fig. 5.19. It consisted of a double-acting compressor cylinder and a separate cylinder containing two identical displacer and valve assemblies in opposed form, separated by a central diaphragm provided with flow passages for the circulation of refrigerating brine. Each displacer cylinder of large diameter was provided at the ambient temperature extremities with a coaxial smaller-diameter cylinder. The displacer, fitting closely in the displacer cylinder, was similarly provided with a coaxial extension fitting the smaller-diameter cylinder. The hollow displacer was equipped with a number of fine-bore tubes serving as regenerative ducts connecting the expansion space, adjacent to the central diaphragm, with the ambient temperature space at the cylinder end. The end of the cylinder was provided with ducts for the cooling water flow.

Inlet and outlet valve seats were provided in the displacer extension to correspond with a double-ended self-acting poppet valve mounted on a valve stem sliding freely in a tube contained within the displacer along the cylinder axis. The valve stem was slightly longer than the displacer overall.

The engine operated as follows. With the element disposed in the arbitrary positions shown, fluid at an increasing pressure entered the valve chamber from the compressor cylinder. The flow of fluid caused the poppet valve to slide to the right closing the inlet valve to the cylinder. The high pressure acting on the displacer extension then moved the displacer inward towards the diaphragm. Near completion of the inward stroke, the end of the valve stem contacted the diaphragm causing the inlet valve to separate

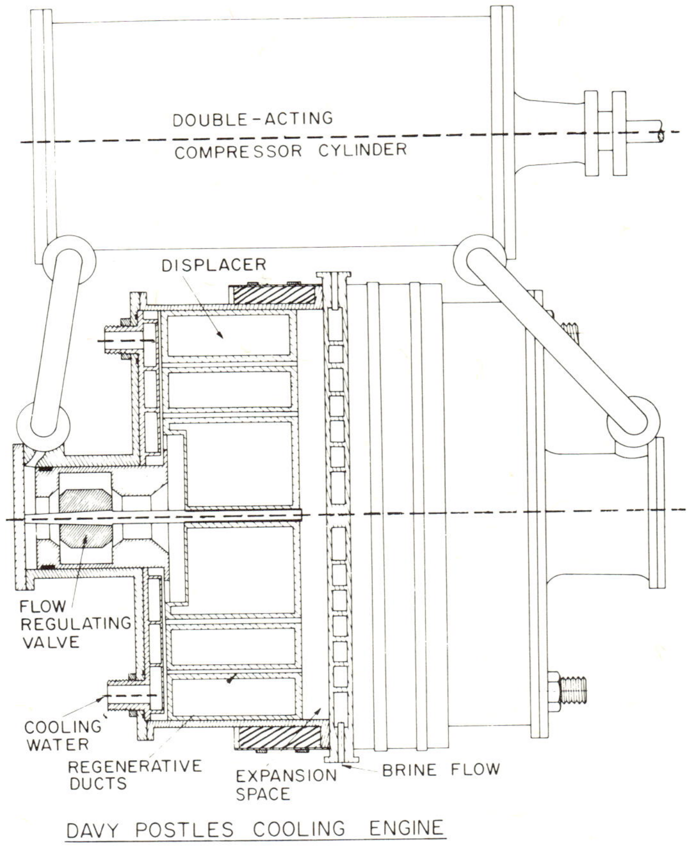

Fig. 5.19. Diagram of the Postle refrigerator.

from the valve seat moving inwards with the displacer. With the inlet valve open, high-pressure fluid filled the cylinder and was cooled by the cooling water heat exchanger on the cylinder end plate.

As the compressor piston reached the end of its stroke and started to return, the pressure began to fall and fluid flowed from the displacer cylinder towards the compressor. The flow acting on the valve caused it to move to the right, thereby closing the exhaust valve. Now the pressure in the cylinder exceeded the decreasing pressure in the valve chamber, thereby inducing the displacer to move from the inward to the outward position. In this movement, fluid was transferred through the regenerative ducts to the central expansion space.

As the displacer approached the end of its outward stroke, the valve stem impacted the cylinder cover plate causing the valve to separate from the exhaust valve seat. Now with the exhaust valve open, fluid passed back to the compressor cylinder all the while expanding in the expansion space and generating refrigeration. When the compressor piston reached the end of its expansion stroke and started to return, the flow of fluid reversed and the operation of the system was repeated as described above.

The novel features of the self-acting valve arrangement of the Postle engine endow the concept with the potential for development as a regenerative cryocooler in the future. An alternative could be equipped with a gas spring to operate at resonant frequency and be supplied with constant pressure fluid from a reservoir.

REFERENCES

Ackermann, R. A., and Gifford, W. E. (1971). "A Heat Balance Analysis of a Gifford–McMahon Cryorefrigerator." *Adv. Cryog. Eng.* **16**, 221–29.

Collins, S. C., and Cannaday, R. L. (1958). *Expansion Machines for Low Temperature Processes.* Oxford University Press, Oxford, England.

Gifford, W. E., and McMahon, H. O. (1960). "A New Low Temperature Gas Expansion Cycle—Part II." *Adv. Cryog. Eng.* **5**, 368–372.

Gifford, W. E., and Hoffman, T. E. (1961). "A New Refrigeration System for 4.2 K." *Adv. Cryog. Eng.* **6**, 82–94.

Gifford, W. E. (1966). "The Gifford–McMahon Cycle." *Adv. Cryog. Eng.* **11**, 152–159.

Higa, W. H., and Wiebe, E. (1978). "One Million Hours at 4.5 Kelvin." Proc. Appl. of Closed-Cycle Cryocoolers to Small Superconducting Devices, pp. 99–108, Boulder, Colorado, October 1977 (Published as NBS Spec. Pub. 508).

McMahon, H. O., and Gifford, W. E. (1960). "A New Low Temperature Gas Expansion Cycle—Part I." *Adv. Cryog. Eng.* **5**, 354–366.

Solvay, E. (1887). Deutsches Reichspatent No. 39,280.

Thomas, W. H. (1873). British Patent No. 709, "Producing Cold for Preserving Animal Food, etc." (Available from Science Museum Library, London).

Chapter 6

Joule–Thomson Cooling Systems

INTRODUCTION

In a Joule–Thomson cryocooler gas expands from high pressure to low pressure at constant enthalpy (often called a throttled or Joule–Thomson expansion) and thereby experiences a decrease in temperature. The system is widely used for gas liquefaction. It may be a simple Linde–Hampson cryocooler or associated with more complex systems involving closed cycle expansion engines to precool the gas before throttled expansion occurs.

Initial experiments on the isenthalpic expansion of a gas were carried out in the early 1850s at Manchester, England by J. P. Joule and William Thomson (later to become Lord Kelvin). The initial experiments were made with gas passing through a nozzle (isentropic expansion). Later a porous plug replaced the nozzle and Joule and Thomson investigated isenthalpic expansion with different gases. They found (Joule, 1852), with initial temperatures near ambient, that many gases (air, CO_2, O_2, N_2) experienced a decrease in temperature but that hydrogen experienced an *increase* in temperature.

The first use of isenthalpic expansion in a refrigerating system is attributed, by Daunt (1956), to Perkins as early as 1834 in the British patent 6662. Sulfuric ether [$(C_2H_5)_2O$] was used by Perkins as the working fluid in his evaporative cooling system and again by Harrison in 1857 for an improved system. Later, in 1876, Karl von Linde in Germany used ammonia as a working fluid for refrigeration. About the same time Pictet in Switzerland used sulfur dioxide as a refrigerant.

The first use of Joule–Thomson isenthalpic expansion for cryocooling was in air liquefiers devised independently but simultaneously by Linde (1897) in Germany and Hampson (1895) in England. Both systems incorporated the recuperative heat exchanger principle invented by Siemens in 1857. Hampson used a close coiled cross flow exchanger of the type now identified as a Hampson heat exchanger. Isenthalpic expansion valves were also incorporated by Claude in the gas liquefaction systems he developed. Nowadays JT expansion valves are used routinely in nearly all modern gas liquefaction and vapor-compression refrigeration systems.

THEORETICAL ASPECTS OF JOULE–THOMSON EXPANSION

Isenthalpic Expansion

Consider the system shown in Fig. 6.1. It consists simply of a duct containing a porous plug, a valve that may be partly "cracked" open, or a metal plate with very fine holes drilled through it. Gas at high pressure enters at station 1 and expands through the constriction to a low pressure at station 2. Fluid passing through the plug experiences high fluid friction which degrades the pressure but no work is done. For the purpose of illustration it is assumed the duct is thermally insulated so that no heat is transferred to or from the fluid. The fluid flows so slowly that the change in kinetic energy in expansion is negligible. The duct is horizontal with no change in potential energy.

Applying the steady flow energy equation to the system shown in Fig. 6.1 and equating changes in kinetic energy, potential energy, work, and heat transfer to zero leaves

$$H_1 = H_2$$

The actual flow in the valve is thermodynamically irreversible and the fluid is not homogeneous. The fluid downstream at 2 can be reasonably assumed to be mixed and homogeneous. It will be found to have the same enthalpy as the inlet fluid.

Outlet conditions of temperature and pressure following isenthalpic expansion from given inlet conditions are plotted on the temperature–pressure plane shown in Fig. 6.2. The curves obtained are lines of constant enthalpy. Starting at the right-hand, high-pressure, side of the figure and

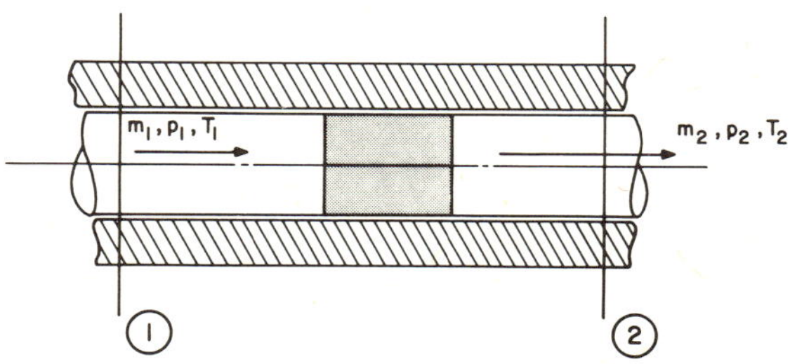

Fig. 6.1. Duct with porous plug for Joule–Thomson expansion of gas.

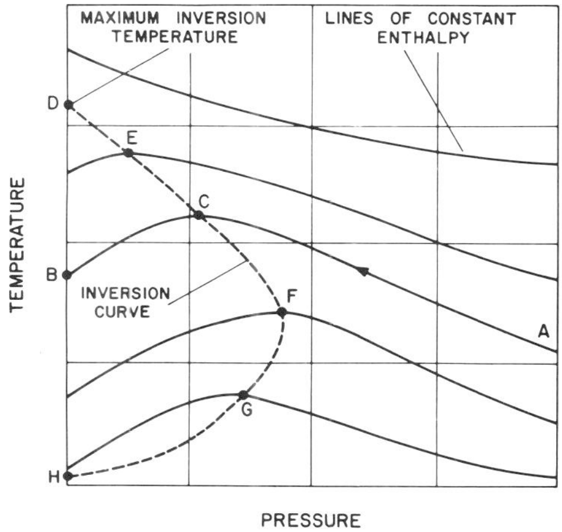

Fig. 6.2. Generalized constant enthalpy curves for a gas displayed on the pressure–temperature plane.

moving along the constant enthalpy curve, *A–B*, the temperature initially increases as the pressure decreases. The increase in temperature reaches maximum at point *C*. Further isenthalpic expansion causes a decrease in temperature.

The Inversion Curve

Curves similar to *A–C–B* are shown at higher and lower initial temperatures of the gas. The loci of the apex points, *D–E–C–F–G–H* are connected to establish the *inversion curve* for the gas. An initial condition of pressure and temperature to the *left* of the inversion curve will result in the gas experiencing a *decrease* in temperature on isenthalpic expansion to a low pressure. Conversely, from an initial condition of pressure and temperature to the *right* of the inversion curve the gas will experience an *increase* in temperature on isenthalpic expansion. If the initial conditions are on the right of the inversion curve and the final pressure is to the left the final temperature may be greater, equal, or less than the initial temperature.

Maximum Inversion Temperature

Point *D* on the inversion curve is called the *maximum inversion temperature*. It occurs where the pressure is zero. Maximum inversion temperatures for different gases are given in Table 6.1.

Gases can be divided into two groups: (a) those for which the maximum inversion temperature is above ambient temperatures, i.e., carbon dioxide,

Table 6.1. Maximum Inversion Temperature for Different Gases[a]

Gas	Maximum inversion temperature, (K)
Above ambient	
Carbon Dioxide	1500
Oxygen	760
Argon	722
Nitrogen	623
Air	602
Below ambient	
Neon	250
Hydrogen	202
Helium	40

[a] After Barron (1966).

oxygen, argon, nitrogen, air, etc., and (b) those for which the maximum inversion temperature is below ambient temperature, i.e., neon, hydrogen, and helium. This difference in maximum inversion temperature, either above or below the ambient value, explains why Joule and Thomson observed the curious phenomena that on expansion through a porous plug hydrogen experienced an *increase* in temperature whereas all the other gases experienced a *decrease*.

Gases with low values of the maximum inversion temperature must be precooled, by an expansion engine or other means, to a temperature within the inversion envelope to ensure a decrease in temperature on isenthalpic expansion.

Joule–Thomson Coefficient

The change in temperature for an isenthalpic pressure drop is defined by the Joule–Thomson coefficient for a fluid: $\mu_H = (\partial T/\partial p)_H$ the change in temperature due to a change in pressure at constant enthalpy. The Joule–Thomson coefficient is the *slope* of the isenthalpic curves shown in Fig. 6.2. The coefficient is zero along the inversion curve (at the peak of the isenthalps) since the slopes of the isenthalps are zero. To the right of the inversion curve the Joule–Thomson coefficient is negative and the temperature increases on expansion. To the left the Joule–Thomson coefficient (slope of the isenthalp) is positive and the temperature decreases on expansion. Figure 6.3 shows the Joule–Thomson coefficient for air as a function of temperature at several different pressures (reproduced from Daunt, 1956).

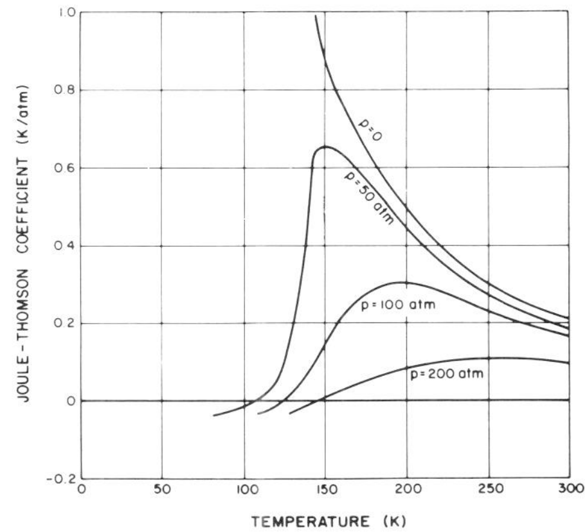

Fig. 6.3. Joule–Thomson coefficient for air as a function of temperature at several different pressures (after Daunt, 1960).

The Joule–Thomson coefficient can be written*

$$\mu_H = \left(\frac{\partial T}{\partial p}\right)_H = -\left(\frac{\partial T}{\partial H}\right)_p \left(\frac{\partial H}{\partial p}\right)_T \tag{6.1}$$

and from elementary thermodynamics

$$dH = \left(\frac{\partial H}{\partial T}\right)_p dT + \left(\frac{\partial H}{\partial p}\right)_T dp \tag{6.2}$$

and

$$dH = C_p\, dT + \left[v - T\left(\frac{\partial v}{\partial T}\right)_p\right] dp \tag{6.3}$$

A comparison of the coefficients in Eqs. (6.2) and (6.3) indicates that

$$C_p = \left(\frac{\partial H}{\partial T}\right)_p \tag{6.4}$$

and

$$\left(\frac{\partial H}{\partial p}\right)_T = v - T\left(\frac{\partial v}{\partial T}\right)_p \tag{6.5}$$

Combining Eq. (6.1), (6.4), and (6.5) we may write

$$\mu_H = \frac{1}{C_p}\left[T\left(\frac{\partial v}{\partial T}\right)_p - v\right] \tag{6.6}$$

* The following is a summary of the excellent treatment of this topic given by Barron (1966).

For a perfect gas: $v = RT/p$ and

$$\left(\frac{\partial v}{\partial T}\right)_p = \frac{R}{p} = \frac{v}{T} \tag{6.7}$$

so that from Eq. (6.6) for a perfect gas

$$\mu_H = \frac{1}{C_p}\left[T\left(\frac{v}{T}\right) - v\right] = 0 \tag{6.8}$$

A perfect gas would experience no change in temperature in a throttled or Joule–Thomson expansion. Therefore a positive or negative Joule–Thomson coefficient must arise from the departure of real gases from perfect gas behavior.

Recall that enthalpy is defined as

$$H = u + pv \tag{6.9}$$

Substituting this in Eq. (6.1) results in

$$\mu_H = -\frac{1}{C_p}\left.\left(\frac{\partial u}{\partial p}\right)_T + \left[\frac{\partial (pv)}{\partial p}\right]_T\right\} \tag{6.10}$$

The first term in Eq. (6.10) is indicative of the departure of a real gas from perfect gas behavior in terms of Joule's law. This states that the internal energy of a gas is a function only of temperature. For real gases the first term is always negative and contributes to a decrease in temperature on expansion. This arises because the molecules of the gas become increasingly separated as expansion proceeds and their microscopic potential energy is increased. No external work or heat transfer occurs, so the increase in potential energy must be offset by a decrease in the microscopic kinetic energy of the gas molecules and is manifested by a decrease in temperature.

The second term in Eq. (6.10) is indicative of the departure of a real gas from perfect gas behavior in terms of Boyle's law. This states that the product of pressure and volume of a perfect gas is a function only of temperature and is illustrated in Fig. 6.4 by the horizontal line. This shows the product of pressure and volume (an isothermal) to be independent of the pressure.

The behavior of real gases is also illustrated in Fig. 6.4. At low pressures and at temperatures near saturation conditions gases are more compressible than Boyle's law predicts since attractive forces are acting to condense the gas. This means the second term in Eq. (6.10) is negative and contributes to the production of a temperature decrease. At high pressures the molecules are closer together and repulsive forces are brought into action. The gas is therefore less compressible than Boyle's law predicts, and the

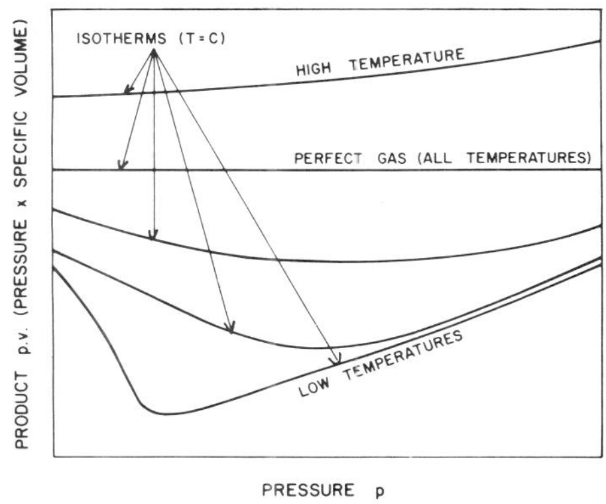

Fig. 6.4. Product of the pressure and volume of a gas as a function of pressure at several different temperatures.

second terms in Eq. (6.10) is positive. This contributes to a temperature increase on expansion.

Increase or decrease in gas temperature on expansion depends upon the relative magnitude of the two terms in Eq. (6.10). Along the inversion curve shown in Fig. 6.2 the two terms are equal in magnitude (and opposite in sign). The second term dominates to the right of the inversion curve and the first term to the left of the inversion curve.

Those wishing to pursue the thermodynamic aspects of Joule–Thomson expansion beyond the elementary level presented here are referred to the paper by Professor Baehr (1963).

CRYOGENIC COOLING AND GAS LIQUEFACTION SYSTEMS USING JOULE–THOMSON EXPANSION

Linde–Hampson Systems

A schematic diagram of an elementary Linde–Hampson gas liquefaction system is shown in Fig. 6.5. The system consists of a compressor, a recuperative heat exchanger, a Joule–Thomson valve, and a liquid/gas separator and liquid reservoir.

The system operates as follows:

(a) Compression (process 1–2″). Gas enters the compressor at condition (1) and is compressed to state (2). Work W is supplied and heat Q rejected during compression of the gas. For the purpose of analysis the compression may be assumed isothermal or isentropic. In practice it would be a polytropic process (1–2″) between the isothermal process 1–2 and the isentropic process 1–2′.

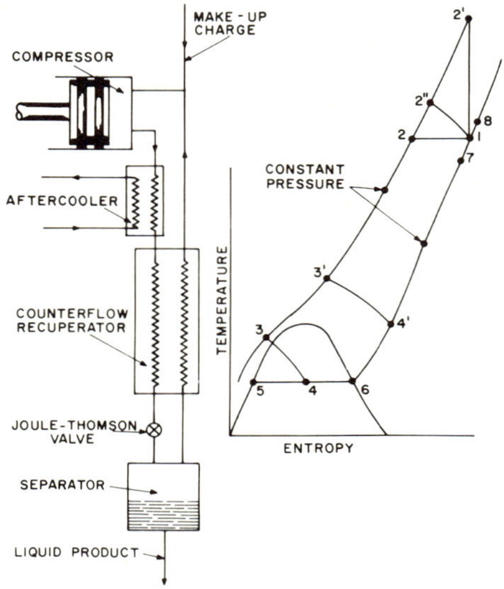

Fig. 6.5. Schematic diagram of simple Linde–Hampson gas liquefaction system.

(b) Aftercooling (process 2″–2). The compressed gas is cooled to ambient temperature in a water or air cooled heat exchanger.

(c) Recuperative cooling (process 2–3). The compressed gas enters a contraflow recuperative heat exchanger and is cooled to condition (3).

(d) Joule–Thomson expansion (process 3–4). The compressed gas expands to low pressure at constant enthalpy. If the gas at (3) is below the inversion temperature it experiences a decrease in temperature. The temperature at (3) may be so low that the final condition (4) is located in the liquid vapor envelope on the *T–S* diagram and the fluid will partially condense during expansion. At (4) it will exist partly as saturated liquid at (5) and partly as saturated vapor at (6). If the temperature at (3′) is not low enough the final condition (4′) will be located outside the liquid–vapor envelope and no saturated liquid will be produced.

(e) Vapor–liquid separation. The saturated liquid at (5) is collected as the useful product of the system and the cold, saturated vapor at (6) is returned to the compressor.

(f) Recuperative heating (process 6–7). Low-pressure saturated vapor at (6) passes through the recuperative heat exchanger *en route* to the compressor and experiences recuperative heating to condition (7). If liquid has been produced the mass of gas flowing in the return line will be less than the mass of compressed gas flowing to the JT valve. The reduced mass flow stream (6–7), will experience a higher temperature differential across the recuperator than the high-pressure stream (2–3) and the exit temperature (7) will approach the compressor inlet temperature (1). To illustrate the make-up/mixing process (7) is shown substantially less than the compressor inlet (1).

(g) Make-up and mixing (process 8/7–1). As liquid gas is produced and leaves at condition (5), new make-up gas must be admitted at the compressor inlet to maintain constant mass flow and continuous operation. The make-up gas is admitted at ambient temperature, (8), and mixes with the return effluent (7) for admission to the compressors, (1).

For the purpose of elementary analysis it is assumed that:

i. the flows take place with no pressure drops,
ii. there are no external heat leaks,
iii. the recuperative heat exchanger is perfect, and
iv. the compression process is isothermal.

Applying the steady-flow energy equation for mass, m, and enthalpy, h, to the recuperative heat exchanger, the expansion valve and the separator we obtain

$$m_7 h_7 + m_5 h_5 = m_2 h_2$$

where

$$m_5 = m_2 - m_7$$

The fraction of gas liquefied:

$$\frac{m_5}{m_2} = \frac{h_7 - h_2}{h_7 - h_5} \tag{6.11}$$

and in the idealized system $h_7 = h_1$. The fraction of gas liquefied therefore depends only on the pressure and temperature at (1) and the pressure ratio p_2/p_1. Many systems for air liquefaction use ambient pressure and temperature for condition (1). This fixes h_1 and h_5 so the only variable is the isothermal compression ratio p_2/p_1 which determines h_2.

The maximum fraction of gas will be liquefied when the optimum pressure p_2 is that at which the enthalpy h_2 is a minimum. Maxima and minima occur when the rates of change or slope of a function are zero. In this case h_2 is a minimum when the rate of enthalpy change as a function of pressure (at constant temperature) is zero, i.e.,

$$\left(\frac{\partial h}{\partial p}\right)_{T=T_1} = 0 \tag{6.12}$$

It can be seen from Eq. (6.1) and Eq. (6.4) that Eq. (6.12) is equivalent to:

$$\left(\frac{\partial h}{\partial p}\right)_{T=T_1} = 0 = (\mu_H \cdot C_p)_{T=T_1} \tag{6.13}$$

The pressure p_2 at which the enthalpy h_2 is a minimum occurs at the apex of the isenthalp, i.e., point C along the enthalp A–B in Fig. 6.2. For air at ambient temperatures of 15 to 20 C the optimum pressure on the inversion curve (point C) is about 41.3 MPa (6000 psi). Unfortunately this is much too high for economic operation, and lower pressures, about half or less of the above figures, are generally used.

It is interesting to note from Eq. (6.11) that the fraction of gas liquefied is negative so long as $h_7(=h_1)$ is less than h_2. This is the case with hydrogen, helium, and neon at atmospheric ambient temperatures and shows that liquids of these gases could never be produced in a simple Linde–Hampson system. Not enough energy can be exchanged in the recuperator to reduce the compressed gas temperature to the point where liquid could be obtained on isenthalpic expansion. The greater limitation with these gases is that the system could never produce any cooling at all from an initial start at ambient temperatures.

Start-Up and Cool-Down of a Simple Linde–Hampson System

The system shown in Fig. 6.5 was assumed to be in a state of steady, continuous flow operation. It is of interest now to consider the conditions prevailing during start-up and cool-down with the whole system initially at the ambient temperature, T_1.

Consider first a fluid having a maximum inversion temperature well above the ambient temperature. The process of cool-down for such a gas is illustrated in the series of T–S diagrams given in Fig. 6.6. The initial operation at time $t=0$ is shown in Fig. 6.6a. After compression (process 1–2″) the gas is cooled in the aftercooler (process 2″–2) and then passes to the recuperative heat exchanger. No cooling is accomplished because the heat exchanger is all at ambient temperature. The gas expands isenthalpically in the Joule–Thomson valve from ambient temperature and high pressure. The temperature at (4′) is lower than the ambient temperature but of course no liquid is produced.

Gas returning to the compressor through the recuperator will be reheated to the initial temperature by the transfer of heat from, and hence cooling, the high-pressure fluid passing to the expansion valve. The system is therefore able to "boot-strap" its way through the initial cool-down process. The situation at some intermediate time is represented in Fig. 6.6b and yet later in Fig. 6.6c when the system is in steady-state operation and producing liquid.

Now consider the other group of fluids given in Table 6.1 where the maximum inversion temperature is below ambient temperature. Such a gas used in the simple Linde–Hampson system would never be able to produce

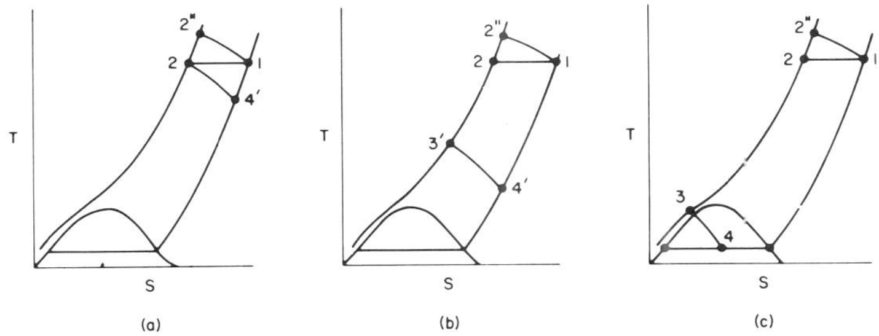

Fig. 6.6. Start-up and cool-down of a simple Linde–Hampson system where the gas has a maximum inversion temperature above ambient.

refrigeration, liquefaction, or cryogenic cooling. This is illustrated in Fig. 6.7.

For the sake of clarity we assume the compression process is isothermal (process 1–2). As before on inital startup there is no effect as the gas passes through the recuperator so that it reaches the Joule–Thomson valve at ambient temperature T_1. On expansion to the lower pressure the gas actually experiences an increase in temperature to (4′). Therefore in returning to the compressor through the recuperator the hot gas will heat rather than cool the incoming high-pressure stream *en route* to the expansion valve. The system could never cool itself down and would simply continue operating at increasingly elevated temperature levels until failure.

Precooled Linde–Hampson System

Precooling is incorporated in the Linde–Hampson system to overcome the problem discussed above in the liquefaction of hydrogen, helium, and neon. One possibility is to cool-down the whole liquefaction circuit to a

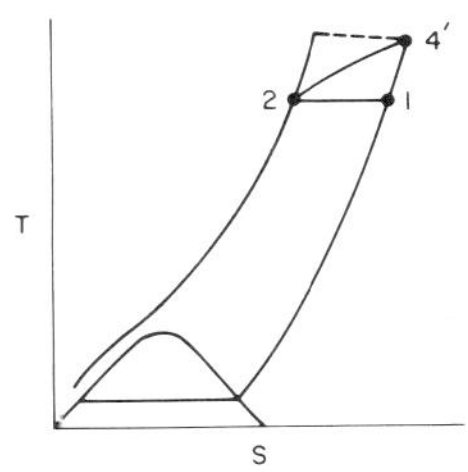

Fig. 6.7. Start-up of a simple Linde–Hampson system where the gas has a maximum inversion temperature less than ambient (hydrogen, helium, and neon).

temperature below the inversion temperature but this is not advantageous thermodynamically. Furthermore, it is not easy to operate compression equipment at temperatures much below ambient. The preferred alternative is to incorporate the precooling at a location downstream of the compressor as shown in the schematic diagram, Fig. 6.8.

The precooler is a cold liquid bath through which the pressurized gas duct passes *en route* to the expansion valve. The precooler can be located downstream of the recuperative heat exchanger and immediately upstream of the Joule–Thomson expansion valve. However, this is less advantageous than the arrangement shown where the recuperative exchanger is divided into two, a high-temperature and low-temperature section. It is most advantageous thermodynamically to arrange the division so the pressurized gas stream enters the precooler heat exchanger at a temperature not much greater than that at which the precooler liqiuid coolant is available.

The precooler liquid coolant can be any liquid with a normal boiling point (at atmospheric pressure) below the maximum inversion temperature of the gas to be liquefied. For neon and hydrogen (maximum inversion temperatures are 250 and 202 K, respectively) the precoolant liquid could

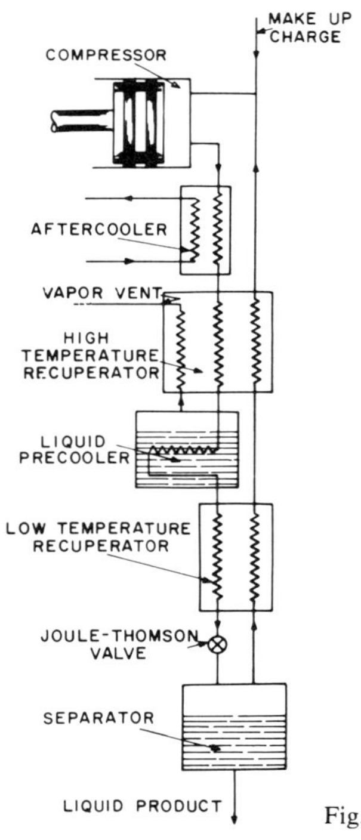

Fig. 6.8. Precooled Linde–Hampson system.

be fluorine (normal boiling point 85 K), oxygen (90 K), air (78 K), methane (113 K), argon (87 K) or nitrogen (77 K). In practice for reasons of economy and safety nitrogen is always used except in unusual circumstances.

Helium has a maximum inversion temperature of only 40 K. Therefore the only liquids suitable for helium precooling are hydrogen (normal boiling point 20 K) or neon (27 K). On grounds of safety neon is the preferred fluid but economic considerations dictate that hydrogen is almost always used.

Cascade System

A helium liquefier operating on the precooled Linde–Hampson cycle consists of a series of separate systems having different fluids at different minimum temperature levels. This is known as a "cascade system." It was used originally for the liquefaction of air. The schematic arrangement of the compound system is shown in Fig. 6.9. The useful product, helium, is available at 4 K. The helium is precooled by a hydrogen bath at 20 K. The hydrogen is precooled in a nitrogen bath at 77 K.

A system of this type was described by Mann *et al.* (1960). The liquefier was 0.5 m diameter and 2 m high and was constructed at the Cryogenic Engineering Laboratory of the National Bureau of Standards, Boulder, Colorado. Mann reports the following performance of the helium liquefier:

i. Liquid helium: 2.4 kg per hour
ii. Helium flow rate: 12.6 kg per hour
iii. Liquid yield $(a/b) = 0.19$
iv. Liquid hydrogen evaporated: 1.2 kg per hour
v. Liquid nitrogen evaporated: 10 kg per hour.

Geist and Lashmet (1960) described a lower capacity (750 mW useful refrigeration at 4.4 K) Joule–Thomson cascade helium liquefier using nitrogen and hydrogen precooler circuits. The system was constructed by Air Products and Chemicals Inc., Allentown, Pennsylvania as a prototype for application with masers, cryotrons, superconducting magnets, and other electronic devices requiring refrigeration at liquid temperatures. The unit was notable for the use of Corblin* diaphragm compressors to eliminate the need for fluid purification equipment. The nitrogen and hydrogen circuit used two-stage compressors having inlet, intermediate and discharge pressures of 0.1, 1.96 (1.69 for H_2) and 24.8 MPa [15, 285 (245 for H_2), and 3600 psi], respectively. The helium compressor operated over the pressure range 0.1 and 1.86 MPa (15 and 270 psi).

* Société des Compresseurs a Membrane Corblin, Paris, France.

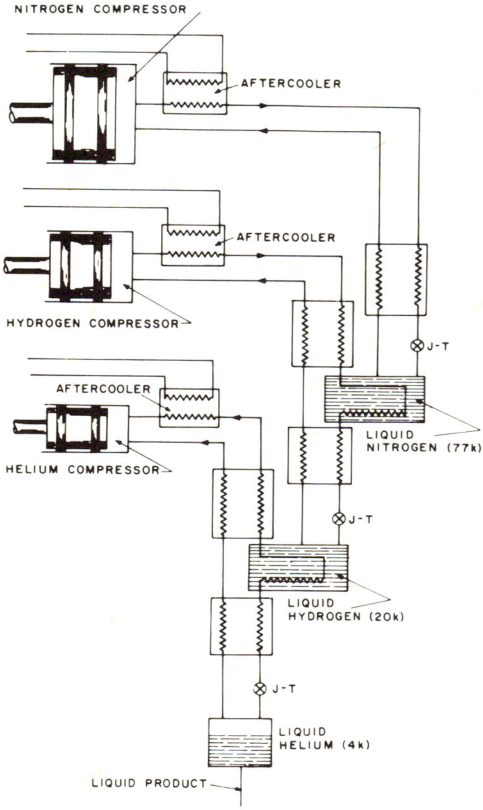

Fig. 6.9. Schematic diagram for helium liquefaction using the external cascade system.

The advantages of precooling and the cascade system are not confined to gases with subambient maximum inversion temperatures (neon, hydrogen, and helium) but apply equally to gases of the other group of gases to improve the liquid yield fraction. Figure 6.10, reproduced from Barron (1966), shows the liquid yield fraction (y) (the fraction of total mass expanded that is saturated liquid after expansion) as a function of the temperature of compression (assumed to be isothermal). The figure refers to a simple idealized Linde–Hampson system using nitrogen. It is evident that the amount of liquid produced is a strong function of the compressor temperature: the lower the compressor temperature the higher the fraction of gas converted to saturated liquid after expansion.

A similar improvement can be obtained using the precooled Linde–Hampson system. An auxiliary refrigerating system cools the compressed gas between the compressor aftercooler and the main recuperative heat

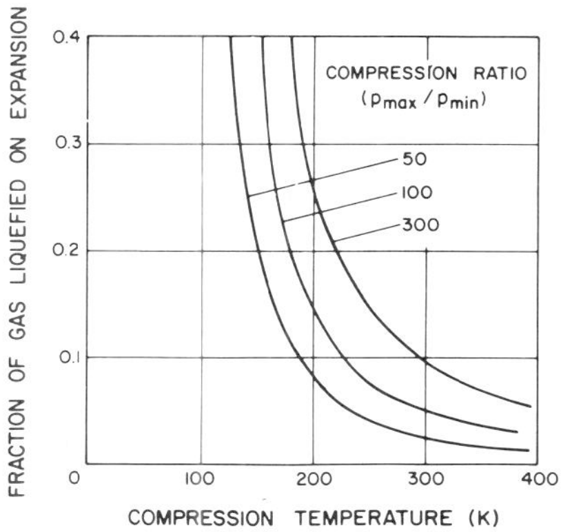

Fig. 6.10. Fraction of gas liquefied in a simple Linde–Hampson nitrogen liquefier as a function of the gas temperature at the end of isothermal compression at several different compression ratios.

exchanger. The auxiliary refrigerator can use the same working fluid but this is not essential since the auxiliary circuit is entirely independent. It is often advantageous to use alternative fluids with more favorable thermodynamic or thermofluid characteristics.

Barron (1966) considered a precooled Linde–Hampson cycle for nitrogen liquefaction with Freon 12 as the refrigerant in the auxiliary precooling circuit. For his study he assumed the nitrogen system to operate between pressures of 0.1 and 20 MPa (15 and 3000 psi) and temperatures of 77 and 294 K (138 and 530 R). The Freon 12 system was assumed to operate between pressure of 103 and 590 kPa (15 and 86 psi) and temperatures of 243 and 469 K (438 and 844 R). He found the addition of the Freon 12 precooler increased the liquid yield by 40% compared with the simple Linde–Hampson nitrogen liquefaction system. At the same time the total work requirement increased only marginally so the work requirement *per unit mass liquefied* decreased by almost 40% compared with the simple cycle. This improvement results from the closer approximation to an ideal reversible system. It can be carried further with multiple refrigeration systems with continuing improvement but at a progressively diminishing rate.

The cascade system was first applied for the liquefaction of air. One system suggested by Keesom (1933) utilized auxiliary refrigeration systems (in order of decreasing temperatures) of (a) ammonia used to liquefy, (b) ethylene used to liquefy, (c) methane used to liquefy nitrogen. The system is known as the "outer cascade cycle."

A later development, known as the "incorporated cascade cycle" uses a compound working fluid having several components which are mixed and processed together in a single compressor. The mixture circulates through a series of heat exchangers and separators so that various fractions of the gas become liquid and are separated at different temperature levels. The various fractions are further expanded and subcooled to provide refrigeration to cool and condense different components of the mixture.

Detailed consideration of this complex system is beyond the level of this book. Suffice to say the incorporated cascade system is used commercially in large installations for the liquefaction of natural gas. It is also used in smaller plants of low cooling capacity. The principal advantage is the high flexibility. Depending on the choice of components and the composition of the gas the system can be used to achieve refrigeration at any temperature level between ambient and cryogenic temperatures. Carbonell *et al.* (1972) have described a development of the incorporated cascade system using a complex working fluid and regenerative rather than recuperative heat exchangers. The regenerators are switched periodically to reverse the direction of fluid flow.

Linde Dual-Pressure System

The Linde dual-pressure system is an alternative to the precooled Linde–Hampson gas liquefier. With the dual-pressure system the work per unit mass of gas liquefied is reduced as in the precooled system but the fraction of gas liquefied *decreases*. The dual-pressure cycle is shown schematically in Fig. 6.11. It consists of high-pressure and low-pressure compressors, a three-way recuperative heat exchanger, intermediate-pressure Joule–Thomson expansion valve and separator, and a low-pressure Joule–Thomson expansion valve and separator.

The cycle operates as follows. The low-pressure return stream and fresh make-up gas combine and enter the low-pressure compressor at (1). After compression to the intermediate pressure the gas passes through the low-pressure aftercooler, mixes with the intermediate-pressure return stream, and enters the high-pressure compressor at (2). After compression the gas passes through the aftercooler, the three-way recuperative heat exchanger and then expands to the intermediate pressure, (5). The vapor, at (7), and intermediate pressure return back through the recuperator as the return stream forming part of the inlet to the high-pressure compressor at (2). The saturated liquid at intermediate pressure, (6), expands through a second Joule–Thomson valve to (8) and, in the low-pressure separator, is separated to the useful product liquid stream at (9), and the low-pressure return stream at (10).

The savings in work gained by the dual-pressure Linde cycle is shown in Fig. 6.12, adapted from Barron (1966). It shows the work to produce a unit mass of liquid as a fraction of the work to produce the same mass in a simple Linde–Hampson cycle. Two curves are given for different ratios

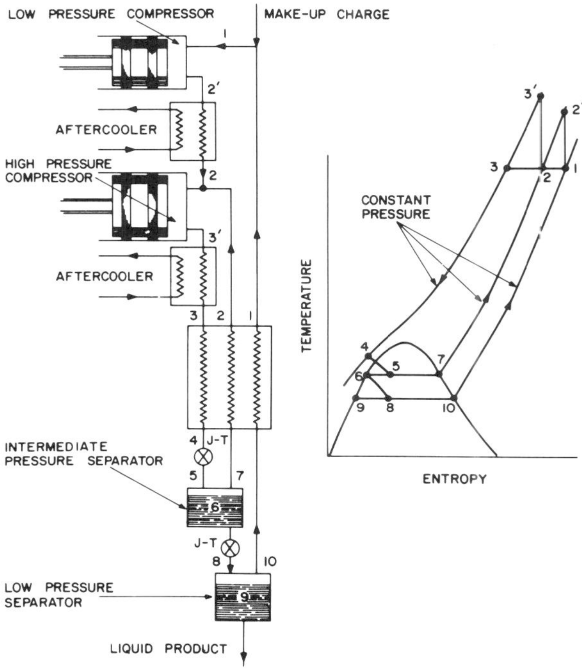

Fig. 6.11. Schematic diagram of gaseous liquefier using the Linde dual-pressure system.

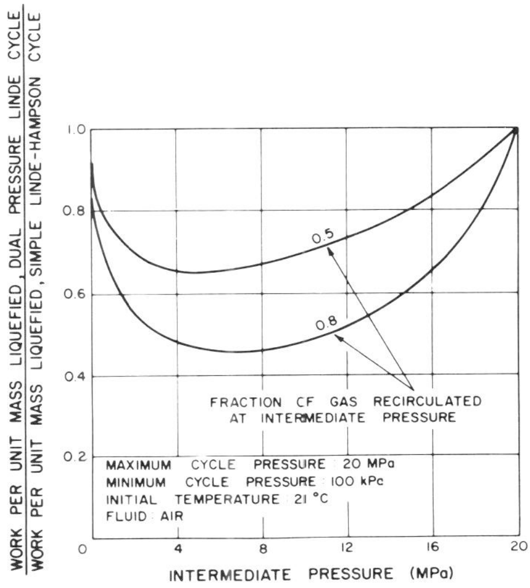

Fig. 6.12. Ratio of the work required per unit mass liquefied in the Linde dual-pressure cycle and the simple Linde–Hampson cycle. The ratio is shown as a function of the intermediate pressure for two different values of the fraction of gas recirculated at the intermediate pressure (abstracted from Barron, 1966).

of gas recirculated at the intermediate pressure. Barron assumed an air liquefaction system operating between pressures of 0.1 and 20 MPa (15 to 3000 psi). With 80% recirculation at the intermediate pressure of 7 MPa over half the work of liquefaction can be saved by use of a dual pressure cycle. Under these conditions the liquid yield per unit mass of gas compressed to the maximum cycle pressure is reduced to about 75% of the yield attained per unit mass of gas compressed in a simple cycle.

The Rietdijk Expansion-Ejector

An interesting variant of the Linde dual-pressure system was explored by Rietdijk (1966) using a single high-pressure compressor and an expansion-ejector to accomplish compression of the low-pressure fluid to the intermediate pressure. A schematic diagram of a liquefaction system incorporating the Rietdijk expansion-ejector is given in Fig. 6.13 reproduced from Haarhuis (1967). The system operates as follows. Fluid enters the compressor at intermediate pressure P_2 and is compressed to P_1. It then enters the recuperative heat exchanger equipped with precooling systems if necessary and is cooled to a low temperature. The fluid expands in a throttled expansion in the expansion-ejector device to the intermediate pressure, P_2, and experiences a reduction in temperature to T_2. The

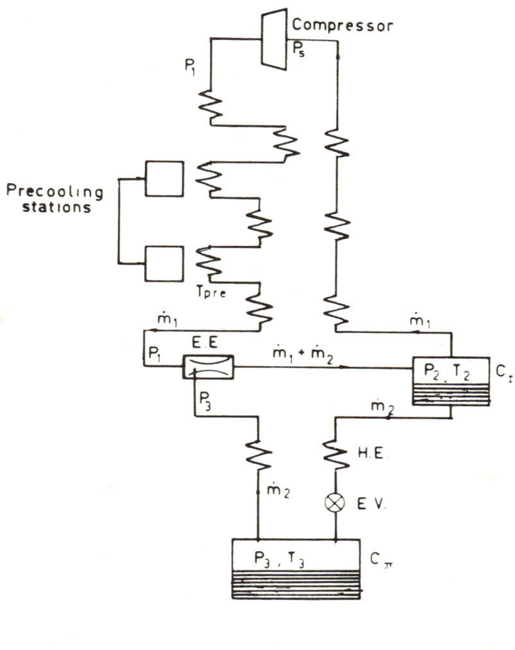

Fig. 6.13. Schematic diagram of a gaseous liquefaction system incorporating a Rietdijk expansion-ejector (after Haarhuis, 1967).

reduction in temperature may be sufficient to cause the fluid to partially condense and collect in the separator identified in Fig. 6.13 as vessel C_I.

The saturated liquid at pressure P_2 passes through a low-temperature recuperative heat exchanger and expands through a Joule–Thomson valve (identified as E.V. in Fig. 6.13) to the minimum cycle pressure P_3 and temperature T_3. The saturated liquid gas accumulating in vessel C_{II} at these conditions is the useful product of the cycle. The saturated vapor at pressure P_3 returns through the low-temperature recuperative heat exchanger and is compressed in the expansion-ejector to the intermediate pressure P_2. Haarhuis (1967) has indicated that a Rietdijk expansion-ejector system results in reduction of the compressor volume of 2.5 times, saving in the size of heat exchangers and the compressor power consumption compared with a conventional Joule–Thomson liquefier.

Haisma and Roozendaal (1967) have given further theoretical and experimental aspects of expansion-ejector performance. Prast (1968) described a helium liquefaction system utilizing a double expansion A-20 Philips Stirling cooling engine and the Rietdijk expansion-ejector.

PERFORMANCE COMPARISON OF GAS LIQUEFACTION SYSTEMS

Figure of Merit

To provide the means of comparison of gas liquefaction systems the *figure of merit* is sometimes used. This is defined as the theoretical minimum work required to liquefy unit mass of gas compared with the actual work requirement of a system: $\text{FOM} = W_{id}/W_{ac}$. The figure of merit can range from zero to unity and indicates how closely a system approaches ideal operation.

Ideal Liquefaction System

Theoretical minimum work to liquefy a gas is estimated assuming the gas is liquefied according to the process shown in Fig. 6.14. This consists of a compressor receiving gas at ambient pressure and temperature, condition (1). The gas is compressed isothermally to (2) and the heat of compression Q_c is rejected from the system. The compressed gas expands isentropically in an engine to (3) where it exists as saturated liquid.

The liquefaction process is the first two processes in the Carnot cycle discussed in Chapter 2. The process of liquefaction is ideal thermodynamically but not suitable for practical use. The principal reason is that the pressure at the end of the isothermal compression is impossibly high (of

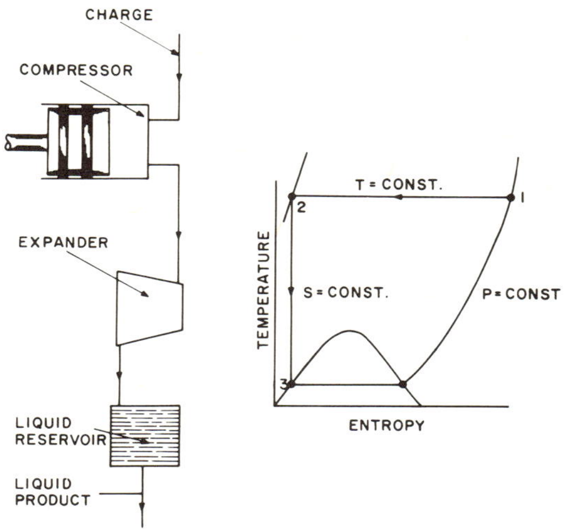

Fig. 6.14. Schematic diagram of ideal gas liquefaction system.

the order of 10^4 MPa for air), and far exceeds the capacity of any compression equipment existing or in prospect. Nevertheless the process provides a useful basis on which to compare theoretical and practical liquefaction systems.

Using the above concept for gas liquefaction Barron (1966) calculated the ideal minimum work requirement for different fluids. His results are reproduced in Table 6.2.

Performance of Linde–Hampson System with Different Fluids

Barron (1966) calculated the performance details of Linde–Hampson gas liquefaction systems for a number of different gases. To provide a common basis of comparison all the cycles were assumed to operate over the pressure range 0.1 to 20 MPa (15 to 3000 psi) with isothermal compression at 294 K (70°F) and no pressure drop or loss in recuperator effectiveness.

The results are reproduced in Table 6.3 arranged in order of successively increasing temperatures. The data include the critical pressure and temperature, the mass of gas liquefied per unit mass of gas compressed the ("liquid yield"), the work required per unit mass of gas compressed and per unit mass of gas liquefied, and the figure of merit for the particular conditions assumed; the figure of merit varied from 0.074 for fluorine to a maximum of 0.600 with carbon tetrafluoride, CF_4. Hydrogen, helium, and neon are not included in Table 6.3 because they cannot be liquefied in the simple Linde–Hampson cycle.

Table 6.2. Ideal-Work Requirements for Liquefaction of Gases Beginning at 294 K and Atmospheric Pressure[a]

Gas	Normal boiling point (K)	Ideal work of liquefaction (kJ/kg)
Helium, He^4	4.2	6650
Hydrogen, H_2	20.3	11,594
Neon, Ne	27.1	1291
Nitrogen, N_2	77.3	742
Air	78.7	710
Carbon monoxide, CO	81.6	742
Fluorine, F_2	85.3	479
Argon, A	87.2	465
Oxygen, O_2	90.2	613
Methane, CH_4	111.6	1044
Carbon tetrafluoride, CF_4	145	171
Ethane, C_2H_6	184.5	330
Ammonia, NH_3	239.1	232

[a] After Barron (1966).

Comparison of Liquefaction Systems for Air

Barron (1966) assembled an interesting tabulation of ideal, theoretical, and observed liquefaction systems for air. Some of the material has been reproduced in Table 6.4. For all the cases the minimum pressure was assumed to be atmospheric (0.1 MPa) and the ambient temperature 294 K (70 F). For the theoretical cases the maximum pressure was taken to be 20 MPa (3000 psi). In some cases the compression process was assumed to be isothermal and the recuperative process to be 100% effective. In other cases, the compressor efficiency was assumed to be 70% and the recuperator effectiveness to be 95%. Conditions for the observed systems were not included by Barron but likely correspond reasonably closely to the conditions of operation of the theoretical systems.

The first entry in Table 6.4 is the ideal reversible liquefaction process discussed above. The second group involves three entries for simple Linde–Hampson systems. The first entry assumes isothermal compression and perfect recuperation. The second assumes nonisothermal compression and nonperfect recuperation. The third gives the observed results of an actual Linde–Hampson air liquefier. The excellent correlation between the theoretical result and the observed result should be noted. The next group of entries refers to the precooled Linde–Hampson cycle. The precooling temperature was assumed to be 227 K. As before the entries included an idealized cycle and another where allowance was made for compressor

Table 6.3. Performance of the Linde–Hampson System Using Different Fluids $p_1 = 0.1$ MPa, $T_1 = 294$ K, $p_2 = 20$ MPa, $T_2 = 294$ K; Heat-Exchanger Effectiveness, $\varepsilon = 1.00$; Compressor Overall Efficiency $\eta_c = 100\%$[a]

Fluid	Normal boiling point (K)	Critical temperature (K)	Critical pressure (MPa)	Fraction of gas liquefied (liquid yield)	Work required per unit mass compressed (kJ/kg)	Work required per unit mass liquefied (kJ/kg)	Figure of merit FOM = W_{id}/W_{ac}
N_2	77.3	126	3.35	0.087	85.1	976	0.141
Air	78.7	133	3.87	0.086	83.0	967	0.137
CO	81.6	132	3.45	0.092	85.5	931	0.148
F_2	85.3	118	5.5	0.053	63.2	1195	0.074
A	87.2	150	4.83	0.124	58.9	474	0.182
O_2	90.2	154.6	5.01	0.110	73.5	670	0.197
CH_4	111.6	190.7	4.58	0.209	141.4	675	0.288
CF_4	145	227.5	3.69	0.303	16.0	53	0.600
C_2H_6	184.5	305.5	4.85	0.505	61.7	122	0.502

[a] After Barron (1966).

Table 6.4. Comparison of Liquefaction Systems Using Air as the Working Fluid $T_1 = 294$ K and p_1 = Atmospheric Pressure, η_c = Overall Compressor Efficiency, ε = Heat Exchanger Effectiveness, i = Fraction of Mass Flow Recirculated[a]

	Air liquefaction system	Fraction of gas liquefied (liquid yield)	Work per unit mass liquefied (kJ/kg)	Figure of merit W_{id}/W_{ac}
1	Ideal reversible system	1.000	132.4	1.000
2	Linde–Hampson system, $p_2 =$ 200 atm, $\eta_c = 100\%$, $\varepsilon = 1.0$	0.086	967.5	0.137
3	Linde–Hampson system, $p_2 =$ 200 atm, $\eta_c = 70\%$, $\varepsilon = 0.95$	0.061	1965	0.068
4	Linde–Hampson system, observed	—	1909	0.070
5	Precooled Linde–Hampson system $p_2 = 200$ atm, $T_3 = -50$ F, $\eta_c =$ 100%, $\varepsilon = 1.00$	0.179	414	0.320
6	Precooled Linde–Hampson system $p_2 = 200$ atm, $T_3 = -50$ F, $\eta_c =$ 70%, $\varepsilon = 0.95$	0.158	683	0.194
7	Precooled Linde–Hampson system observed	—	1032	0.129
8	Linde dual-pressure system $p_3 =$ 200 atm, $p_2 = 60$ atm, $i = 0.8$, $\eta_c = 100\%$, $\varepsilon = 1.00$	0.060	507	0.261
9	Linde dual-pressure system, $p_3 =$ 200 atm, $p_2 = 60$ atm, $i = 0.8$, $\eta_c = 70\%$, $\varepsilon = 0.95$	0.032	1479	0.090
10	Linde dual-pressure system, observed	—	1173	0.113
11	Linde dual-pressure system precooled to -50 F, observed	—	662	0.201
12	Cascade system, observed	—	602	0.221

[a] After Barron (1966).

efficiency and recuperator ineffectiveness. Agreement between the calculated and observed results was not as good as the case of the precooled cycle.

The next group refers to the Linde dual-pressure cycle. Again three entries are included, arranged as before. There was poor correlation between the observed and calculated results.

The final entries present the observed results in terms of liquid yield and work per unit mass liquefied for a Linde dual-pressure cycle with precooling to 227 K, and a cascade, presumably of the "outer cascade cycle" type although the details of operation and the fluids used were not included by Barron. It is interesting to compare the observed figures of merit for the various systems. The simple Linde–Hampson system had the worst performance with a figure of merit of only 0.070. With precooling this improved to 0.129. The Linde dual-pressure system had a figure of merit of 0.113. The combination of precooling and dual pressure improved the figure of merit to 0.201. This compared favorably with the cascade system having the best figure of merit, 0.221. The simple Linde-Hampson system therefore consumed about three times the work per unit mass liquefied of the thermodynamically superior but very much more complicated cascade or precooled, dual-pressure Linde cycle.

MINIATURE JOULE–THOMSON COOLERS

Hymatic Coolers

Miniatnre Joule–Thomson cooling systems have been developed for electronic applications, particularly infrared thermal detectors. The development was carried out in Great Britain by the Hymatic Engineering Co. Ltd., Redditch, Worcester with Government sponsorship through the Royal Radar Establishment, Malvern, Worcester. Accounts of this work have been given by McInroy (1967), Parkinson (1959, 1967), and Nicholds (1968, 1970).

McInroy described the construction and gave the performance data for both liquid air and liquid hydrogen/liquid nitrogen Joule–Thomson miniature liquefiers. Figure 6.15 is a cross section of the miniature air liquefier described by McInroy as follows:

> The miniature liquefier is enclosed by a representative cell for producing temperatures of 79 K (air) or 77 K (N_2) in the region of the detecting element of the cell. Liquid is produced in the collecting chamber by the Joule–Thomson effect from high pressure air (60 to 400 atm) applied to the inlet.
>
> The weight of the liquefier is 5.3 g and has a heat exchanger of 7.2 mm diam × 25 mm long mounted on a hollow mandrel, or former, of Monel the lower end of which is closed by a brass cap to which the nozzle assembly is fitted. The cavity thus produced is filled with expanded perlite to reduce heat leaks by convection. A sintered bronze filter of 2 micron pore size is placed in the inlet cavity of the former to exclude harmful dust particles.

The original prototype liquefier built at the Royal Radar Establishment had no nozzle assembly or inlet filter. Instead the end of the finned tube heat exchanger was squeezed together to a fish-tail shape which restricted the flow to the desired volume. This operated satisfactorily but required very pure air for its operation as the fine slit became easily blocked. Contaminant gases such as CO_2 at low vapor pressure sublimed at the cold end to form minute particles which blocked the exist orifice. The addition of an inlet filter and the porous nozzle effected a substantial improvement in reliability, but the requirement for a very high degree of gas purity remained. Indeed the problem remains today as perhaps the single greatest impediment to the wider use of miniature Joule–Thomson coolers.

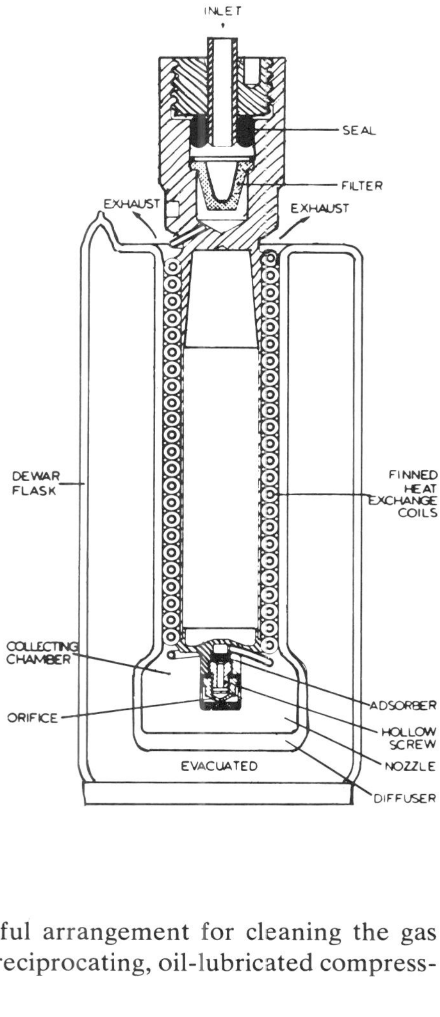

Fig. 6.15. Cross section of miniature Joule–Thompson air or nitrogen liquefier (after McInroy, 1967).

McInroy described one successful arrangement for cleaning the gas compressed to 40 MPa (6000 psi) in reciprocating, oil-lubricated compressors as follows:

1. Air compressor, oil lubricated crankcase and upper cylinders, 400 atm, water cooled cylinders and intercoolers (max. cooling water temperature 50°C), maximum compression ratio in any one stage 4.9, minimum distillation temperature of lubricating oil 230°C.
2. Mechanical oil and water trap for collection of oil and water droplets.
3. Soda-lime bed for chemical absorption of CO_2 and SO_2 traces. This must be first sorbent bed because water is formed.
4. Activated charcoal for bulk hydrocarbon removal in wet state.
5. Activated alumina for water adsorption. Preferably arranged to be reactivated *in situ*.

6. Activated charcoal for remaining hydrocarbon removal in dry state.
7. Molecular sieves, 13X for remnants of H_2O, CO_2, SO_2 and hydrocarbons.
8. 5 micron filter.

Air produced by this process has been called "Pure Air" and conforms to the following specification:

	Maximum allowable	Minimum obtainable
Water content	1 wpm (Dew Point minus 77°C)	0.1 wpm (Dew point minus 90°C)
CO_2 content	1.2 wpm	0.3 wpm
Hydrocarbon content	0.045 wpm	0.03 wpm

Performance of Miniature Joule–Thomson Coolers

Typical performance characteristics for both air and hydrogen liquefiers were given by McInroy and are reproduced in Figs. 6.16 and 6.17. Each series of three diagrams per figure shows the gas flow, time to achieve liquefaction, and the gross cooling power achieved. The heat leak to the system must be subtracted to obtain the net available cooling power. McInroy gave the heat leak in the configuration shown as 0.4 W at 22 C ambient temperature for the air liquefier, and 0.13 W for the hydrogen liquefier when the nitrogen/air precooler was operating.

When argon was used rather than air as the working fluid in the system shown in Fig. 6.15 an appreciable improvement in performance was noted. At an ambient temperature of 22°C the time to achieve liquefaction decreased to 0.7 of the time with air. The cooling power was 50% greater than with air. With an increase in the ambient temperature to 75°C the improvement in cooling power was even greater to 2.2 times the cooling power with air. Such a result as this might of course be anticipated from Table 6.3, which clearly shows the advantage of argon compared with air as a working fluid in Joule–Thomson systems.

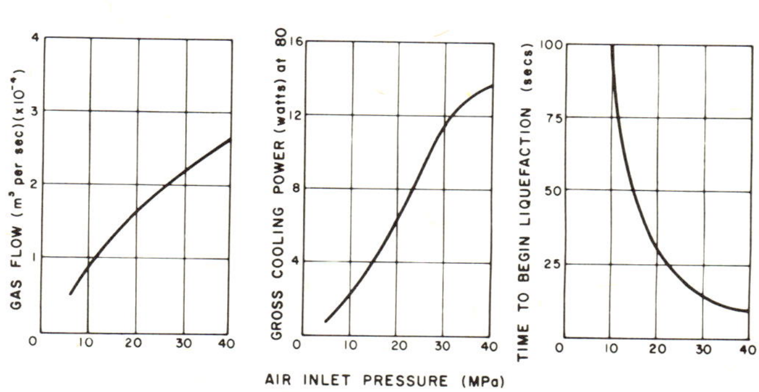

Fig. 6.16. Performance of miniature Joule–Thomson nitrogen liquefier (after McInroy, 1967).

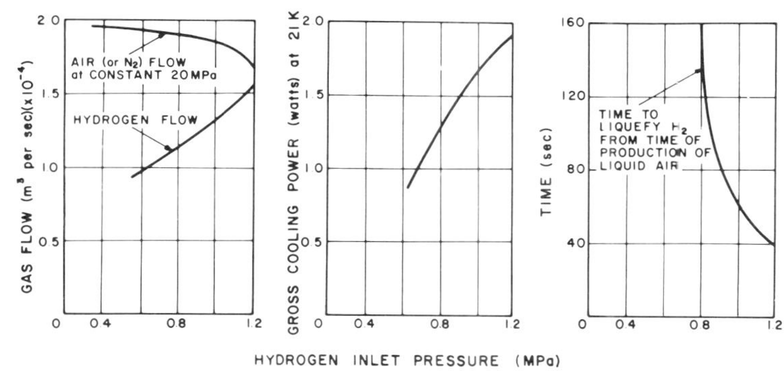

Fig. 6.17. Performance of miniature Joule–Thomson hydrogen liquefier (after McInroy, 1967).

Self-Regulating Miniature Joule–Thomson Coolers

The development of fixed orifice miniature coolers of the type described by McInroy above was followed by the introduction of an important development, the self-regulating miniature Joule–Thomson cooler. In the self-regulating system the expansion orifice is throttled when sufficient liquid has been generated to provide a stable pool of liquid to cool the detector. With a cooler supplied from compressed gas bottles rather than directly from a compressor the self-regulating system gave an order of magnitude reduction in the total system weight because of reduced gas flow.

The self-regulating cooler can cool down more rapidly since the initial gas flow can be very much higher than normal. Another important advantage is that the problem of contamination is much reduced. Solid contaminants at the expansion nozzle tending to obstruct the flow simply cause the self-regulating device to open so as to maintain the flow necessary to achieve the desired cooling. The self-regulating cooler has significantly less noise than the fixed orifice Joule–Thomson coolers which are themselves greatly advantageous in this regard compared with cooling engines of the integral Stirling type.

Nicholds (1970) gave performance details and the test protocol for self-regulating Joule–Thomson coolers. Stephens (1970) has given further details of the self-regulating system including the operation of the regulatory flow orifice needle valve activated by a vapor pressure thermometer and metal bellows arrangement. Buller (1971) has presented interesting data on the performance of similar self-regulating Joule–Thomson miniature coolers constructed at the Hughes Santa Barbara Research Center, Goleta, California.

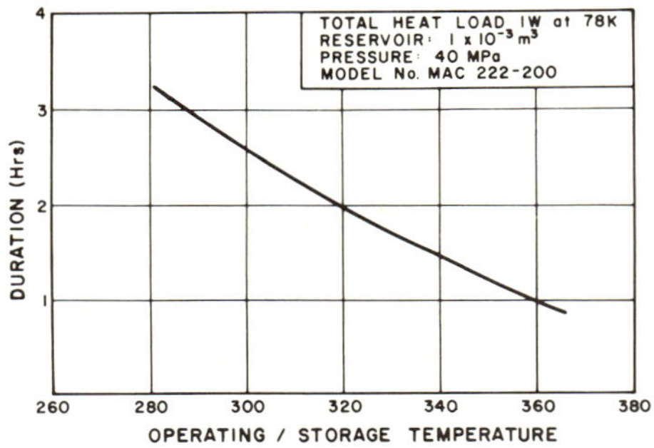

Fig. 6.18. Operating time for miniature Joule–Thomson self-regulating cooler (Hymatic Model No. MAC 222-200) as a function of ambient temperature (after Stephens, 1970).

Complete Refrigeration System

Little work on miniature Joule–Thomson coolers has been reported from United States sources but one paper by Stephens (1970) is of interest. In this he disclosed that the Bendix Corporation of Davenport, Iowa had a licensing agreement with the British Hymatic Engineering Company to manufacture Hymatic coolers in North America. One interesting characteristic given by Stephens (1970) is reproduced in Fig. 6.18. This shows the duration of operation of a nominal 1 W at 77 K self-regulating Hymatic Joule–Thomson cooler (Model No. MAC 222-200), energized from a compressed gas bottle having a capacity of 1×10^{-3} m^3 and containing nitrogen compressed, at the start, to 40 MPa (6000 psi). The duration of operation is shown as a function of the ambient operating temperature. A

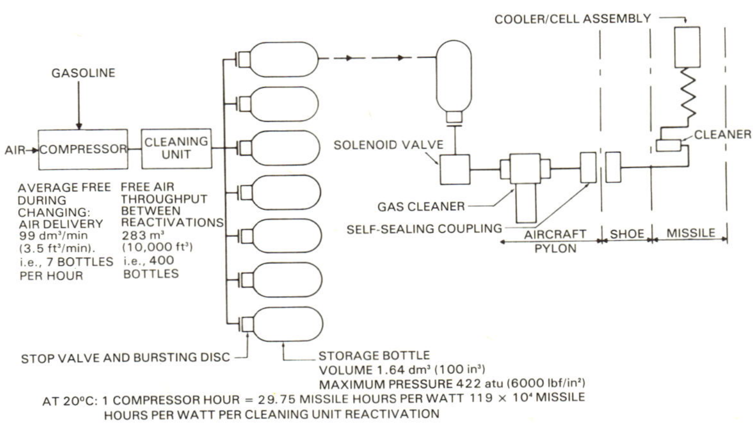

Fig. 6.19. Schematic arrangement for complete refrigerating system to provide miniature Joule–Thomson cooler system for heat-seeking air-to-air missiles aboard aircraft (after Stephens, 1970).

reduction of 40 K from 360 to 320 K doubled the duration from 1 to 2 hr. Further reduction of 30 to 290 K increased the duration to 3 hr. This result might be anticipated from our previous acquaintance with the precooled Linde–Hampson cycle.

Stephens (1970) included an interesting description of a complete refrigeration system based on miniature Joule–Thomson coolers suitable for heat-seeking, air-to-air missiles on board aircraft. A schematic arrangement of the system in Fig., 6.19 is reproduced from Stephens (1970). The following operational advantages were listed:

Operating temperature −40°C to +70°C,
Rapid cooldown, less than 30 s,
Minimal microphony,
Minimal mass on guidance gimbals,
Long standby time,
Instant readiness,
Short logistic supply train,
Simple servicing,
Transportable support equipment,
Simple power supplies.

Air Products Coolers

Geist and Lashmet (1960, 1961) have presented data on the design and performance of miniature Joule–Thomson coolers developed by Air Products Inc. of Allentown, Pennsylvania. The systems described included open cycle systems with hydrogen to provide $\frac{1}{2}$ to 2 W of useful refrigeration in the range 15 to 60 K. Refrigeration to cool the hydrogen below the inversion temperature was obtained by an auxiliary Joule–Thomson system using compressed nitrogen or by liquid nitrogen. Closed cycle hydrogen and neon systems were also described. These were intended for relatively long-term operation. The closed cycle systems incorporated a two-stage nonlubricated enclosed compressor with electric drive.

Later Currie (1967) described a small open cycle Joule–Thomson cooler using helium as the working fluid to generate a useful refrigerating capacity of 0.5 W at 4 K. The unit was a cascade system using gaseous hydrogen Joule–Thomson expansion to liquefy hydrogen for precooling the helium. Liquid nitrogen was used to precool both the high-pressure hydrogen and helium gases *en route* to the expansion nozzles through compound recuperative heat exchangers. Kessler (1970) of Deutsches Elektrones—Synchrotron, Hamburg, West Germay described a refrigeration system incorporating an Air Products Joule–Thomson unit to provide refrigeration for a small hydrogen "target" for use with nuclear research.

JOULE–THOMSON EXPANSION COMBINED WITH COOLING ENGINES

In all of the above we have explored various ways and means to utilize Joule–Thomson expansion alone to produce low-temperature refrigeration and gas liquefaction. Many other systems exist which incorporate both cooling engines and a Joule–Thomson expansion system. One such combination mentioned in discussion of the Reitdijk expansion-ejector system was described by Prast (1968). This combines a double-expansion Stirling engine as the two-stage precooler of a helium liquefaction system. Haarhuis (1967) described a similar system incorporating two Philips double-expansion Stirling engines.

Chapter 5 includes the description of a helium liquefier built by A. D. Little Inc. combining a three-stage Gifford–McMahon cooling engine with a Joule–Thomson expansion process. Further development of that concept was described by Stuart and Hogan (1965).

Chapter 7 describes many Claude cycle cryocoolers that include Joule–Thomson expansion valves particularly where the fluid is to be liquefied. It is customary for the fluid to be expanded partially in the engine and partially in the Joule–Thomson valve. Expansion in the engine occurs isentropically and achieves a greater temperature difference than isenthalpic (JT) expansion over a given pressure ratio. However, expansion of two-phase fluids in engines presents problems, so that the preferred combination is partial expansion in an engine to approach the saturated vapor condition and then Joule–Thomson expansion to achieve liquefaction.

REFERENCES

Baehr, H. D. (1963). "On the Thermodynamics of the Cold-Air Cycle with Throttling." Proc. XI Int. Congr. of Ref./Prog. in *Ref. Sci. Technol.* **1**, 319–328, Munich, Germany (Pubs. Pergamon Press, Oxford and Vertag Muller, Karlsruhe).

Barron, R. (1966). *Cryogenic Systems.* McGraw-Hill Book Co., Toronto.

Buller, J. S. (1971). "A Miniature Self-Regulating Rapid-Cooling Joule–Thompson Cryostat." *Adv. Cryog. Eng.* **16**, 205–213.

Carbonnell, E., Chovet, P., Johannes, C., Marinet, D., and Solente, P. (1972). "Refrigerator without Moving Parts at Low Temperature Able to Cool Down to 90–100 K." *Proc. 4th Int. Cryo. Eng. Conf.*, Eindhoven, pp. 68–70.

Currie, R. B. (1967). "A Joule–Thomson Laboratory Expander." *Adv. Cryo. Eng.* **12**, 557–563.

Daunt, J. G. (1956). *The Production of Low Temperatures down to Hydrogen Temperatures.* Handbuch der Physik Bd IV, Springer Verlag, Berlin.

Geist, J. M., and Lashmet, P. K. (1960). "Miniature Joule–Thomson Refrigeration Systems." *Adv. Cryog. Eng.* **5**.

Geist, J. M., and Lashmet, P. K. (1961). "Compact Joule-Thomson Refrigeration Systems 15–60 K." *Adv. Cryog. Eng.* **6**.

Haarhuis, G. J. (1967). New Type Helium Liquefier." Proc. XII Int. Cong. of Ref./Prog., in *Ref. Sci. Technol.* **1**, 121, Madrid (11R Paris).

Haisma, J., and Roozendaal, K. (1967). "Investigation of the Behaviour of an Expansion-Ejector in the Low Temperature Region beyond the λ-Transition of Helium." Proc. XII Int. Cong. of Ref./Prog., in *Ref. Sci. Technol.* **1**, 111–210, Madrid (11R Paris).

Hampson, W. (1895). British Patent 10165.

Joule, J. P. (1852). *Sci. Pap.* **2**, p. 216.

Joule, J. P., and Thomson, W. (1852). *Phil. Mag.* **4**, 481.

Keesom, W. H. (1933). *Commun. Phys. Lab. Univ. Leiden*, suppl. 76a.

Kessler, G. (1970). "Joule–Thomson Hydrogen Refrigerator Target." *Adv. Cryog. Eng.* **15**, 443–446.

Linde, H. V. (1897). German Patent 88824 [see also *Z. ges Kälteind* **4**, 23 (1897); and the *Engineer*, Nov. 13 and 20 (1896)].

Mann, D. B., Bjorkland, W. R., Macinko, J., and Hiza, M. J. (1960). "Design, Construction and Performance of a Laboratory Size Helium Liquefier." *Adv. Cryog. Eng.* **5**, 346–353.

McInroy, J. (1967). Miniature Joule–Thomson Coolers." Proc. XII Int. Cong. of Ref./Prog. in *Ref. Sci. Technol.* **1**, 59–68, Madrid (11R, Paris).

Nicholds, K. E. (1968). "Low Temperature Devices for Laboratory Operation." Proc. 2nd Int. Cryo. Eng. Conf., pp. 65–66, Brighton Iliffe, Sci. and Tech. Publ., Guildford, U.K.

Nicholds, K. E. (1970). (a) "Performance of Self-Regulating Joule–Thomson Minicoolers," pp. 277–282. (b) "Miniature Cryogenic Cooling Systems for an Upper Atmosphere Infrared Research Programme," pp. 283–286, Proc. 3rd Int. Cryog. Eng. Conf., Berlin, Iliffe Sci. and Tech. Publ., Guildford, U.K.

Parkinson, D. H. (1959). "Some Problems in the Design of Helium Liquefiers Based on the Joule–Thomson Effect." Proc. X Int. Cong. of Ref./Prog. in *Ref. Sci. Technol.* **1**, 53–57 (Publ. Pergamon Press, Oxford).

Parkinson, D. H. (1967). "Miniature Refrigeration Systems—A Review." Proc. XII Int. Cong. of Ref./Prog. in *Ref. Sci. and Tech.*, *Vol.* **1**, 69–77, Madrid (11R, Paris).

Prast, G. (1968). "A 3.5 K Refrigerator Based on the Three-Space Stirling Refrigerator." Proc. 2nd Int. Cryo. Eng. Conf., Brighton, Iliffe Sci. and Tech. Pubs. Ltd., Guildford, U.K.

Rietdijk, J. A. (1966). "The Expansion-Ejector—A New Device for Liquefaction and Refrigeration at 4 K and Lower." Included in *Liquid Helium Technology*, Bull. IIR, Annexe 1966–65, p. 241.

Stephens, S. (1970). "A Self-Regulating Miniature Joule–Thomson Refrigerator." *Appl. Cryog. Technol.* **3**.

Stuart, R. W., and Hogan, W. H. (1965). "A Small Helium Liquefier." *Adv. Cryog. Eng.* **10**, 62–68.

Chapter 7

Claude and Joule–Brayton Systems

INTRODUCTION

Here we consider a family of cryocoolers using recuperative heat exchangers and achieving cryogenic temperatures by expanding gas in an engine thereby producing work. When the gaseous working fluid is to be liquefied it is customary for the final stage of expansion to be done with a Joule–Thomson expansion valve. Systems with expansion engines and a final Joule–Thomson expansion are called *Claude cycle liquefiers*. When the system is used as a refrigerator rather than a liquefier the final Joule–Thomson stage is not included and all the expansion occurs in expansion engines. These are called *Joule–Brayton cycle refrigerators*.

This family exists in endless variety. It is widely used for gas liquefaction, separation, and processing systems and less widely for simple refrigeration applications where regenerative systems are preferred. Proper consideration of Claude systems has been deferred to a companion volume devoted to systems for gaseous liquefaction, separation, processing and refrigeration in large and intermediate capacities involving turbomachinery.

Here the field of interest has been limited to cryocoolers of intermediate or less capacity using reciprocating expanders. The important areas of gaseous separation and processing associated with large-scale gaseous liquefaction systems have been ignored. Such systems are fundamental to the production of industrial gases, principally oxygen for welding, steelmaking, and hospital use and nitrogen for food-processing. The liquefaction and processing of natural gas is necessary to the separation of helium and the intercontinental transport of liquid natural gas as an energy fuel.

ISENTROPIC VS. ISENTHALPIC EXPANSION

Practically all cryocoolers involve expansion of the working fluid from a high pressure to a low pressure. The working fluid is usually a gas but

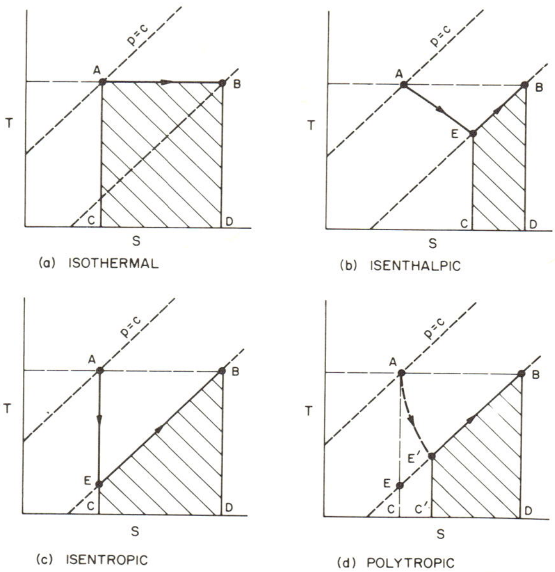

Fig. 7.1. Expansion processes represented on the T–S plane.

can be a liquid (gas previously liquefied) or wet vapor (gas cooled below the critical temperature with liquid droplets entrained in it). The expansion can approximate various idealized thermodynamic processes, namely, isothermal, isenthalpic, isentropic and polytropic. These are represented on the temperature–entropy diagram (T–S plane) as shown in Fig. 7.1.

Assume first the gas is at some high pressure and low temperature T_A. The gas expands to a low pressure and heat is supplied to bring the temperature back to the initial temperature T_A. An apparatus to carry out such comparative work is shown in Fig. 7.2. It consists of a piston reciprocat-

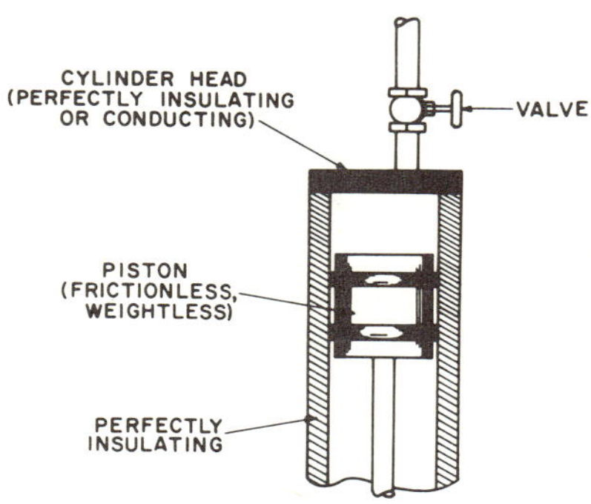

Fig. 7.2. Idealized expansion engine.

ing in a cylinder closed at one end. A valve is connected to the cylinder through which fluid is admitted or expelled from the cylinder. The cylinder head is assumed to be perfectly insulating or perfectly heat conducting as required. The walls of the cylinder are perfectly insulated and the piston is assumed to be without mass and to experience no friction in moving about in the cylinder.

For the *isothermal* process represented in Fig. 7.1a assume that the cylinder is filled with gas at condition A. The filling process occurs when the piston is at the top of its stroke (volume between the piston and cylinder head is least). When the piston moves down the cylinder the gas expands to a low pressure. The cylinder head is assumed to be perfectly conducting and as the gas expands the temperature tends to decrease so that heat is transferred through the cylinder surroundings to maintain the temperature constant. This heat abstracted, the refrigeration effect, is represented by the shaded area A–B–C–D in Fig. 7.1a. Work is done on the piston by the gas. The amount of work is exactly equal to the heat abstracted from the surroundings.

For the *isenthalpic* process represented in Fig. 7.1b assume that, with the cylinder empty of gas, the piston is moved to the bottom of the stroke so the volume between the piston and the cylinder head is a maximum. The inlet valve is opened slightly and working fluid slowly enters the cylinder. The pressure increases and eventually reaches the final required pressure, point E, Fig. 7.1b. Gas passing the inlet valve is said to be throttled. A throttled expansion occurs at constant enthalpy and is categorized as a Joule–Thomson expansion. If the temperature T_A before expansion is below the inversion temperature (see Chapter 6) then on expansion a decrease in temperature (to T_E, Fig. 7.1b) will occur. Assume the cylinder head becomes perfectly conducting and the fluid is heated to the final condition, T_B, by heat transfer through the cylinder head from the surroundings. The heat abstracted is represented by the area E–B–C–D on Fig. 7.1b, which is less than area A–B–C–D on Fig. 7.1a. For gas experiencing Joule–Thomson expansion the refrigerating effect is about one third of that with isothermal expansion even though the fluid started and finished at the same states. In the isenthalpic process no work was done; the piston remained stationary at the lower dead-point and the cylinder volume was constant.

For the *isentropic* process, Fig. 7.1c, we assume the same sequence as the isothermal case. The cylinder is filled with gas at state A and expanded to a low pressure. The only difference is that the cylinder head is perfectly insulated during expansion and no heat is transferred during expansion so the entropy remains constant ($\Delta S = \int \Delta Q/T = 0$). Work is done on the piston and the temperature is reduced as shown by the process A–E in

Fig. 7.1c. The cylinder head then becomes perfectly conducting and heat is transferred from the cylinder surroundings to heat the gas to the final temperature, T_B. The shaded area E–B–C–D in Fig. 7.1c is larger than the isenthalpic process and smaller than the isothermal process.

The polytropic process of Fig. 7.1d follows a similar pattern to the isentropic with the difference that the temperature at the end of the expansion T'_E, is greater than at the end of isentropic expansion T_E. This difference may arise from two causes, limited heat transfer during the expansion (usually the case with reciprocating engines but not turbines), and irreversibilities (principally friction) which occur during expansion (particularly in turbines) to increase entropy and, hence, the fluid temperatures. The effect of limited heat transfer and irreversibilities during expansion is to decrease the potential refrigerating effect compared with the isentropic process. The potential refrigerating effect is the shaded area E'–B–C'–D in Fig. 7.1d, smaller than the area for the isentropic process by area E–E'–C'–C.

Figure 7.1 shows clearly that in terms of refrigerating effect the order of preference of the four processes is isothermal, isentropic, polytropic, and isenthalpic.

In practice isothermal expansion cannot be achieved. It would require infinite rates of heat transfer or the expansion to proceed infinitely slowly. It is customary to adopt an isentropic process as the reference for expansion engines. This cannot be achieved in practice and realistic cycle analyses adopt an arbitrary value (60% to 85%) for the isentropic efficiency to account for the polytropic process.

The same work is necessary to compress the gas regardless of the subsequent expansion. Since the Joule–Thomson process produces so little refrigerating effect compared with isentropic expansion it is to be expected that Joule–Thomson cryocoolers will be less efficient than those incorporating expansion engines. The principal reason for using Joule–Thomson expansion is simplicity. It is much easier and cheaper to provide a simple throttle valve than an expansion engine with means to absorb the work produced. All small domestic refrigeration systems operating on the vapor-compression cycle have JT expansion valves.

In larger refrigerating and cryogenic cooling systems the additional expense of expansion engines can be justified by the improved efficiencies. Even there Joule–Thomson expansion systems are invariably used for the final stage of expansion when there is partial liquefaction of the working fluid. Two-phase working fluids are difficult to handle in expansion engines.

Daunt (1956) discussed isenthalpic and isentropic expansion in more formal thermodynamic terms than the intuitive-descriptive method followed above. The following is abstracted from Daunt's presentation:

... Writing the First and Second Laws of thermodynamics as:

$$T\,dS = C_p\,dT - T\left(\frac{\partial v}{\partial T}\right)_p dp \tag{7.1}$$

we obtain for the case of isentropic expansion ($dS = 0$) an immediate evaluation for the coefficient of cooling,

$$\alpha_S = \left(\frac{\partial T}{\partial p}\right)_S = \frac{T}{C_p}\left(\frac{\partial v}{\partial T}\right)_p \tag{7.2}$$

This compares with the Joule–Thomson coefficient of cooling:

$$\alpha_H = \left(\frac{\partial T}{\partial p}\right)_H = \frac{T}{C_p}\left(\frac{\partial v}{\partial T}\right)_p - \frac{v}{C_p} \tag{7.3}$$

whence

$$\alpha_H = \alpha_S - \frac{\nu}{C_p} \tag{7.4}$$

As p is increased or T decreased, thereby diminishing v the Joule–Thomson coefficient α_H approaches the reversible isentropic cooling coefficient α_S. Moreover as the critical temperature is approached C_p tends to an infinite value; hence in regions near critical, isenthalpic expansions also must be highly efficient. ...

The favorable environment for Joule–Thomson expansion near the critical point is evident on reference to a temperature–entropy diagram. Figure 7.3, a *T–S* plane for helium, shows two constant-pressure lines, the liquid vapor envelope, the critical point *C*, and three isenthalpic curves *A–B*, *D–E*, and *F–G*. The isenthalp *A–B*, starting at the lowest initial temperature, T_A, has a high slope not too significantly different from the isentrope *A–H*. At higher temperatures expansion *over the same pressure*

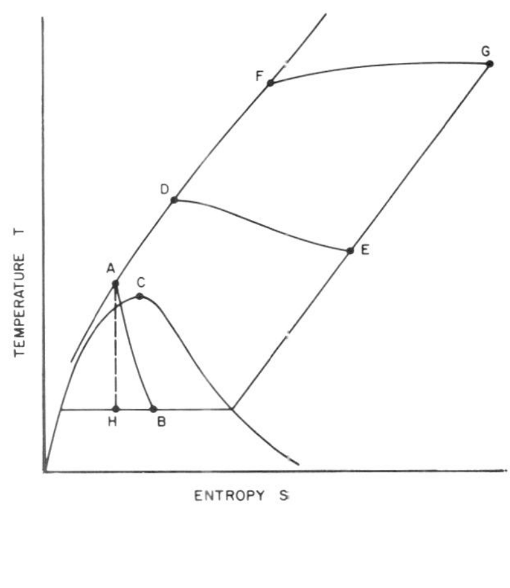

Fig. 7.3. Comparative slope of isenthalps on the *T–S* plane.

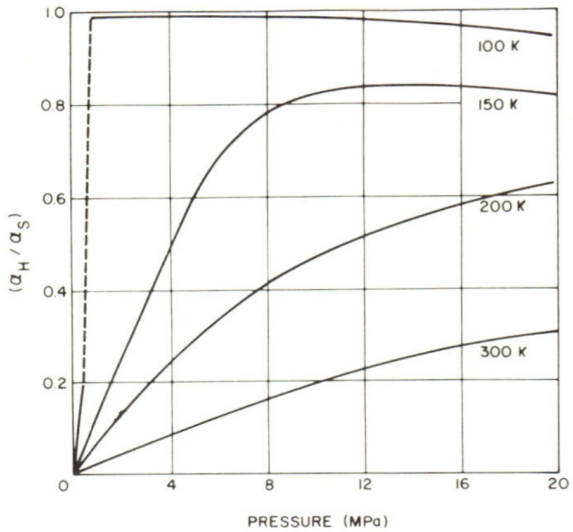

Fig. 7.4. Ratio of the isenthalpic–isentropic coefficients of cooling as a function of pressure at the start of expansion (for air) (after Daunt, 1956).

range results in a much smaller temperature decrease (isenthalpe D–E). If the initial temperature, T_F, is above the inversion curve the temperature will actually increase on isenthalpic expansion.

The relationship between α_H and α_S can be written [from Eqs. (7.2) and (7.3)] as

$$(\alpha_H/\alpha_S) = 1 - \frac{v}{T}\left(\frac{\partial T}{\partial v}\right)_p \tag{7.5}$$

This relationship for air as a function of pressure for different values of temperature T (start of expansion) is given in Fig. 7.4 (after Daunt, 1956). As the temperature reduces, (α_H/α_S) increases and at 100 K approaches unity. At higher temperatures the ratio (α_H/α_S) is a strong function of pressure, increasing progressively with increase in pressure. This helps to explain why the Linde dual-pressure cycle is so much more efficient than the simple Linde–Hampson cycle. In the dual-pressure cycle the principal mass of fluid is maintained at the intermediate pressure (about 4 to 5 kPa).

CLAUDE CYCLE SYSTEMS

Claude Cycle

A Claude cycle system for gas liquefaction, shown in Fig. 7.5, includes (a) compressor equipped with an aftercooler, air or water cooled, (b) recuperative heat exchangers, designated as high- (ambient), intermediate-, and low-temperature exchangers, or alternatively upper-, intermediate-,

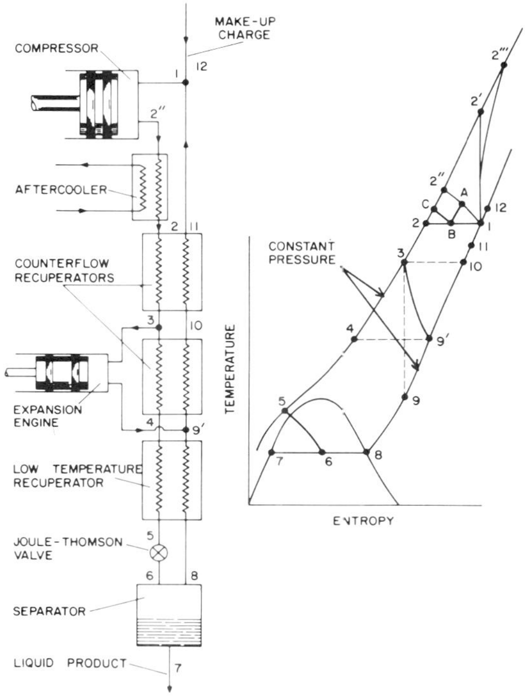

Fig. 7.5. Schematic diagram of typical Claude cycle cryocooler and liquefier.

and lower-stage exchangers, (c) expansion engine, (d) Joule–Thomson expansion valve, and (e) separator and liquid product reservoir. The system operates as follows:

i. Recirculated vapor at condition (11) mixes with fresh charge at condition (12) for the compressor inlet charge at condition (1).
ii. The vapor is compressed to condition (2″). Frequently the process is assumed to be isothermal [process (1–2)] or isentropic [process (1–2′)]. In practice the process would correspond to the process (1–2″) with reciprocating compressors, or the process (1–2‴) with a rotary compressor. The compression process may be arranged to occur in two or more stages with intercooling back to ambient temperature between stages. In Fig. 7.5, the compression process is shown in two stages, process (1–*A*) and process (*B*–*C*) with intercooling, process (*A*–*B*) between stages. Multistage compression with intercooling reduces the compression work. The work of compression decreases as the number of stages increases but at a progressively diminishing rate. It is rare to find more than

four stages of compression in reciprocating systems. Multistage compression is used for rotary and reciprocating compressors.

iii. Fluid leaving the compressor is cooled to ambient temperature in the watercooled aftercooler and then passes to:
iv. The high- (ambient) temperature recuperator and is cooled to condition (3).
v. A fraction of the fluid then enters the expansion engine and expands. The process of expansion may be isothermal (3–10), isentropic (3–9), or an intermediate, polytropic process (3–9′).
vi. The remaining fluid at high pressure passes through the intermediate recuperator, is cooled to condition (4), and then enters:
vii. The low-temperature recuperator and is further cooled to condition (5).
viii. Finally, the fluid expands through the Joule–Thomson valve and expands isenthalpically to condition (6), a two-phase, saturated liquid, saturated vapor mixture.
ix. The mixture is separated into the fluid and vapor phases in the separator.
x. The saturated liquid remains in the separator reservoir and is drawn off at condition (7) as the liquid product of the system. Sometimes the saturated liquid, perhaps liquid air or nitrogen, is not drawn off but remains in the separator reservoir as the precooler bath for a second Claude, or Joule–Thomson system for helium, hydrogen, or neon liquefaction.
xi. Saturated vapor leaves the separator at state (8) and enters the low-temperature recuperator and is warmed to condition (9) by heat exchange with the downcoming high-pressure stream.
xii. The effluent stream mixes with the exhaust from the expansion engine to condition (9′). The temperature of the fluid leaving the low-temperature recuperator will not be exactly equal to that of the exhaust stream from the expansion engine. Condition (9′) is the state of the mixed streams.
xiii. The combined low-pressure streams now pass through the two recuperative heat exchangers being heated to a final temperature, condition (11), and subsequently mix with incoming make-up charge to provide inlet gas to the compressor.

Strictly speaking, the two separate streams: (a) process (3–9′), and (b) process (3–4–5–6–8–9′) cannot be represented on the *T–S* plane because the diagram is constructed for *unit mass* of the working fluid. It is useful to represent the processes as followed here, providing it is always recalled that per unit mass of working fluid compressed, fraction $(1-x)$ passes through the expansion engine and only x is involved in the actual liquefaction process.

Analysis of Claude Cycles

Claude systems are analyzed by assuming steady state continuous operation and applying the steady flow energy equation to determine the fluid enthalpy at critical points in the cycle. Per unit mass of working fluid entering the compressor:

(1) The compressor work is

$$W_c = (H_2' - H_1) \tag{7.6}$$

(2) The heat transferred in the aftercooler is

$$(H_2' - H_2) \tag{7.7}$$

(3) The heat transferred in the high- (ambient) temperature recuperator is

$$(H_2 - H_3) = (1-y)(H_{11} - H_{10}) \tag{7.8}$$

where y is the fraction of gas liquefied per unit mass of gas compressed.

(4) The heat transferred in the intermediate temperature recuperator is

$$x(H_3 - H_4) = (1-y)(H_{10} - H_{11}) \tag{7.9}$$

(5) The work done by the gas in the expansion engine is

$$W_E = (1-x)(H_3 - H_9) \tag{7.10}$$

(6) The heat transferred in the low temperature recuperator is

$$x(H_4 - H_5) = (x-y)(H_{9'} - H_8) \tag{7.11}$$

(7) The enthalpy after expansion in the JT valve, H_6 is equal to the enthalpy before expansion H_5.

(8) The fraction of gas liquefied [condition (7)] is

$$y = x\left(\frac{H_6 - H_7}{H_8 - H_7}\right) \tag{7.12}$$

(9) The fraction of gas not liquefied but existing at state (8) is

$$(x-y) \tag{7.13}$$

The work done by the expansion engine may be used to partially offset the work of compression. In that case the net cycle work is

$$(W_C - W_E) = (H_2' - H_1) - (1-x)(H_3 - H_9) \tag{7.14}$$

The work generated by the expansion engine is relatively small (about 10% of the compression work). Accordingly the expansion engine and compressor are generally not coupled. The expansion work can be absorbed in useful ways to drive pumps or generators or simply dissipated in hydraulic or electrical dynamometers.

Optimum Recirculation Fraction

It is of interest to determine how the pressurized stream should be split at condition (3) between the fraction (x) going to the Joule–Thomson expansion valve and the remainder ($1-x$) expanding to provide refrigeration to cool the x stream from T_3 to T_4. Resolution of the optimum division is a complex problem best approached by numerical optimization procedures. Daunt (1956) reproduced data given in the literature for air liquefiers. The calculations, based on typical heat exchanger effectiveness parameters, thermal losses, and irreversible effects in expansion, approximate practical conditions.

Figure 7.6 (after Daunt, 1956) shows the compressor power input per unit mass of liquid air produced (kW hr/kg liquid) as a function of the fraction of gas compressed passing to the Joule–Thomson expansion valve. Several curves are included for different pressures at the end of compression. The important points to note are that (a) the different curves have well-defined minima representing the optimum fraction (x) for most

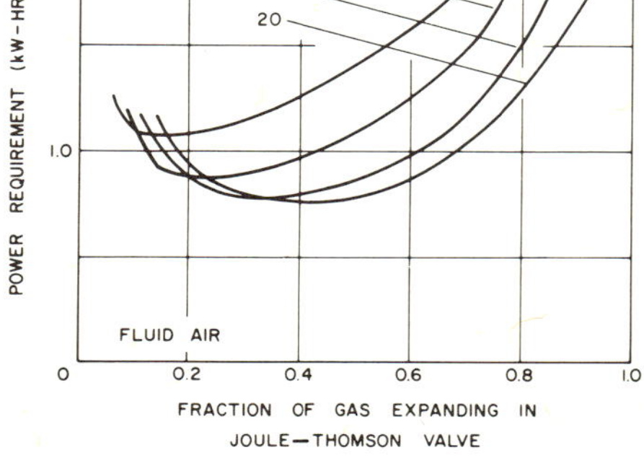

Fig. 7.6. Power requirement per unit mass of gas liquefied as a function of the fraction of gas expanded in the Joule–Thomson valve (for air) (after Daunt, 1956).

efficient operation, and (b) as the pressure increases the maximum efficiency and the fraction to be expanded in the Joule–Thomson expansion increase.

The conditions of maximum efficiency for operation of an air liquefier were provided by Davies (1949) in the data reproduced in Table 7.1. This shows, for several different compressor pressures, the fraction of gas passing through the expansion engine, i.e., $(1-x)$, the optimum temperature for gas to enter the expansion engine, the theoretical power required in kW hr per liter of liquid produced and, finally, the actual power required (kW hr per liter of liquid produced). These data confirm, as in Fig. 7.6, that the optimum fraction of gas expanded in the engine decreases and the system efficiency improves as the pressure increases. The temperature for admission of the gas to the expansion engine increases as the pressure increases. At a maximum cycle pressure of 200 MPa the temperature has increased to 293 K, approximately the ambient temperature.

Heylandt Cycle

Heylandt was the first to appreciate that the optimum temperature for admission to the expansion engine increased as the cycle pressure and system efficiency increased. Heylandt introduced a modified Claude cycle by eliminating the upper-stage recuperative heat exchanger of Fig. 7.5 and elevated pressure level (for an air liquefier) from 4 MPa (used by Claude) to 20 MPa. He further increased the fraction of mass flow passing to the Joule–Thomson valve from 0.2 (at 4 MPa) to 0.4 (at 20 MPa).

As evident from Table 7.1 there is little difference in the theoretical power requirement for the Heylandt–Claude cycle but there are substantial practical advantages. The actual power required per unit mass liquefied with the Heylandt system is only 75% of the power with the original Claude system. Air is admitted to the expansion engine at ambient temperature immediately following the aftercooling process on leaving the compressor. About 60% of the air passes through the expansion engine and exits from the engine at a temperature and pressure of about 150 K and 0.1 MPa (condition 9′ in Fig. 7.5). The relatively high temperature of operation of the expansion engine has important practical advantages. The high-temperature recuperator is eliminated and appreciable savings gained thereby. The lubrication problems of the piston in the cylinder and the thermal insulation or isolation problems are greatly minimized. The efficiency of the Heylandt cycle is the highest of all simple Claude systems and was much favored for air liquefaction systems particularly in the United States.

Gardner and Smith (1960) have given a comprehensive summary of the thermodynamics of the Heylandt cycle from the aspect of power

Table 7.1. Conditions for Operation of Claude Cycle Air Liquefier at Maximum Efficiency[a]

	Turbo expansion machine (Kapitza)			Original Claude system			Heylandt cycle
Maximum pressure (MPa)	0.65	2	3	4	4	10	20
Fraction $(1-x)$ expanding in the expansion engine	—	0.88	0.84	0.80	0.75	0.68	0.60
Temperature at inlet to expansion engine (K)	115	155	173	191	215	241	293
Power required (theoretical) (kW hr per liter liquid)	—	—	—	0.33	—	—	0.32
Power required (practically) (kW hr per liter liquid)	1.5	1.08	1.00	0.90	0.83	0.78	0.70

[a] After Davies (1949).

consumption and thermodynamic reversibility associated with gas liquefaction and separation.

Low-Pressure Air Liquefiers

The Heylandt cycle represents one limiting case of the Claude system, when the cycle pressure is so high that the optimum temperature for admission to the expansion engine has increased to ambient temperature. The fraction of fluid to be recirculated through the engine is a minimum 60%.

The other limiting case occurs when the maximum cycle pressure is so low that a very high fraction of the gas must be recirculated through the expansion engine (>90%). This system is used for the large-scale production of liquid air or industrial gases. The combination of low pressure and high flow rate favors the use of turbines and rotary compressors and permits appreciable savings in plant costs with increased reliability. Another advantage is that the use of regenerative rather than recuperative heat exchangers facilitates the removal of impurities (water vapor and carbon dioxide) in the air to be liquefied.

The low-pressure turbo air liquefier was introduced in the early 1930s using the reversing regenerator concept described somewhat earlier by Frankl. Hence the term Linde–Frankl is frequently used to describe the whole system for gaseous separation rather than simply liquefaction. The Russian cryogenics pioneer P. Kapitza is also frequently associated with the low-pressure turbo air liquefier. In 1939 he provided the first detailed description of the system including the results achieved in a small experimental prototype unit. The applicability of the low-pressure system to large-scale processing with turbo equipment puts the Linde–Frankl system beyond our field of interest here except to note in passing. Daunt (1956) has given a good account of the system and references the source documents.

Collins Low-Pressure Air Liquefier

A small scale air liquefier involving relatively low-pressure and reciprocating machinery was described by Collins *et al.* (1958). The system is of interest because it involves the use of reversing recuperative heat exchangers to achieve the same result as the Frankl regenerative heat exchangers with periodic flow switching, that is, the eliminating of special equipment to remove water vapor and carbon dioxide from the air to be liquefied.

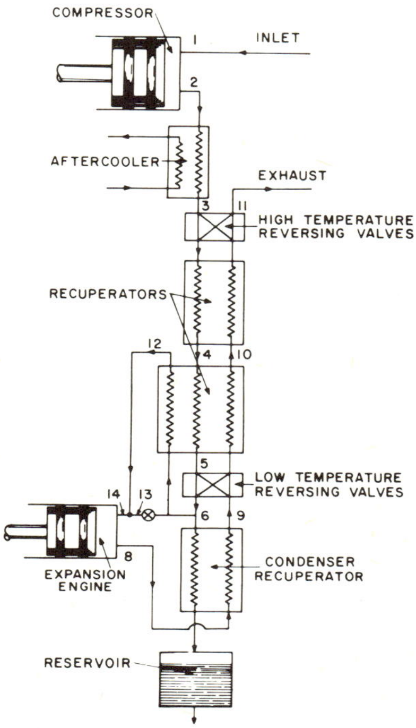

Fig. 7.7. Collins low-pressure air liquefier.

A schematic of the flow system for the Collins air liquefier is shown in Fig. 7.7. The system operates as follows:

(i) Inlet air (at atmospheric conditions) enters the compressor at condition (1).

(ii) The air is compressed to 13 MPa and passes to a watercooled aftercooler.

(iii) The compressed fluid then passes through the reversing valves, the high- and intermediate-temperature recuperators and the low temperature reversing valves. At condition (6), downstream of the low-temperature reversing valve, the pressure and temperature is 13 MPa and 125 K. By the time this temperature is achieved all the significant water vapor, carbon dioxide, and other miscellaneous impurities in the inlet air have precipitated on the walls of the recuperative heat exchangers. If operated continuously the flow passages in the recuperators would soon be blocked by accumulations of these precipitated solids. However, the reversing valves operate at approximately 3-min intervals to cause the return flow to pass through the passages containing the precipitated contaminants. As the return flow warms in passing through the recuperators the contaminants reevaporate from the walls of the flow passages, are entrained in the return flow and exhaust from the system at approximately atmospheric conditions. In this system there is no recirculation of the return effluent as in the other cases considered earlier.

(iv) At condition (6), pressure 13 MPa, temperature 125 K, the flow separates into two streams. The minor stream enters the condenser/recuperator where it condenses under pressure to the liquid phase, condition (7), and hence passes to the liquid reservoir and subsequently to a rectification column for separation into the liquid-nitrogen useful product and gaseous oxygen for release to the atmosphere.

(v) The major stream at condition (6) passes to the expansion engine where it expands to a final pressure and temperature of approximately 0.15 MPa and 87 K.

(vi) The low-pressure, low-temperature effluent enters the condenser/recuperator to accomplish the condensation of the liquid air in (d) above and hence passes to exhaust through the upper stage recuperators. In passing the effluent reevaporates and absorbs the contaminants precipitated in the course of the initial cooling.

(vii) Prior to fluid actually entering the expansion engine the opportunity is taken to provide auxiliary cooling to the intermediate stage recuperator. In passage to the expansion engine the fluid at condition (6), pressure 1.3 MPa, temperature 125 K, is further subdivided. A small amount of the gas is fed through a third set of flow passages in the intermediate stage recuperator and reheated to about 200 K, condition (12). The major stream is partially throttled to about 1.25 MPa, condition (13), to provide sufficient pressure potential to induce flow through the intermediate recuperator.

(viii) The two streams at condition (12) and (13) are mixed to provide inlet pressure and temperature, condition (14), of 1.25 MPa and 146 K to the expansion engine.

Auxiliary cooling in the intermediate recuperator is necessary to cool the pressurized incoming stream to 125 K (at condition (5)), where all the carbon dioxide will have precipitated out.

Collins (1956) conceived the use of reversing recuperative exchangers as an alternative to Frankl regenerative exchangers for relatively low-pressure systems. Application in systems with high-pressure streams would pose formidable sealing problems in the low-temperature reversing valves. On flow reversal the high-pressure fluid contained in the void volume of the heat exchangers simply passes to exhaust and serves no useful purpose. This so-called "carry-over loss" would incur a serious decline in the system efficiency at high pressures.

Precooled Claude System

Figure 7.8 is a schematic diagram of a Claude cycle system incorporating a precooler heat exchanger. The precooler is shown downstream of the upper stage recuperator but could also be located elsewhere depending on the temperature regime. If necessary two stages of precooling at different temperatures might be employed, one downstream of the aftercooler and the other as shown.

Precooling improves the performance of Claude cycle systems to the same degree as we saw in Chapter 6 with Linde–Hampson systems. Gases with low inversion temperature—neon, hydrogen, and helium—need to be precooled in Linde–Hampson systems for liquefaction to be achieved. The same need for precooling does not exist with Claude systems since the cooling effect arises from the more effective isentropic expansion as well as Joule–Thomson isenthalpic expansion. All gases, even helium can be

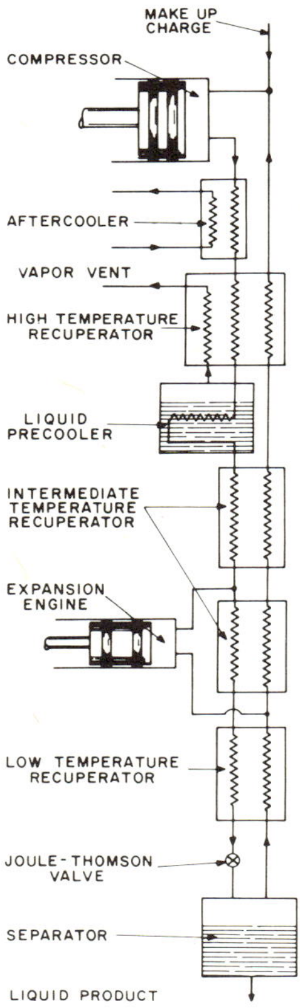

Fig. 7.8. Schematic diagram of Claude cycle with liquid precooler.

liquefied in Claude cycle systems with no precooling. Nevertheless, precooling confers advantageous benefits to the cooling performance and should always be considered. For air the precooler can be a cold liquid refrigerant, Freon, propane, ammonia, etc., generated by a vapor compression refrigeration system. For gases with a low inversion temperature—neon, hydrogen, or helium—the precooler will probably contain liquid nitrogen. Barron (1966) asserts that with liquid-nitrogen precooling a Claude system for hydrogen production has a figure of merit (see Chapter 6) 50% to 75% higher than that of the Linde–Hampson system.

More sophisticated systems for hydrogen liquefaction have been developed. Figure 7.9 is a schematic diagram for a Claude cycle hydrogen liquefaction plant incorporating a liquid-nitrogen precooler and a further stage of low-temperature precooling generated by an auxiliary refrigeration system using helium as the working fluid. The helium system is itself precooled in the liquid nitrogen bath and remains in the gaseous state throughout. The helium system does *not* include a Joule–Thomson

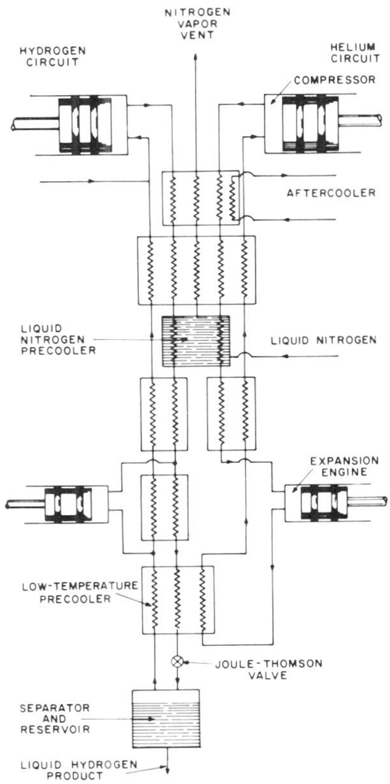

Fig. 7.9. Schematic diagram of hydrogen liquefier operating on the Claude cycle with a helium precooler circuit operating on the Joule–Brayton cycle.

expansion downstream of the expansion engine and may be characterized as a precooled Joule–Brayton cycle.

A preferred arrangement to that shown in Fig. 7.9 would be to eliminate the hydrogen expansion engine and Joule–Thomson expansion valve so as to produce liquid hydrogen by direct condensation in a low-temperature precooler using the refrigeration produced by expansion in the helium expansion engine. This allows the use of relatively low pressures, for the hydrogen need be compressed only to the extent required to maintain the flow through the heat exchangers (less than 1 MPa). It relaxes both the compressor work requirement and the difficult sealing problems with high-pressure hydrogen, a highly inflammable gas.

The helium gas-refrigerated, hydrogen liquefaction system is relatively insensitive to helium pressures permitting the use of relatively low values. Barron (1966) claims that for helium gas pressures of 1 MPa a system figure of merit of 0.11 can be achieved.

Dual-Pressure Claude Cycle System

Another modification to the basic Claude cycle is the dual-pressure system shown schematically in Fig. 7.10. In this cycle the principal mass of working fluid circulates in the low to intermediate pressure fluid circuit embracing the low-pressure compressor and aftercooler, the expansion engine, and the upper and intermediate stage recuperators. The cycle is encompassed with the state points 1–2′–2–12–10–11–1 shown on the *T–S* plane included in Fig. 7.10.

A small fraction of the fluid at condition (2), the intermediate pressure, is further compressed to high pressure and cooled in recuperators to condition (6). It is then expanded in the Joule–Thomson valve to the system minimum pressure and passed to the liquid/vapor separator so that saturated liquid [condition (8)] and saturated vapor [condition (9)] are produced. In a brief description of the dual-pressure cycle, Barron (1966) claims the work requirement per unit mass of gas liquefied is reduced compared with the basic Claude cycle system, but this does not accord with other data given by Barron, reproduced in Table 7.2.

Comparison of Claude Cycle Liquefaction Systems for Air

The performance of different Claude cycle systems for air liquefaction was compared by Barron (1966). His data are reproduced in Table 7.2. The table contains a brief specification of the system, the liquid yield per unit mass of gas compressed, the cycle work input required per unit mass of gas liquefied, and the figure of merit. The table is comparable with a

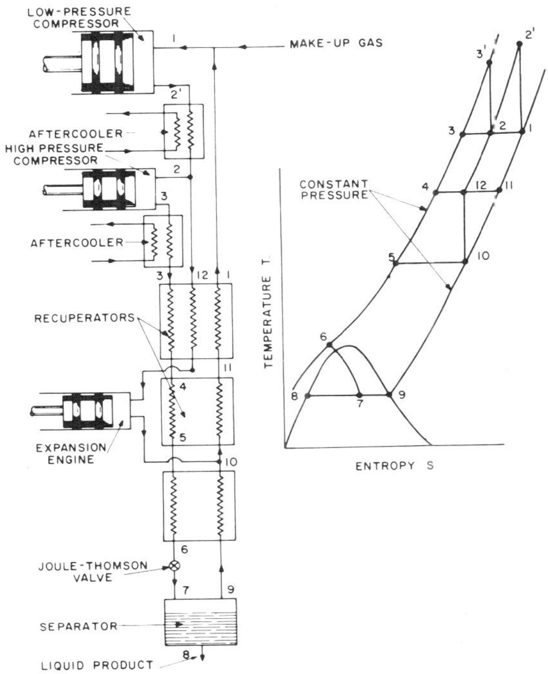

Fig. 7.10. Schematic diagram of the dual-pressure Claude cycle cooling system.

similar table reproduced in Chapter 6 for variants of the Linde–Hampson cycle. In Table 7.2 the first three entries refer to basic Claude cycle systems. The first entry is idealized by the assumption of compressor and expander isentropic efficiencies of 100% and perfect recuperation. The second entry refers to a more realistic theoretical cycle with assumed values for machine efficiencies and with imperfect recuperation. The third entry is a record of an actual Claude system in operation. It is of interest to note the observed figure of merit (0.201) is about half the predicted "realistic" cycle of the second entry (0.356) and one quarter the predicted idealized cycle (0.808).

The next three entries in the same format as above refer to Heylandt systems where the cycle pressure is increased from 4 MPa to 20 MPa. All three entries have figures of merit somewhat better than the corresponding entry for the basic Claude cycle.

The final two entries in the table refer to dual-pressure Claude cycle systems. Barron did not include observed values for the dual-pressure Claude cycle system. The figure of merit for the dual-pressure Claude cycle was substantially less than that of the corresponding entry for the basic Claude cycle.

Table 7.2. Comparison of Liquefaction Systems for Air as the Working Fluid[a]

	Liquid yield, y	Work per unit mass liquefied kJ/kg	Figure of merit
1. Claude system, $p_2 = 4$ MPa, $(1-x) = 0.7$, $\eta_c = \eta_e = 100\%$, $\varepsilon = 1.0$	0.260	164.2	0.808
2. Claude system, $p_2 = 4$ MPa, $(1-x) = 0.7$, $\eta_c = 70\%$, $\eta_e = 73$, $\varepsilon = 0.95$	0.189	372.7	0.356
3. Claude system, observed	—	662.08	0.201
4. Heylandt system, $p_2 = 20$ MPa, $(1-x) = 0.6$, $\eta_c = \eta_e = 100$, $\varepsilon = 1.0$	0.373	157.7	0.821
5. Heylandt system, $p_2 = 20$ MPa, $(1-x) = 0.6$, $\eta_c = 70\%$, $\eta_e = 72\%$, $\varepsilon = 0.95$	0.301	345.2	0.384
6. Heylandt system, observed	—	614.8	0.216
7. Dual-pressure Claude system, $p_2 = 0.65$ MPa, $p_3 = 4$ MPa, $(1-x) = 0.7$, $\eta_c = \eta_e = 100\%$, $E = 1.0$	0.119	219.7	0.604
8. Dual-pressure Claude system, $p_2 = 0.65$ MPa, $p_3 = 4$ MPa, $(1-x) = 0.7$, $\eta_c = 70\%$, $\eta_e = 72\%$, $\varepsilon = 0.95$	0.073	503.0	0.264

[a] $T_1 = 294$ K, $p_1 = 0.1$ MPa, η_c is the compressor overall efficiency, η_e is the expander overall efficiency, ε is the heat exchanger effectiveness (after Barron, 1966).

MULTIPLE-EXPANSION ENGINES

Claude Stepped Piston Two-Stage Expander

So far we have considered only systems with a single stage of expansion. Significant advantage can be gained by multistage expansion, usually arranged to occur at progressively decreasing temperatures and pressures.

The earliest use of multiple expansion was described by Claude (1913) with reference to an air liquefier. The system is shown schematically in Fig. 7.11. A notable feature was the use of a single *stepped piston* with the high-pressure and low-pressure cylinders arranged coaxially and operating as follows: Air compressed to about 4 MPa passed through the watercooled aftercooler and a column containing potassium hydroxide to dry the air and remove the carbon dioxide. The clean compressed air passed through the upper stage recuperator and was cooled to 140 K. Thereafter the flow divided into two streams. The principal stream passed through the high pressure expander emerging at some intermediate pressure and temperature. It then passed through the intermediate-stage recuperator and was reheated before final expansion in the low-pressure expander to the

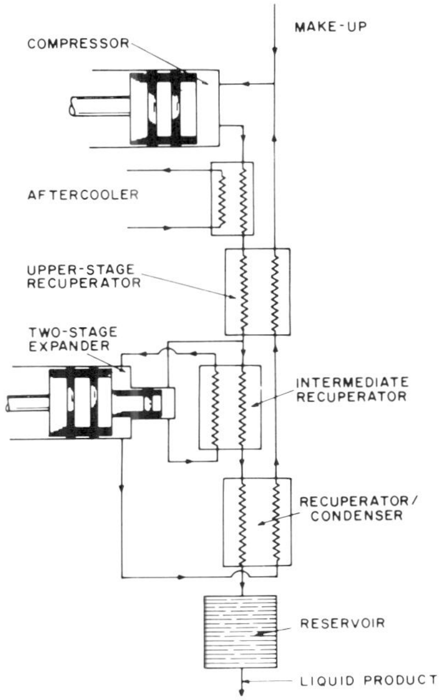

Fig. 7.11. Schematic diagram of the Claude multiple-expansion, stepped-piston, air liquefier.

lowest cycle temperature. The low-pressure, low-temperature exhaust then entered the third-stage recuperative condenser and thence to the upper-stage recuperator back to the compressor.

The minor stream continued at progressively diminishing temperature through the intermediate- and lower-stage recuperators. The flow condensed in the lower-stage recuperator and the liquid product collected under pressure in the reservoir for subsequent separation into oxygen and nitrogen.

The use of stepped pistons for multiple-stage expansion in Claude systems was discussed by Collins *et al.* (1968) with regard to his helium liquefier. About the same time Kapitza and Danilov (1968) advocated the use of stepped pistons to facilitate sealing of the low-temperature stage in small expansion engines. Later Danilov and Kovatchev (1972) discussed the construction and performance of a small helium expander with a two-stage reciprocating piston having a relatively high adiabatic efficiency. Stepped pistons, sometimes called differential pistons, are of course used elsewhere. They are found in reciprocating compressors, and in multiple-stage Gifford–McMahon cooling engines. Zimmerman used a multiple-step piston in the split Stirling system shown in Fig. 1.2.

Collins Multiple-Expansion Engine

One of the most significant developments of cryogenic cooling systems occurred in the late 1940s when Collins (1947) made a multiple-expansion Claude cycle system capable of liquefying helium. Claude cycle helium liquefiers had been constructed earlier. The first to use a reciprocating engine was described by Kapitza (1934). This was a single-stage, precooled Claude cycle system. The Kapitza engine was further notable as the first to use a hydrodynamic gas lubricated piston to minimize frictional heat generation at low temperature.

Collins incorporated some of the ideas exemplified in the Kapitza engine and introduced other innovative features including the use of multiple-expansion engines. The system developed was capable of liquefying any gas, air, hydrogen, or helium without the benefit of precooling although performance was improved by precooling.

A schematic diagram of the Collins helium liquefier is shown in Fig. 7.12. For the sake of simplicity, only two engines are shown, but, depending on the helium inlet pressure, systems with up to five expansion engines have been used. The expansion engines can be physically separated but coupled to a common crankshaft or can be the stepped (differential) piston type discussed above. A Joule–Thomson valve is always incorporated as the final stage of expansion in the liquefier flow circuit.

Typical performance details for a two-stage Collins liquefier were given by Weinstock (1968). Approximately 25% of the helium flow is expanded in the first expansion engine to provide cooling in the 30–60 K range. About half the helium flow is expanded in the second expansion engine to provide cooling in the 8–15 K range. The remainder is expanded in the Joule–Thomson valve. Liquid nitrogen precooling of the helium before expansion in the engines increases the fraction of helium liquefied by two or even three times the production with no precooling. Some important parametric effects of multiexpander systems were explored by Class *et al.* (1960) with particular reference to large-scale hydrogen liquefaction.

Arrangements in Multiple-Expansion Systems

Systems with multiple expansion can be in parallel or series arrangement of the expansion engines. The two alternatives are shown in Fig. 7.13. In both cases the expansions take place at significantly different temperatures separated by a recuperative heat exchanger.*

* This should not be confused with the alternative situation where an expansion proceeds at more or less the same temperature, either (a) in parallel where gas expands over the same pressure range; The cylinders are supplied from a common source and may exhaust to a common duct; or (b) in series where the gas expands in a number of stages at successively lower pressure as for example in a multistage turbine or reciprocating engine.

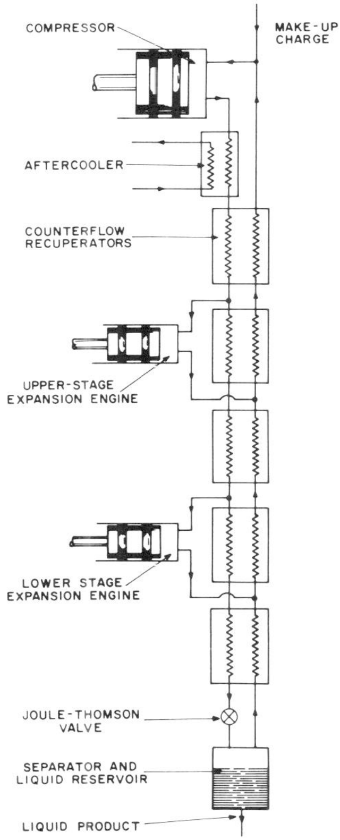

Fig. 7.12. Schematic diagram of the Collins–Claude cycle multiple-expansion gas liquefier.

There is perhaps an elegant thermodynamic argument to demonstrate the superiority of the parallel arrangement, but the argument is not obvious. However, it is difficult to avoid a strong intuitive feeling that the parallel arrangement is best. The question, basically, is which is most advantageous: (a) to expand two small masses over the whole pressure range, or (b) to expand a single mass (the sum of the two masses) in two partial expansions.

The dilemma is illustrated in Fig. 7.14. Idealized work diagrams are shown for the upper-stage and lower-stage expansions of the parallel arrangement [diagrams (a) and (b) on the left side of the figure] and the series arrangement [diagrams (c) and (d) on the right side of the figure]. On all four diagrams the isotherms are drawn, (a–b) for temperature T_1 and (c–d) for temperature T_2. It is assumed these temperatures are the gas temperatures at the start of the expansion process, T_1 for the upper stage, T_2 for the lower stage.

The isotherms are located on the P–V planes so that on diagrams (a) and (b) the isotherm for T_1 passes through the point where the volume is

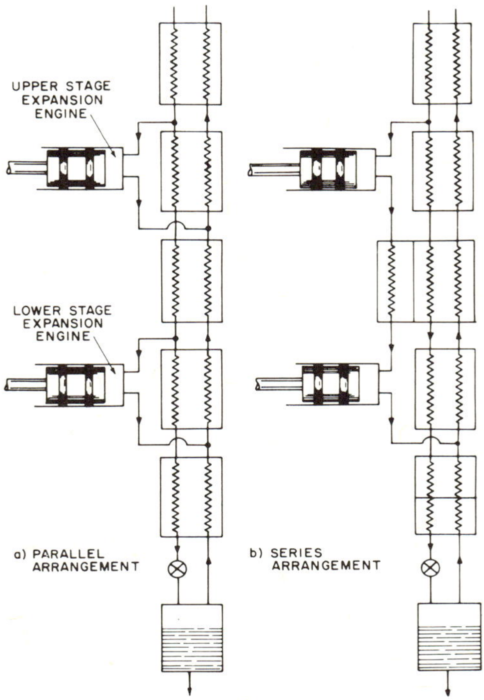

Fig. 7.13. Parallel and series arrangement of multiple-expansion engines in Claude cryocoolers.

unity and the pressure 3, in arbitrary units. The isotherm is drawn for a perfect gas according to the equation $PV = \text{const}$. The isotherm for T_2 passes through the point where $P = 3$ and $V = 0.5$ units. This is equivalent to specifying the temperature as T_2 is half the temperature at T_1. Assume an equal mass of gas expands in both the upper and lower expansion stages. Then the volume of a given mass of gas at temperature T_2 will be half the volume of the same mass of gas at temperature T_1.

Diagrams (a) and (b) have been drawn for the above situation. In both cases the expansion process was assumed to be an arbitrary polytropic process so that the temperature is somewhat less at the end of expansion than at the beginning. To a first approximation the work produced in the upper stage engine is twice the work produced in the lower stage engine (comparing the areas of the two diagrams).

Now consider the series arrangement represented in the diagrams (c) and (d) of Fig. 7.14. In this case the sum of the two separate masses experiences the expansion but in two stages. The isotherms for T_1 and T_2 have been drawn for the situation where double the mass of gas was present for the diagrams (c) and (d) than was assumed for diagrams (a) and (b).

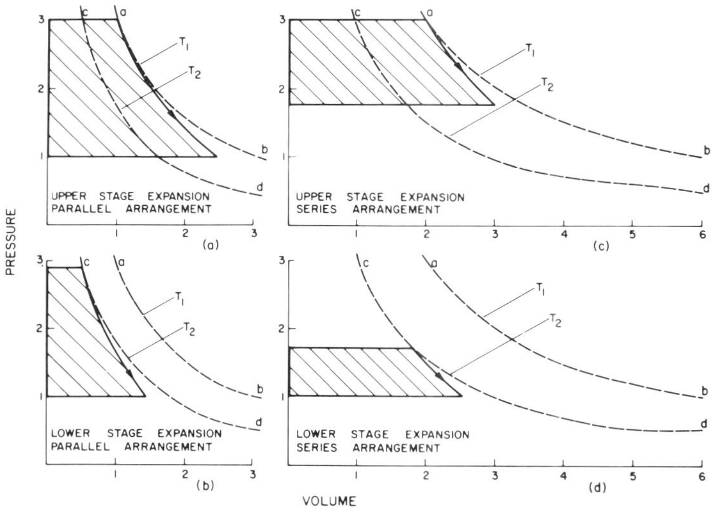

Fig. 7.14. Work diagrams for the upper and lower stage expansion engines in parallel (diagrams a and b) and series arrangements (diagrams c and d).

This simply means that the isotherms pass through points where, when $P = 3$, the volume is 2 for T_1 and unity for T_2.

The arbitrary decision was made to select the expansion ratio to be equal in both stages, i.e., $p_{r(\text{upper})} = p_{r(\text{lower})} = \sqrt{3} = 1.73$. The work diagrams then correspond to those shown in diagrams (c) and (d) of Fig. 7.14. As before an arbitrary polytropic expansion was assumed where the temperature at the end of the expansion was somewhat less than the temperature at the start.

The sums of the areas of diagrams (a) and (b) are greater by approximately 10% than the areas of the diagrams for (c) and (d). This indicates that for this arbitrary idealized example the parallel arrangement will produce more refrigeration (evidenced by the greater work output) than the series arrangement. Confirmation of this view was provided by Carbonnell (1974) in a discussion of the relative merits of parallel and series arrangements. He confirmed the advantage of the parallel arrangement in terms of a 10% increased refrigeration capacity for given constant adiabatic efficiency and pressure expansion ratio.

Carbonnell was concerned with large-scale refrigeration using turbines and pointed out that the series arrangement was characterized by high flow rates over low expansion ratios. These factors contribute to increased turbine adiabatic efficiency great enough to favor series coupling for turbines. Reciprocating engines, on the other hand, can handle high expansion ratios at relatively low mass flow rates. This favors the parallel arrangement where reciprocating expanders are used. Confirmation was provided by the eminent Russian cryogenic engineer I. B. Danilov in the discussion to a paper concerned with multiple-expansion helium liquefiers (Kapitza and Danilov, 1968).

RECIPROCATING EXPANSION ENGINES

Claude Expansion Engines

The expansion engines used by Claude in his original air liquefier were, in the words of Collins (1958), "simply a judicious adaptation of the steam engine of the day to the task in hand." Figure 7.15 reproduced from Collins (1958) is a cross-section view of a Claude-type engine. The resemblance to a steam engine is striking. Claude's first intent was to induce liquefaction of air directly to the engine cylinder to overcome the problem of low-temperature lubrication of the piston by using the liquid air as a lubricant. On initial startup a light petroleum distillate was used as the lubricant.

Claude met with early success in his experimental work. In his book Claude (1913) relates in entertaining fashion his early struggles. It was . . .

> a cold snowy night in November (1898) within the somewhat forbidding tramway depot of the Bastille-Charenton Tramways, a totality of conditions of doubtful comfort . . .

that a temperature thought to be less than −150°C was attained. It was not until years later that Claude realized there was an instrumentation error and that the true temperature was about −100°C, scarcely better than had been achieved by Solvay in his expansion engine a dozen years earlier. In ignorance of this but with much enthusiasm and incredible endurance, Claude persisted with his experiments, moving in 1900 with ". . . new apparatus installed at the compressed air works of the omnibus company at La Villette."

Further disappointments awaited him until one Sunday in May 1901, Claude conceived the notion of liquefying the air under pressure *outside* the cylinder. This was at the eleventh hour, for the shareholders, their patience exhausted, had arranged a meeting the following day to receive his proposals to terminate the work. In desperate straits Claude feverishly modified his apparatus that day and was rewarded:

> . . . at last to see a thin thread of liquid escaping from the cock and filling little by little one of the three litre liquid air holders which had so long awaited its appearance.

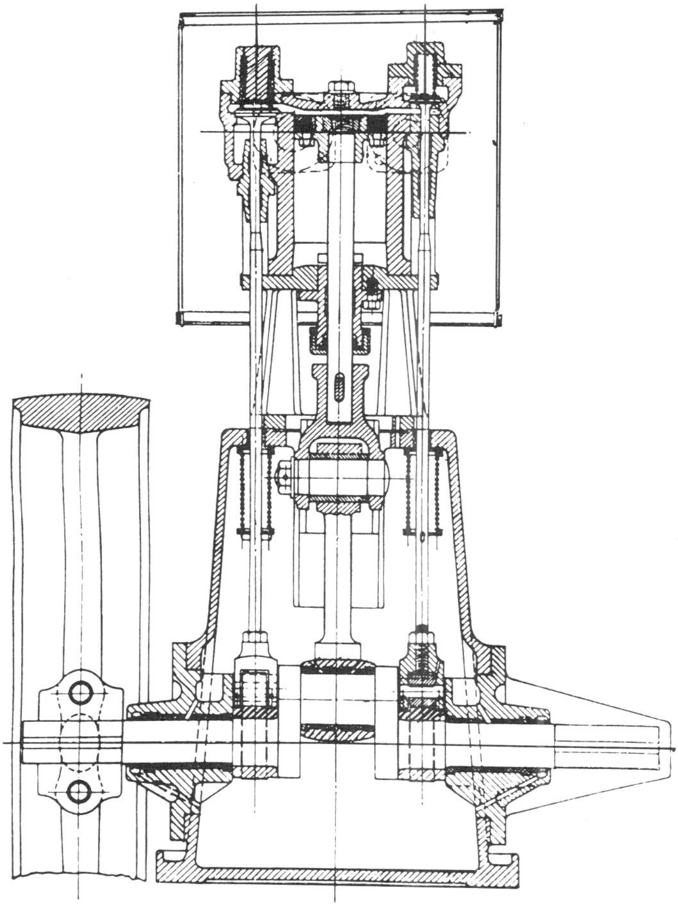

Fig. 7.15. Sectional view of Claude-type expansion engine.

Thereafter was laid the foundation of the great French industrial syndicate L'Air Liquide, Ltée.

Claude was much concerned about piston lubrication with metal rings but eventually (1912) adopted the leather cup as a piston seal. It is surprising this took so long, for leather had long been used on air expansion engines for ordinary refrigeration and Claude must have been aware of this.

In 1920, Claude constructed hydrogen expansion engines to obtain temperatures below the boiling point of liquid nitrogen. These engines were used to separate hydrogen from coke oven gas. Collins (1958) described hydrogen expansion engines "currently in use" constructed in the United States by E. I. duPont de Nemours and Co. to Claude design as

> ... single cylinder, double-acting machines of 10 inch bore and 12 inch stroke. The pistons are fitted with "bakelite" rings with bronze expanders. Segmental carbon-"bakelite" piston rod packing is employed. Wear on cylinder and rings is said to be inappreciable in a year of essentially continuous operation. The inlet valve is balanced and is mechanically actuated; the discharge valve, also mechanically actuated, is of the poppet or mushroom type.
>
> In normal operation for recovery of hydrogen from coke oven gas, the engine inlet pressure is about 25 atm. and the discharge pressure about 1.6 atm. The inlet temperature is about 113 K and the discharge temperature about 68 K. Much lower temperatures could be obtained easily but are not wanted because of the resultant solidification of methane and other impurities. The energy output of the hydrogen engines is used for compressing air.

Land (1960) has given a brief but useful review of contemporary expansion engine practice, including piston and turbine engines from the aspect of lubrication, efficiency losses, power absorption, and control.

Many modern reciprocating expansion engines favor the hydrodynamic gas lubricated piston concept introduced by Kapitza (1934) and other innovations introduced by Collins (1958) including tensile connecting rods. Some designs still prefer the more traditional approach. Witter (1966) of the Union Carbide Corp., Linde Division, in the description of a small closed cycle helium liquefier operating on the Claude cycle with two stages of expansion, gave an interesting critique of different design features. He was in favor of the compressive connecting rod and carbon-filled fluorocarbon piston rings. The principal advantage seen by Witter was elimination of the seals around the connecting rod and valve rods. An interesting feature mentioned by Witter was the use of a metal bellows seal around the valve rods operating at ambient temperatures.

The use of dry-rubbing piston rings as an alternative to hydrodynamic gas lubrication appears to be popular for larger plants, although a clear trend to increasing use of turboexpanders is evident. Aberle and Westbrook (1963) described large helium and nitrogen liquefaction systems with dry-rubbing reciprocating expansion engines. These were constructed by the Linde Company, Union Carbide Corp. of Tonawanda, New York for space simulators at the U.S. Air Force Arnold Engineering Development Center, Tullahoma, Tennessee and at the NASA Lewis Research Center, Cleveland, Ohio. Similarly, Sato (1968) has confirmed a preference for expansion engines with piston rings (of filled PTFE material) in the discussion of large-scale helium liquefiers constructed by the Tokyo Oxygen Co. of Japan.

An important and comprehensive discussion of a large reciprocating expansion engine for tonnage hydrogen liquefaction was given by Morain *et al.* (1963). Various aspects relating to the mechanical design of the engine were given with estimates of the various losses due to heat leaks and ring friction. Comparative data were presented for the predicted and actual engine performance on test.

Heylandt Crowned Piston

A significant improvement in expander technology occurred with the introduction by Heylandt (1912) of the crowned piston concept shown in Fig. 7.16. The normal piston carried a long extension of low thermal conductivity and the cylinder was similarly extended, again with low-conductivity material. The piston extension or piston crown was a clearance fit in the cylinder and moved without friction. The main piston was equipped with sealing rings or other device to prevent leakage from the cylinder past the piston. These operated at substantially ambient temperatures with a water cooled heat exchanger provided around the cylinder to carry off the frictional heat generated. Low-temperature expansion of the gas occurs in the cylinder above the piston crown thermally isolated by the low-conductivity materials of the piston crown and cylinder extension from the sealing device on the piston operating at ambient temperatures.

It will be recalled that Heylandt was also responsible for a modification of the Claude cycle where gas is admitted to the expander at ambient temperature downstream of the compressor aftercooler and at high pressure. Engines of the Heylandt type have been extensively used for the manufacture of industrial gases involving liquefaction and separation processes.

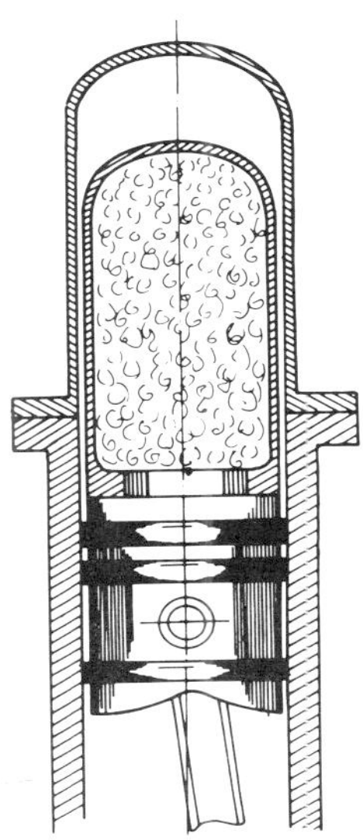

Fig. 7.16. Heylandt crowned piston.

Kapitza Hydrodynamic Lubricated Piston

Kapitza (1934), then a professor at the University of Cambridge, described an apparatus for the liquefaction of helium, essentially a pre-cooled Claude cycle system with a reciprocating expansion engine. Figure 7.17 is a reproduction of the sectional view of the system given by Kapitza.

The engine included some novel features, principally the provision of hydrodynamic gas lubrication of the piston. The piston of stainless steel with no sealing rings or other sealing device was made to be a loose fit in the cylinder of bronze. In operation the leakage past the piston was high during the early part of the power stroke when the pressure in the cylinder was maximum. Therefore a hydraulic work-absorbing system was used instead of a conventional crank-connecting rod system coupled to a dynamometer. The hydraulic mechanism permitted the power (expansion) stroke to be accomplished rapidly to minimize leakage past the piston with a relatively slow return. The piston diameter was 3.12 cm (1.25 in.) and the stroke varied from 3.5 to 5.0 cm (1.4 to 2.0 in.). Clearance of the piston in the cylinder was 0.05 mm (0.002 in.). It was found necessary to cut grooves in the piston along its length at intervals of 6.25 mm (0.25 in.) to prevent seizure of the engine. The grooves were 0.25 mm (0.01 in.) deep and wide.*

The engine operated at 2 Hz (120 strokes per minute) with an inlet pressure of 3 MPa and a discharge pressure of 0.22 MPa. The temperature change of helium in the engine was from about 20 K to 10 K. Interesting details of the subsequent development of this device were given by Malkov *et al.* (1968) and by Kapitza *et al.* (1968).

Collins Expansion Engine

The Collins system has become the virtual standard machine for use in the Western world. In a recent technical memorandum (CT1-1978) the manufacturers claimed that over 125 of the smaller model 1410 liquefier were in use in over 20 countries. A decade earlier Weinstock (1968), quoting a private communication, indicated that 500 Collins liquefiers were sold by A. D. Little in the twenty years since the prototype development in 1946. Whatever the number it is certainly true that up to 1950 the number of laboratories dealing with liquid helium was limited to a handful of specialist operations. The development of the Collins machine put the

* The piston would not be naturally centered in the cylinder as one might expect. Instead the high-pressure gas flowing past the piston from the high-pressure to the low-pressure region causes the piston to move to one side of the cylinder and provide the flow path with least restraint.

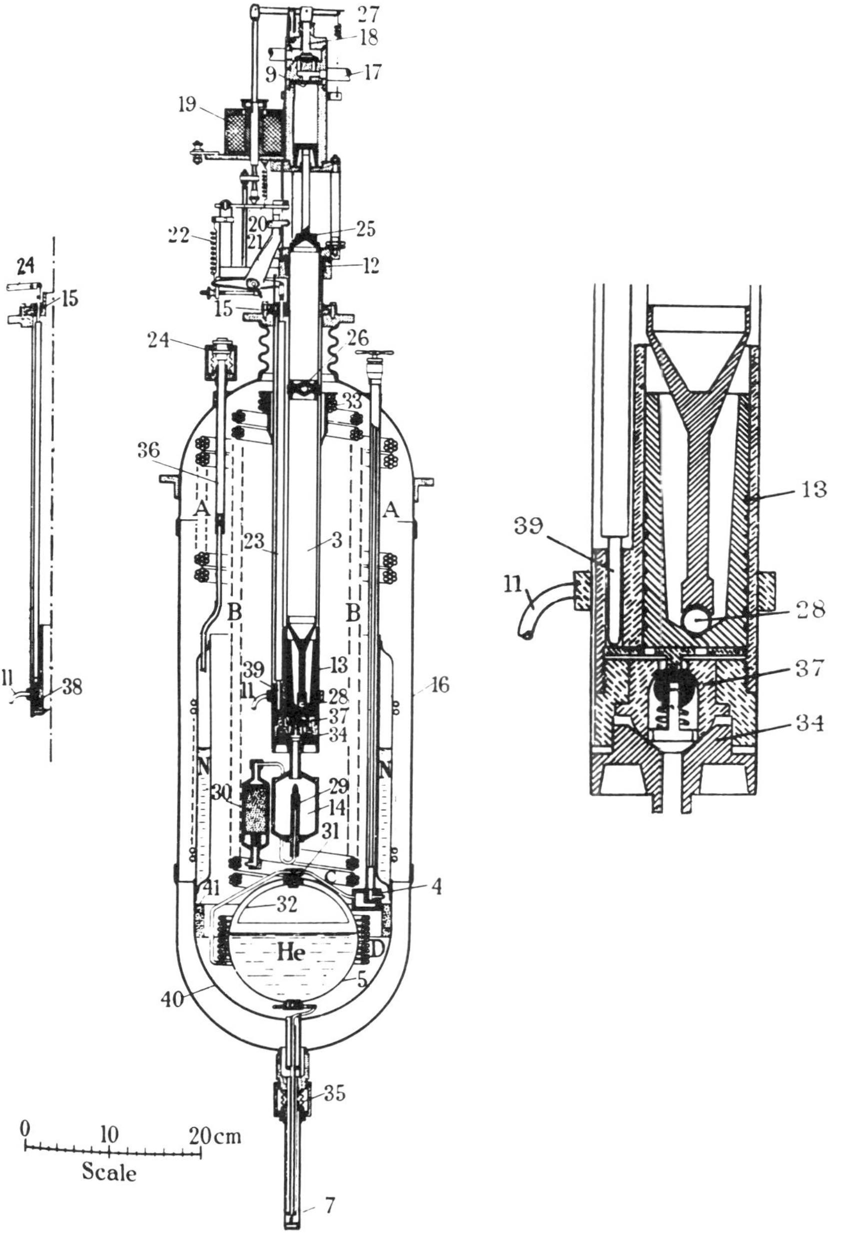

Fig. 7.17. Sectional view of the Kapitza helium liquefier including an expansion engine with hydrodynamic lubrication (after Kapitza, 1934).

production of liquid helium onto a routine push-button basis and thereby transformed the capabilities and potential of low-temperature research.

Collins evolved the design of the helium liquefier through a variety of systems. In his book Collins (1958) describes in detail an early diaphragm engine completed in 1938 capable of attaining temperatures of 10 K. The engine experienced minor leakage problems which had a highly deleterious effect on the vacuum insulation surrounding the engine. The Second World

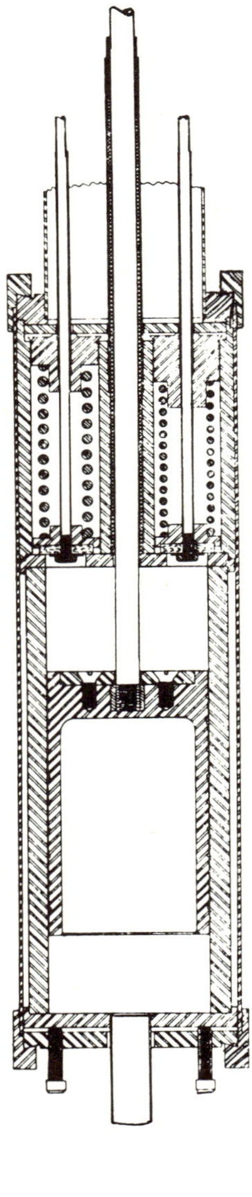

Fig. 7.18. Section diagram of low-pressure Collins expansion engine (after Collins, 1958).

War intervened and Collins concentrated his attention on the development of the low pressure air liquefier (shown in Fig. 7.7) with reversing recuperative heat exchangers used for oxygen production. The system incorporated expansion engines for relatively low inlet pressure of 0.6 to 1.5 MPa.

After the war Collins did not resume work on the diaphragm engines but incorporated his wartime experience in a multiple-expansion helium liquefier using reciprocating expansion engines. Figure 7.18 is a section view of a Collins expansion engine cylinder and valve chamber. Particular features that led to successful operation include the use of a very close fitting piston (of soft material over a steel core) sliding in a cylinder with hardened surface, a highly efficient continuous recuperative heat exchanger (see Chapter 8), and the design innovation of putting the piston and valve rods in *tension* rather than the customary compression mode. This latter feature permitted the use of thin flexible rods reducing the conductive heat leak and facilitating free movement of the piston in the cylinder.

Another important feature was to contain the engines and heat exchangers in a stainless steel Dewar vessel containing the liquid-helium product. The external environment of the engines and heat exchangers was low-pressure helium, therefore minor leaks from the pressurized engine/recuperator system were of no significance compared with a similar leak to a vacuum enclosure.

Further construction and operating details of the system were given by Collins (1947, 1952, 1956, 1958, and 1966). The following is a description of the chief characteristics of the helium liquefier by Collins (1958):

Rate of liquefaction	25–32 l/hr
Power required	45 kW
Helium circulated	215 g mol/min (185 cfm)
Operating pressure	12.5 atm
Refrigeration	Expansion engine plus liquid nitrogen
Liquid nitrogen	(If no liquid nitrogen is used, rate of helium liquefaction is 10 l/hr)
Heat exchanger	Hampson type
Actual work expended (N_2 plant included)	3.1 kW hr/l
Computed requirement (actual liquefier but with N_2 plant and helium compressor assumed reversible)	0.87 kW hr/l
Computed requirement (entire process reversible)	0.24 kW hr/l

Description of Liquefier

A flow diagram of the actual cycle is given in Figure [7.19]. It differs from the idealized cycle, not only because the heat exchangers and expansion engines are necessarily imperfect, but also because practical considerations have influenced the choice of apparatus and procedures. For the sake of compactness of the liquefier and greater production of liquid from available compressed helium, liquid nitrogen is employed to the extent of its utility. The transfer of heat from gaseous helium to liquid nitrogen evaporating at a constant temperature is irreversible, of course, and a net increase of entropy is incurred. A final difference lies in the substitution of a throttle valve for the fourth engine of the cycle.

In the flow diagram shown [Figure 7.19], compressed helium (about 12 atm) from the compressor is treated for entrained oil in an oil trap and for vaporized oil in a refrigerated heat exchanger 3. Thereafter the stream divides, about 8 percent going to heat exchanger 4, in which it is cooled to 80 K by means of liquid nitrogen and then expanded in engine E_1, the remainder going to the principal heat exchanger 5. The temperature of the gas in heat exchanger 5 ranges from room temperature at the upper end to 15 K at the bottom. At the zone of exchanger 5 where the temperature is 40–45 K, a second fraction (about 15 percent of the whole) of the compressed helium is led off for expansion in the second engine, E_2. At the lower end

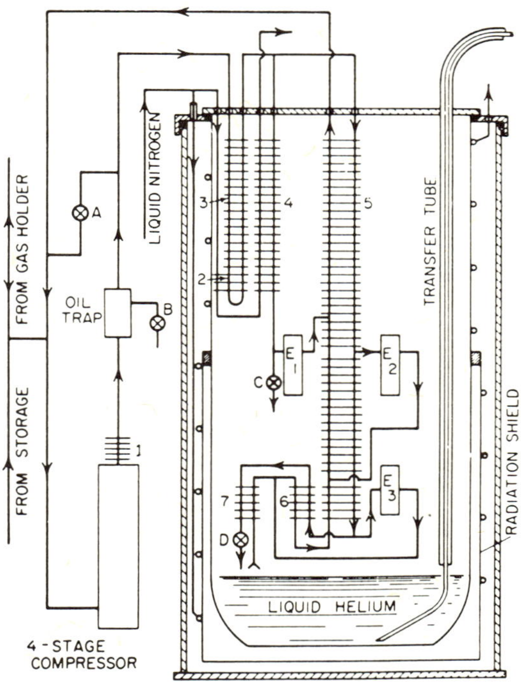

Fig. 7.19. Flow diagram of relatively large-scale Collins helium liquefier (after Collins, 1952).

of exchanger 5 a final division of the stream occurs. About 52 percent is used in expansion engine E_2 and 25 percent flows through the small exchangers 6 and 7 to the expansion valve D.

All the helium that enters exchanger 4 passes through engine E_1, the pressure falling to substantially one atmosphere and the temperature falling from 80 K to about 45 K. The expanded helium joins the low pressure stream in exchanger 5. All three engines are served by a single crankshaft and are identical in size. The bore is 2 inches, and the stroke is 2 inches. Each engine is comprised of a single cylinder. Although the three engines receive compressed helium from the same supply line and discharge their spent gas into the same low-pressure conduit, they operate at different temperature levels and embrace different sections of the heat-exchange system. Helium enters E_2 at about 45 K and is discharged at about 25 K. The inlet temperature of E_3 is about 15 K and the outlet temperature is estimated to be about 9 K.

Liquid helium forming at the throttle valve D drops into the bottom of the Dewar vessel in which the heat exchangers and engines are suspended. Space for 30 liters is provided.

Certain features of the earlier helium cryostat have been retained. The heat exchangers and engines, which hang from a steel plate, are surrounded by an atmosphere of helium rather than by the insulating vacuum. Minor leaks from the high-pressure stream can be tolerated. The helium atmosphere is contained by a large metal Dewar vessel, the vacuum jacket of which is continuously pumped. The lower half of the inner wall of the vacuum jacket is enclosed by a nitrogen-cooled radiation shield.

The engine cylinders and pistons are made of nitrided nitralloy and are so closely fitted that piston rings are unnecessary. The piston rods are relatively long and slender, and are made to operate in tension to promote perfection of alignment of the piston within the cylinder. There exists a thermal gradient in the helium atmosphere surrounding the heat exchangers and engines, the temperature being approximately 295 K at the top and 4.2 K at the bottom. As far as practicable, the cylinders of the engines are located at the proper elevation for matching temperatures inside with outside in order to reduce convection to a minimum. The stuffing boxes for the piston rods and valve pull rods and the running gear of the engines are placed on top of the lid of the Dewar so that heat generated in these parts can be kept out of the cold region.

The piston rods and valve pull rods are attached to the ends of horizontal walking beams, as in ancient steam engines. The beams are 2 feet in length, are pivoted at one end, and at the point immediately above the crankshaft are fitted with ball bearings to act as cam followers. With a stroke of only 2 inches the end of the piston rod travelling in an arc of 2-foot radius is not pulled away from the vertical by an appreciable angle. Speed control is achieved by centrifugal action of a split flywheel within a brake drum. The two halves of the flywheel are fitted with brake shoes. Speed is adjusted by changing the compression of a spring. This adjustment may be made while the engine is running.

The Collins engine became the prototype for a family of helium liquefier systems in series production up to the present time. Figure 7.20 shows the model 1430 helium liquefier system currently in production at

Fig. 7.20. Collins–Claude cycle helium liquefier (CTI Model No. 1430).

CTI, Waltham, Massachusetts. The system has the capacity to produce 5 to 40 liters per hour of liquid helium or 19 to 100 W of refrigeration at 4.6 K depending on the compressor option and mode of operation chosen. For maximum output the helium refrigerator module requires 40 liters per hour of liquid nitrogen and 2.5 kW of electric power. The liquefier module weighs 770 kg and is contained in an envelope $1.27 \times 0.375 \times 0.69$ m. The associated helium compressor module requires 100 kW of electric power and 70 kg per min of cooling water at 24°C and 0.3 MPa. The mass of the compressor module is 1455 kg and is contained in an envelope $0.415 \times 0.475 \times 0.715$ m. The constituents and general arrangement of a typical CTI/Collins helium liquefier are shown in Fig. 7.21.

Following the pioneer work by Collins at the Massachusetts Institute of Technology and the initial development of the commercial version at A. D. Little successive improvements and variants appeared. Pastuhov

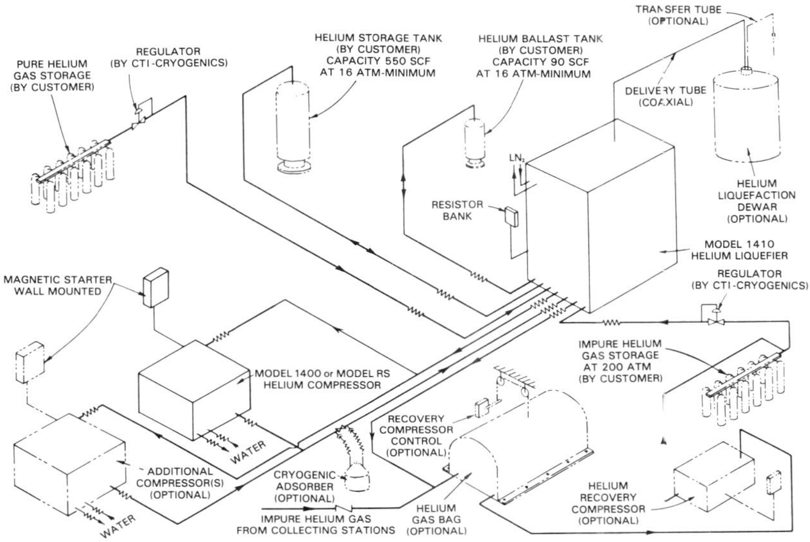

Fig. 7.21. Typical liquid helium installation utilizing Collins–Claude cryocooler.

(1960) of A. D. Little Inc. discussed application of a Joule–Brayton cycle variant of the Collins–Claude engine for refrigeration over the temperature range 20 to 100 K. Later Collins and Streeter (1968) described a variant which by the addition of a very low-pressure Joule–Thomson stage was able to generate refrigeration at 1.8 K. Robinson (1965) described a large-capacity (120 l/h) Collins helium liquefier constructed by A. D. Little. Further details of a large-scale helium refrigerator installed at the Brookhaven National Laboratory to cool a superconducting magnet were given by Robinson and Hogan (1968).

Subsequently, Johnson, Collins, and Smith (1971) described a modification of the large-scale liquefier where the Joule–Thomson expansion valve was replaced by an expansion engine with condensation of the helium during the expansion stroke. The engine was driven by a hydraulic plunger mechanism independent of the upper stage expansion engine. The valves were also operated by a hydraulic–pneumatic system. Use of the expansion engine instead of a Joule–Thomson valve increased the liquid yield by 33%.

Development of Collins engines stimulated the construction of similar machines of different sizes and capacities. Hood *et al.* (1964) discussed the performance of Joule–Brayton helium refrigerators for operation in the 10–30 K range that have been clearly inspired by Collins work. These units were constructed by Cryovac Inc. of Columbus, Ohio for the U.S. Air Force Arnold Engineering Development Center.

Zeitz (1963) described a closed cycle helium refrigerator for 2.5 K constructed by Air Products and Chemicals Inc., Allentown, Pennsylvania. This unit was a compound device consisting of a high-pressure (0.2–2 MPa) double expander Joule–Brayton refrigerator generating refrigeration at 10 K to precool a low-pressure (0.009–0.5 MPa) Linde–Hampson system generating refrigeration at 2.5 K. The expansion engines of the high-pressure system are acknowledged to be "similar to the machines developed by Collins and Kapitza."

Furihara *et al.* (1968) described what they whimsically called "a new type of closed cycle helium refrigerator for laser applications." From the description given it appeared to be a carbon copy of the Collins–Claude cycle double expander with final-stage Joule–Thomson expansion. A Collins-type expansion engine was also incorporated for helium expansion in a compound helium–hydrogen–neon refrigerating system described by Zeitz *et al.* (1965). It was constructed by Air Products and Chemicals Inc. for the NASA Lewis Research Center to cool a cryogenic magnet used for plasma heating and acceleration experiments.

Another large-scale (1.4 kW at 4.5 K) helium refrigerator incorporating two-stage Collins expansion engines was described by Winters and Snow (1966). The unit was constructed by Air Products and Chemicals Inc., Allentown, Pennsylvania for the NASA Langley Research Center. The same company also constructed the lower-capacity Claude cycle system using a single Collins expansion engine with no precooler described by Meier and Currie (1968). These units were built to provide 1 W of useful refrigeration at 4 K with additional refrigeration at 30 K. The units were designed for long-term reliable automatic operation with redundant compressor and expansion engines provided with automatic switches following component failure. At the time of reporting (1968) 10 systems in field use had accumulated 32,000 operating hours with only one machinery failure.

Doll–Eder Valveless Expansion Engines

Two researchers at the Bavarian Academy of Sciences in Munich, West Germany, R. Doll and F. X. Eder (1964) introduced an important new concept in expansion engine technology: the valveless engine with hydrodynamic gas lubricated piston. A simplified cross section of the engine is shown in Fig. 7.22 reproduced from Kneuer *et al.* (1968). The engine cylinder (1) has two annular channels (2) and (3) for the admission of high-pressure gas and the exhaust of low-pressure gas. The piston (4) has a central bore (5) connected via radial channels 6 to a ring groove (7) on

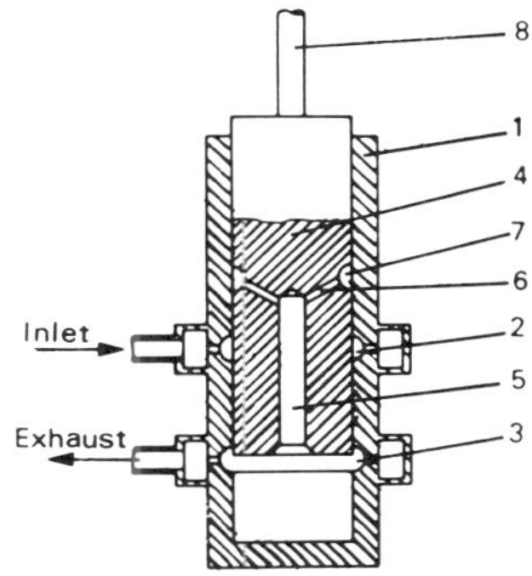

Fig. 7.22. Simplified cross section of Doll–Eder valveless, gas lubricated piston expansion engine.

the piston surface. Work is transferred via the piston rod to the crank-gear and hence to a dynamometer.

Gas flow and the resultant work diagram for the engine are illustrated in Fig. 7.23 (after Kneuer *et al.*, 1968). At condition 1, with the piston in the bottom dead center position, the ring groove on the piston is aligned with the inlet annular channel in the cylinder. High-pressure gas enters the engine filling the clearance space, volume V_1, principally the central bore of the piston. As the piston rises the piston ring groove and the cylinder inlet annular channel move out of alignment thereby closing off admission of further fluid at high inlet pressure. The fluid contained in the cylinder expands to a lower pressure and work is produced.

At condition (2) the lower edge of the piston clears the annular exhaust channel in the cylinder releasing expanded fluid to exhaust. The pressure drops rapidly to the exhaust pressure [condition (3)] as the piston ascends to top dead center. Finally the piston descends closing off the exhaust annular channel and compressing the gas retained in the cylinder to condition (4). Near the piston top dead center position the inlet port is again uncovered and fresh charge is admitted to the cylinder to repeat the cycle.

There is a leakage of high-pressure gas to exhaust in the annulus formed by the piston in the cylinder. The leakage is indicated, by the arrows on Fig. 7.23, as a flow from (a) the inlet to the exhaust annular channel, (b) the underside of the piston to the exhaust channel, and (c) the inlet channel to the space above the piston. This leakage flow is the hydrodynamic gas lubricant for the piston. Kneuer *et al.* (1968) identify the flow as a loss that is unavoidable with valveless engines and give the following equation for the magnitude of the flow loss:

$$\dot{v} = (p_1^2 - p_2^2)a^3 U / 24 p_1 L \beta \tag{7.15}$$

where $\dot{v}$ is the volumetric rate of flow, p_1 is the inlet pressure, p_2 is the

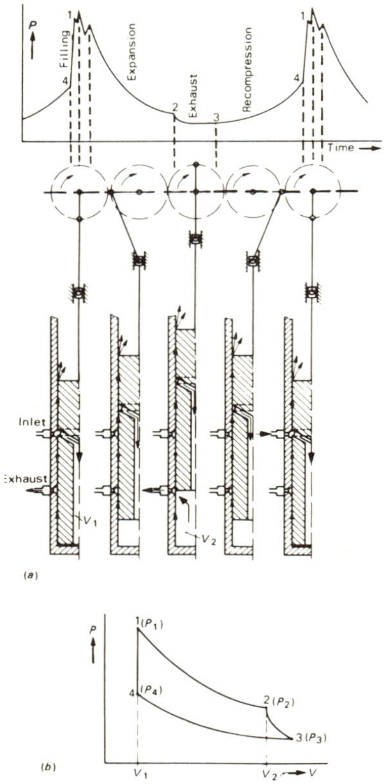

Fig. 7.23. Operational cycle and work diagram for Doll–Eder valveless engine.

exhaust pressure, a is half the clearance of the piston in the cylinder, U is the circumference of the piston, L is the length of the leakage path, and β is the viscosity of the working fluid. The flow is a function of the *cube* of the clearance in the cylinder and compels close tolerance clearance fitting of the piston in the cylinder. Kneuer *et al.* (1968) mention a clearance of less than 2 and 3 μ with piston diameters of 28 and 38 mm.

Becker *et al.* (1968) in their discussion of the leakage loss suggest that with 2 μ clearance the loss is 5% to 10% of the total gas flow with pressures in the range 0.3 to 3 MPa. This flow is said to be sufficient to stabilize the piston and provide complete gas lubrication:

> . . . by means of a large number of grooves on the piston surface. The clearance between successive grooves has a proper profile and achieves high stabilizing forces.

Doll–Eder valveless expansion engines have been used extensively by the German company Linde Aktiengesellschaft of Munich. In 1968 Kneuer *et al.* wrote that over 50 valveless engines had been incorporated in helium liquefiers in the previous four years. Baldus and Sellmaier (1968) described a complicated system to generate temperatures of 1.8 K in a helium II refrigerator utilizing a Doll–Eder valveless expansion engine in a Claude cycle system with a low-pressure Joule–Thomson expansion circuit generating 30 W of refrigeration at 1.8 K.

Publication of the original paper in 1964 stimulated considerable research on valveless expanders. Leger *et al.* (1971) described, in fascinating detail, the design and development of a Claude cycle helium liquefier with two valveless expansion engines, in a parallel arrangement. One interesting feature of the Leger engine was the use of different materials for the cylinder and the piston. The piston was made of material having a lower thermal expansion coefficient than the cylinder. On starting, the clearance of the piston in the cylinder was as high as 11 μ and reduced, due to differential contraction, to 2.5 μ at the normal running stage. The use of different materials is contrary to conventional practice where the same materials are used for piston and cylinder to maintain a constant clearance during cooldown. In early engines with the piston and cylinder made of the same steel, Leger found the piston cooled more slowly than the cylinder and resulted in premature wear contributing to reduced performance at low temperatures. In some cases the differential rate of cooling caused the engine to seize and resulted in deformation of the cylinder.

Matsubara *et al.* (1968) have described a hybrid expander with the inlet high pressure fluid entering through valveless ports as in the Doll–Eder engine but exhausting through a crank-operated valve of conventional design. This allows independent selection of exhaust valve timing but has no other advantage over the Doll–Eder engine and introduces the complexities of mechanically operated valves.

Another valveless, gas-lubricated expansion engine of exceptional interest was described, albeit briefly, by Clarke *et al.* (1968). The engine was a free-piston machine taking the form of a double-ended Doll–Eder engine with a single reciprocating element, the piston, operating in both cylinders. One end operated as a conventional Doll–Eder expansion engine at low temperature. The other end operated as an auxiliary compressor absorbing the work of the expansion engine to compress helium. The engine had a bore and stroke of 31.75 mm ($1\frac{1}{4}$ in.) and operated at a frequency of 45 Hz with an inlet pressure and temperature of 2.5 MPa and 24 K. It was said that

> . . . many hundreds of hours of development running carried out on prototype engines of this type confirm the original concept of simplicity and reliability. Depending on the design and operating requirements the efficiency on isentropic heat drop basis can be well over 70 percent . . .

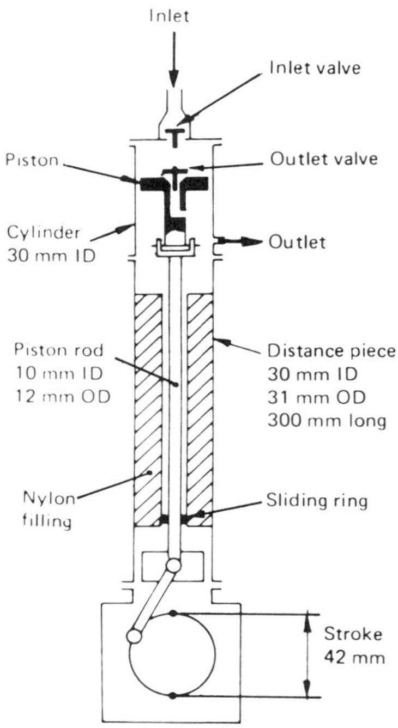

Fig. 7.24. Sketch of expansion engine with valves operated by piston motion (after Claudet *et al.*, 1972).

Claudet and Verdier (1972) described an interesting prototype expansion engine for Claude cycle systems with valves operated by the piston motion. Figure 7.24 is a sketch of the engine (after Claudet *et al.*, 1972) and Fig. 7.25 shows the relative piston and valve positions during operation.

The cycle of events in the engine is as follows:

(1) Exhaust. With the piston at bottom dead center and beginning its ascent the inlet valve is kept closed by the high pressure of the inlet fluid. The outlet valve located in the piston is opened by a spring F located under the valve. Gas is released from the cylinder to exhaust.

(2) Compression. Exhaust gas leaves the cylinder as the piston ascends towards top dead center. As the piston approaches the end of the upward stroke the spring F above the exhaust valve is compressed and closes the exhaust valve. Shortly thereafter the spring F is compressed sufficiently to cause the inlet valve to lift from its seat (alternatively an extension of the exhaust valve impinges on the underside of the inlet valve causing it to lift) and new charge is admitted to the engine.

(3) Induction. The piston passes the top dead center position with the inlet valve open and the exhaust valve closed. As the piston descends the

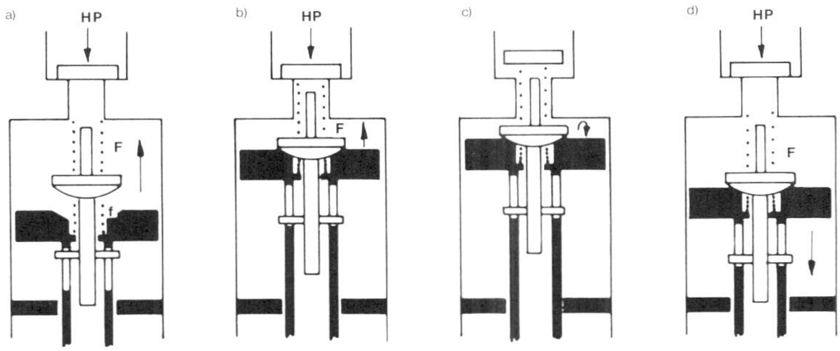

Fig. 7.25. Piston and valve positions in Claudet–Verdier engine operation (after Claudet *et al.*, 1972).

spring force F holding the inlet valve open decreases and the inlet valve closes.

(4) Expansion. The piston descends with both valves closed. The pressure decreases as the expansion proceeds.

The performance of the prototype Claudet/Verdier engine was most encouraging. With inlet–outlet pressures of 0.9 and 0.14 MPa they observed isentropic efficiencies in the range 50% to 70%. The temperature difference across the engine ranged from 5 K with an inlet temperature of 20 K to 26 K with an inlet temperature of 76 K.

Russian work on valveless piston expansion engines was described by Dobrov and Orlov (1974). Their engine utilized a six-phase gas distribution cycle in an attempt to overcome some of the deficiencies of the conventional four-phase Doll–Eder engine. The Dobrov–Orlov concept required mutual but out-of-phase reciprocation of both the piston and cylinder, but it is not clear how this was accomplished. One's feeling is that the added complexity obviates the principal attraction of simplicity in the Doll–Eder valveless engine.

Parulekar and Narayankhedkar (1972) and (1976) discussed the design and performance of a valveless expansion engine with novel features including a regenerative heat exchanger contained within the piston. The piston was activated by alternating fluid pressure in the "warm" and cold, expansion, part of the cylinder. Fluid pressure in the fluid drive region was regulated by solenoid valves activated by the piston motion. There seems every possibility this system could be made to operate at resonant frequency as a free piston machine of the same family as that described by Clarke *et al.* (1968). It could also be classified as an Ericsson engine of the types described in Chapter 5.

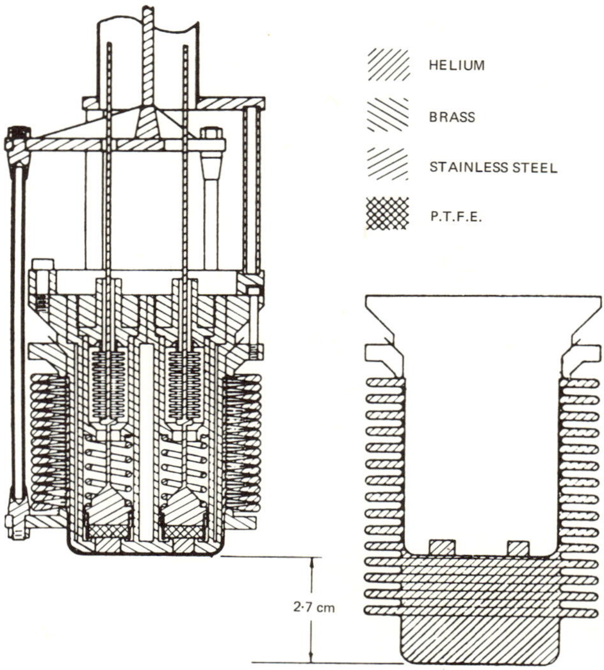

Fig. 7.26. Sectional view of formed bellows engine (after Long and Simon, 1954).

Bellows Expansion Engines

Replacement of the conventional cylinder and piston of an expansion engine by a flexible metal bellows was suggested by Simon as early as 1928. The earliest known account of a practical engine of this type was given by Long and Simon (1954). A simplified cross section of their engine is given in Fig. 7.26. The internal diameter of the bellows was 4 cm and the stroke of the engine was 2.7 cm with 15 bellows convolutions. The inlet and outlet pressures were 0.55 and 0.13 MPa and the frequency of operation was 2 to 3 Hz. Figure 7.26 shows the admission and discharge valves were contained in a central valve block within the bellows cavity. The valve rods operated in tension to open the valves against the force of stainless steel springs attempting to close the valves. Expansion of the bellows was restrained by (a) a thin-wall stainless-steel tube, in compression, supporting the valve block, and (b) a connecting rod in tension attached to the bottom flange of the bellows. Long and Simon reported they were able to liquefy helium with the bellows engine.

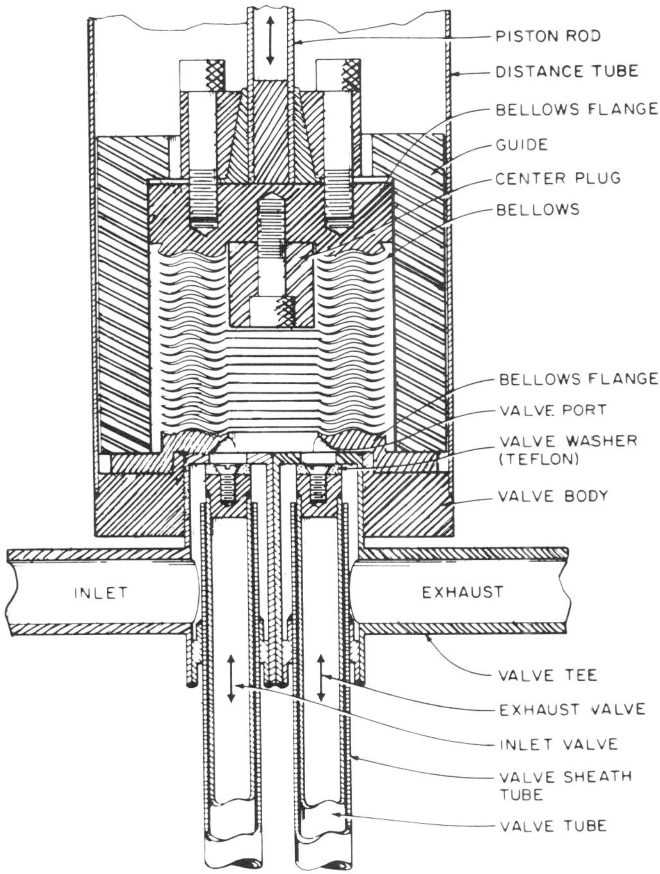

Fig. 7.27. Sectional view of welded bellows engine (after Smith, 1967).

The advantage of a bellows engine is elimination of the piston leakage and friction losses of conventional engines. In combination with a diaphragm compressor a completely sealed system free of all contaminants can be obtained. The principal disadvantage was expected to be the limited life of the bellows. The operating experience showed at low temperatures the fatigue life improved so much that by limiting the pressure a virtually unlimited life could be obtained.

Long and Simon used formed bronze bellows. They have a fairly high volume of retained gas when the bellows are in the compressed state. This

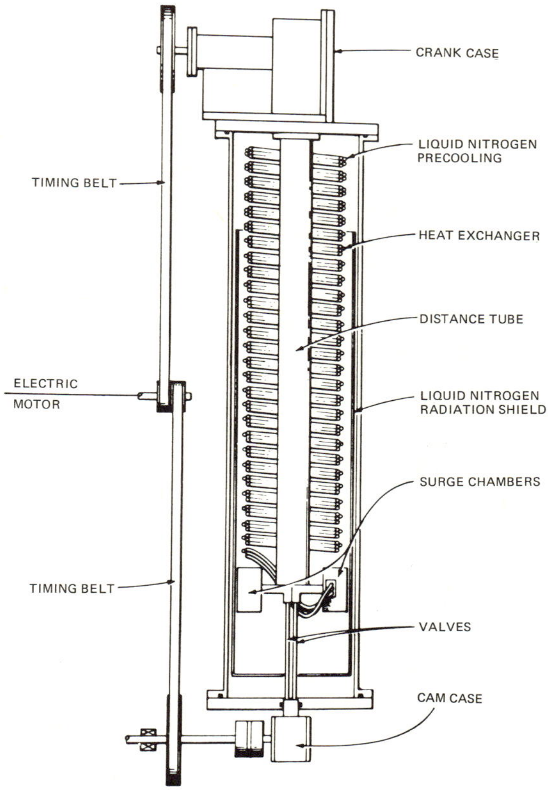

Fig. 7.28. Bellows engine test apparatus (after Smith, 1967).

corresponds to a conventional expander with a high clearance volume involving substantial recompression of expanded gas.

A later development of a bellows expansion engine was reported by Smith (1967). This work was carried out at the Massachusetts Institute of Technology for the U.S. Air Force. A cross section of Smith's bellows engine is shown in Fig. 7.27 and the engine test apparatus in Fig. 7.28. Smith utilized welded metal bellows of stainless steel in place of the formed bronze bellows of the earlier engine. The welded metal bellows were superior from two aspects. First, the bellows may be more tightly collapsed. They were formed by edge welding rings of the sheet metal so that only the weld beads prevent the convolutions from closing tight. Second, the bellows were made of high-tensile-strength stainless steel free from the effects of the severe deformations required to produce a formed bellows.

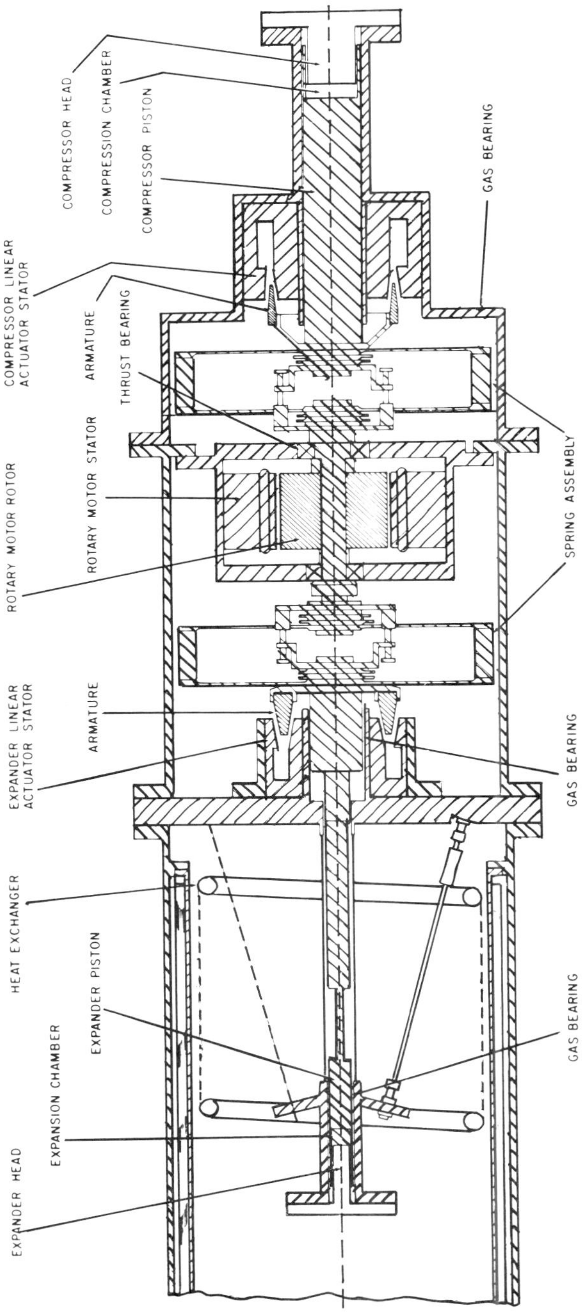

Fig. 7.29. Diagrammatic view of rotary-stroking engine (after Schulte *et al.*, 1965).

Smith used bellows having an outer diameter of 3.17 cm (1.25 in.) and an inner diameter of 1.27 cm (0.5 in.). Nineteen convolutions of 0.1 mm (0.004 in.) thickness provided a stroke of 1.9 cm (0.75 in.) and a displacement of 1.1×10^{-4} m^3 (0.44 in.^{3}). The inlet and discharge pressures were 0.51 and 0.11 MPa (75 and 16 psi). Two bellows to this design failed at 1.3×10^6 cycles (43 hrs) and 1.15×10^6 (38 hrs). The design was then modified as detailed by Smith to a bellows of 35 convolutions and a diaphragm thickness of 0.2 mm (0.008 in.) with the stress at the inner weld always in compression. At the time of reporting this assembly had operated for 7.7×10^6 cycles (258 hrs) without failure at temperatures of 28 and 18 K at inlet and exhaust, respectively.

Rotary Stroking Engine

Schulte *et al.* (1965) described a novel Joule–Brayton engine with reciprocating compressor and expander tuned to operate at a relatively high resonant frequency (60 Hz). The system is shown in Fig. 7.29 (reproduced from Schulte *et al.*, 1965). In addition to the reciprocating motion the dynamic elements were also driven with rotary motion so as to generate the hydrodynamic gas lubrication to maintain the piston "floating" in the cylinders.

Later Breckenridge (1969) described a development of the Schulte engine now expanded to a Claude cycle system with two stages of compression, two stages of engine expansion and a Joule–Thomson expansion stage. Subsequent developments of the rotary stroking engine have been reported regularly by Breckenridge. Many references under this name may be found listed in the Bibliography in both the open literature and the list of government reports.

Free-piston Stirling engines for power generation have been developed and are in series production (see Walker, 1980, Chapter 11). These have demonstrated the feasibility of resonant frequency systems with reciprocating elements and self-induced hydrodynamic gas lubrication. Elimination of the need to rotate as well as reciprocate the compressor–expander elements must lead to a simpler, and therefore more reliable, system. Given this parallel development it is hard to justify a sanguine view for the future of the Schulte rotary stroking engine.

ROTARY EXPANSION ENGINES

It is tempting and appropriate at this stage to venture forth into a review of rotary expanders (turbines) applied to Claude cycle systems. We shall resist the temptation. The subject is simply too big for this already over-long chapter to contain. Instead we consign turbines to a separate folio, companion to this volume, concerned primarily with large-scale refrigeration, liquefaction, and gas processing. Such division is in the natural order of things. Rotary machines, turbines, and compressors, are characteristically best suited for high flow rate, low-pressure ratio systems as compared with reciprocating devices which favor low-flow, high-pressure ratio systems.

Much work has been done on miniature turbo expanders and compressors for both Claude and Joule–Brayton cycle systems. It is appropriate to briefly review that work below since our interest is limited to small- and intermediate-capacity systems. However, a detailed consideration has been left to the companion volume. Many papers on miniature turbine systems may be found in the extensive Bibliography included here under the names of Colyer and Sixsmith. Some of these are briefly discussed below.

MINIATURE CLAUDE CYCLE SYSTEMS

There are few references to miniature Claude cycle systems. Most work has been done with miniature refrigerators for flight systems operating on aircraft or spacecraft. In most cases these operate without a Joule–Thomson valve for the final stage expansion and so are Joule–Brayton rather than Claude cycle systems. A miniature Claude or Joule–Brayton system was the rotary stroking engine reviewed above and described by Schulte *et al.* (1965) and later by Breckenridge (1969).

Much work has been done for the Air Force by the General Electric Co. Schenectady, New York on miniature rotary Joule–Brayton and Claude cycle systems. Gessner and Colyer (1968) have reviewed the early progress of this work and many subsequent reports have been made by Colyer (see Bibliography). Other miniature turbomachinery Claude and Joule–Brayton cycle systems have been described by Maddocks (1968). This work was done by the AiResearch Manufacturing Co., Los Angeles for the U.S. Army Satellite Communications Agency. Campbell (1967) reported on the development and performance of a "miniature Claude cycle" refrigerator. The system included a reciprocating engine but no Joule–Thomson expansion valve, so it was a Joule–Brayton rather than a Claude system.

It is not difficult to understand the paucity of literature. Claude systems are not well suited for miniaturization due to valves and recuperative heat exchangers. A better choice for miniature applications with recuperative heat exchangers is the Linde–Hampson system, basically a Claude cycle system with the expansion engine (a bulky, expensive unit) removed. The principal objective of miniaturization—an ultracompact design—is more readily achieved, and in small capacities there is no difficulty in using reciprocating compressors to achieve the high pressures characteristic of Linde–Hampson systems. Despite these advantages for miniaturized applications little development of Linde–Hampson systems has occurred. The reason is that the low flow rates of a miniature system require extremely fine passages in the recuperative heat exchangers and the Joule–Thomson valve. They are susceptible to contamination and blockage for the flow is continuous and one way so accretion of contaminants is continuous and cumulative. The need for uncontaminated fluid requires filtering and absorbing devices that make it difficult to achieve the low mass and volume objectives of miniaturized equipment.

Regenerative heat exchangers are the preferred alternative to recuperative heat exchangers. They tend to be self-cleaning and less susceptible to blockage by contaminants in the working fluid. They are easy to fabricate at low cost compared with recuperative heat exchangers and can be made very small and of low mass. Regenerative systems can be pressurized to a high level so as to keep the system volume and mass within acceptable bounds. They exist in enough variations, Stirling (integral or split), Vuilleumier, and Gifford–McMahon, to meet demand limitations better than any miniature Claude system yet in prospect.

TERBOT MIXED REFRIGERANT CYCLE

An interesting modification to the basic Claude cycle was described by Terbot (1976). The flow diagram shown in Fig. 7.30 is reproduced from Terbot (1976). In the low-temperature region the cycle utilizes helium (or any other gas) and is identical to the conventional Claude cycle with two stages of expansion engine and Joule–Thomson expansion valve.

The difference between the Terbot and conventional Claude cycle is that at the inlet to the compressor another refrigerant (Terbot suggests dichlorodifluoro-methane, refrigerant 12) is mixed with the helium and compressed with it in the compressor. After passing through the aftercooler and the upper-stage recuperative heat exchanger the mixture, cooled by this time to 120 K, consists of liquid R12 and gaseous H_2. The liquid R12 is separated from the helium and recirculated back through the upper stage recuperator mixed with helium *en route* to the compressor inlet. Addition of the refrigerant R12 to helium increases the molecular weight of the mixture to be compressed and allows use of "compact, reliable inexpensive standard centrifugal compressor[s]."

Terbot presented comparative data for conventional Claude cycle systems of intermediate and large capacity using reciprocating and oil-

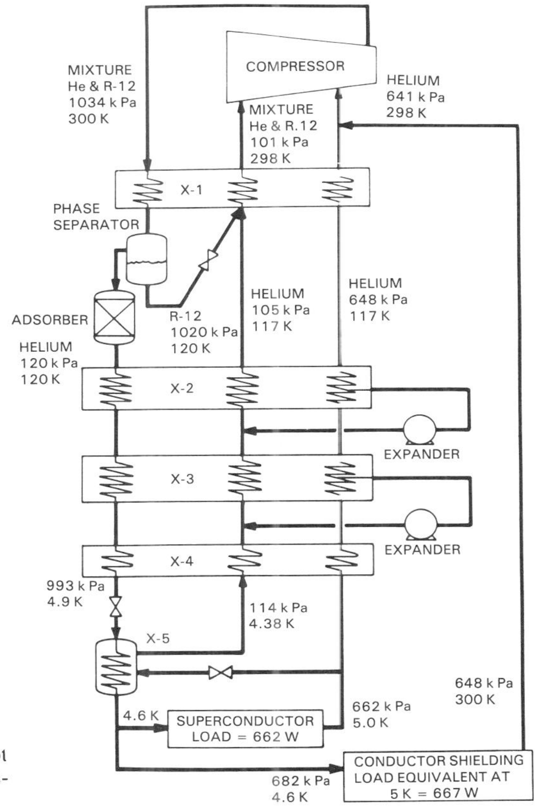

Fig. 7.30. Flow diagram for Terbot mixed refrigerant Claude cycle system (after Terbot, 1976).

flooded screw compressors and for the new mixed refrigerant cycle. The Claude cycles included liquid-nitrogen precooled cycles in some cases. The mixed refrigerant cycle emerged from this comparison indicating appreciable (15%) savings in the capital and operating cost.

No studies are known of the Terbot mixed refrigerant cycle in smaller systems using reciprocating compressors or with auxiliary refrigerants other than R12.

COMPRESSORS FOR CLAUDE CYCLE SYSTEMS

Compressors are a necessary component of all Claude, Joule–Brayton and Joule–Thomson systems. In some systems (Joule–Thomson) very high pressures (up to 40 MPa) are used to achieve an effective compact system. In other cases (Claude and Joule–Brayton), more moderate pressures can be utilized. In all, the compressor is vital and frequently a significant fraction of the overall cost. The main input of energy is power to drive the compressor—normally, the prime operating cost. The compressor is the element most likely to require maintenance attention and is the key component affecting reliability.

Improved reliability is often gained by the use of redundant (spare) units. In many cases the provision of a redundant compressor rather than duplication of the whole system dramatically improves the reliability of a system. Provision can be made for automatic start-up of the auxiliary or redundant compressor in the event of an unplanned shut-down of the principal compressor. Compressors can generally be classified into two types: (a) reciprocating compressors, and (b) rotary compressors.

Reciprocating Compressors

Reciprocating compressors are massive, bulky, slow running (up to 50 Hz) machines having a high isentropic efficiency able to handle a virtually unrestricted range of pressure ratios but comparatively small mass flow. They are easy to start and to control. They are insensitive to sudden and large changes in load. They have a high capital cost and, because they require relatively frequent maintenance, have a high operating cost.

Their efficiency can be improved with multistage compression processes having interstage cooling in air or water cooled heat exchangers. Aftercoolers downstream of the compressor cool the compressed gas to near ambient temperatures. Two stages of compression are commonly used for the low-pressure Claude cycles (e.g. pressure ratios of up to 8 or 10). Above this pressure ratio an increased number of stages may be used. It is rarely worth going beyond four stages for the degree of improvement occurs at a progressively diminishing rate.

Reciprocating compressors are found in a variety of shapes, sizes, and arrangements. They may be double or single-acting crank-connecting rod machines with automatic spring control or cam regulated valves. They may be air-cooled or water-cooled, with a variety of drive systems, electric motors, steam turbines, diesel engines. They exist in many mechanical arrangements: inline multiple cylinder, vee-cylinder, radial cylinder, or parallel cylinder with swash-plate drive.

An important aspect of compressors is the method of lubricating the rubbing surfaces, the reciprocating piston rings, the rotating crankshaft main bearings and the oscillatory big and little end bearings of the connecting rods. Rubbing surfaces properly lubricated with oil have superior friction and wear characteristics to any other known combination. The science of establishing and maintaining the hydrodynamic oil film of a self-energized metal piston ring sliding in a metal cylinder, developed to a high level for combustion engines, can be directly carried over to gas compressors. The act of providing an oil film on the cylinder wall must inevitably lead to contamination of the compressed gas by oil vapor and droplets. While of no concern to an internal combustion engine it is the principal drawback to oil lubrication in refrigeration compressors. Provision must be made downstream of the compressor to absorb or freeze out the entrained oil and other contaminants, such as water and carbon dioxide, that may be present in the compressed gas. Collins (1956) states that alternate layers of fine steel wool and fibre-glass have been found to be "especially effective in removing atomized oil."

To avoid the contamination problem many compressors are "dry-lubricated." No oil is provided for lubrication. Instead dry piston rings rub on cylinder walls and shafts run in bearings that are completely dry. This has long been done for air compressors using carbon piston rings and bearings, sometimes filled with a metal powder reinforcement. When the gases are very dry, carbon is unsuitable; apparently, a trace of water is an active participant in the local surface chemistry of a carbon bearing. A vapor pressure of −30°C is sometimes quoted as the criterion for use of a "dry-rubbing" carbon bearing. Below this the carbon behaves as an abrasive and the rubbing friction is much increased.

In these cases and many where carbon was previously used, another material is available, the plastic polytetrafluoroethylene (PTFE, Teflon, Fluon). It has a low coefficient of friction when rubbed on another surface coated with the same material. When a piston ring or bearing of PTFE material is rubbed on a metal cylinder wall or shaft the frictional resistance and wear are initially high but very quickly the metal wall is "plated" with an extremely thin layer of the wear debris torn from the PTFE component. Thereafter the friction and wear is reduced so low that many compressors now have PTFE-based piston rings. PTFE journal bearings are also used but to a lesser extent for prelubricated sealed roller or ball bearings can be used in these applications with very high reliability.

To improve the wear characteristics of PTFE a hard filler is added of metal powders, brass, copper, silver, etc., chopped glass, or carbon powder. The glass-filled PTFE material known as Rulon-A, dark red in color, enjoys a high reputation.

Successful designs of compressors could be engineered with hydrodynamic gas lubrication of pistons in the same way used by Kapitza and Collins in expansion engines. Gas-lubricated pistons are used routinely in free-piston Stirling engines (see Walker, 1980) which operate in both the compressive and expansive mode. The only compressor known to have gas-lubricated pistons is that included in the Joule–Brayton rotary stroking engine discussed above (Schulte *et al.*, 1965). This machine was complicated by rotating the reciprocating elements to energize the hydrodynamic bearing film. This is not acceptable for widespread application and has been demonstrated to be unnecessary by contemporary free-piston Stirling developments.

Reciprocating compressors are not particularly sensitive to the physical characteristics of the gas to be compressed. Moreover, the commercial demand for cryogenic systems has been relatively limited. As a consequence most of the reciprocating compressors found in cryogenic service are not recognizably different from the gas compressors found in other applications. Many compressors used are simply air compressors suitably modified for the gas to be compressed. Modification usually takes the form of an effort to reduce leakage and contamination by air or lubricant. For single-acting compressors the entire crankcase is sometimes made airtight. For double-acting compressors the provision of a specially packed piston rod gland is all that is necessary. With oil-lubricated machines a lubricant having a low vapor pressure is selected. Modifications to the intercooler and aftercooler may be necessary and supplementary filters and absorbers will likely be added.

Rotary Compressors

Rotary compressors may be characterized as small, light-weight, fast-running machines having a relatively low isentropic efficiency able to handle a limited range of pressure ratio with very high rates of mass flow. They are easy to start and control but have a limited range of operation and are acutely sensitive to sudden large changes in load. They are comparatively inexpensive and require little maintenance with an excellent record of reliability. There are three principal types of rotary compressor: axial flow, centrifugal flow, and rotary screw.

Centrifugal flow compressors are the simplest type with a single rotating disc carrying vanes. They are low cost and reliabile but have a relatively low efficiency and are capable of handling only a limited pressure ratio.

Axial flow compressors comprise several stages of fixed and moving blades in which the gas is successfully compressed. The compression ratio per stage is very small but there may be many stages and a high overall compression may be gained. Axial flow machines are expensive but can have a higher efficiency than centrifugal machines. They are sensitive to the physical characteristics of the fluid being compressed.

Rotary screw compressors appear to be finding increased application in cryogenic systems. They consist of two or more cylindrical rotors with multiple helical lobes or grooves meshing so that as the rotors contrarotate the space between the lobes decreases and the fluid trapped therein is compressed. Oil is sprayed into the rotors at the low pressure end and hence the term "oil-flooded screw compressor" is frequently used to describe the system.

Gas-lubricated bearings are used increasingly with the turbomachinery, compressors and expanders, in cryogenic systems over the entire range of refrigeration capacity from miniature systems to large-scale commercial gas liquefaction.

REFERENCES

Aberle, J. L., and Westbook, A. J. (1963). "Liquid Helium and Nitrogen Supply Systems for Space Simulators." *Adv. Cryog. Eng.* **8**, 190–198.

Anon.—CTI (1978). "Standard Helium Liquefier Model No. 1410." Tech. Mem. 91477, CTI, Waltham, Massachusetts.

Baldus, W., and Sellmaier, A. (1968). "A Continuous Helium II Refrigerator." *Adv. Cryog. Eng.* **18**, 434–440.

Barron, R. (1966). *Cryogenic Systems.* McGraw-Hill Book Co., Toronto.

Becker, H., Doll, R., and Eder, F. X. (1968). "Measurement of the Efficiency of the Valveless Doll–Eder Expansion Engine." *Proc. Second Int. Cryog. Eng. Conf.*, Iliffe Sci. and Tech. Pub., Guildford, U.K., pp. 9–11.

Breckenridge, R. W. (1969). "A 3.6 K Reciprocating Refrigerator." *Adv. Cryog. Eng.* **14**, 387–393.

Campbell, D. N. (1967). "Miniature Closed Circuit Single Stage Claude Cycle Refrigerator for 28 K." *Proc. First Cryog. Eng. Conf., Tokyo*, pp. 197–198, Heywood-Temple Ind. Pub., London.

Carbonnell, E. (1974). "Large Scale Refrigerators with Turbines." *Proc. Fifth Int. Cryog. Eng. Conf., Kyoto*, pp. 353–356, IPC Sci. and Tech., Guildford, U.K.

Clarke, M. E., and Gardner, J. B. (1968). "New Developments in Expansion Machinery for Low Temperature Refrigerators." *Proc. First Cryog. Eng. Conf., Tokyo*, pp. 273–275, Heywood-Temple Ind. Pub., London.

Class, C. R., Spero, R. P., and McIntosh, G. E. (1960). "Efficient Utilization of Ortho-Para Catalyst." *Adv. Cryog. Eng.* **3**, 64–72.

Claude, G. (1913). *Liquid Air Oxygen and Nitrogen.* P. Blakiston Son and Co., Phila, Pennsylvania.

Claudet, G., and Verdier, J. (1972). "Simplified Cryogenic Reciprocating Expansion Engine." *Proc. Fourth Int. Cryog. Conf.*, pp. 80–82, Eindhoven, IPC Sci. and Tech. Press Ltd., Guildford, U.K.

Collins, S. C. (1947). "A Helium Cryostat." *Rev. Sci. Instrum.* **18**, 157.

Collins, S. C. (1952). "Helium Liquefiers." *Science* **116**, 289.

Collins, S. C. (1956). *Helium Liquefiers and Carriers. Handbuch der Physik* Vol. **XIV**, 113–135.

Collins, S. C. (1966). "Helium Refrigerator and Liquefier." *Adv. Cryog. Eng.* **11**, 11–15.

Collins, S. C., and Cannaday, R. L. (1958). *Expansion Machines for Low Temperature Processes.* Oxford University Press, Oxford.

Collins, S. C., and Streeter, M. H. (1968). "Refrigerator for 1.8 K." *Proc. First Cryog. Eng. Conf., Tokyo*, pp. 215–217, Heywood-Temple Ind. Pub., London.

Cowans, K. W., and Walsh, P. J. (1965). "Continuous Cryogenic Refrigeration for Three to Five Micron Infrared Systems." *Adv. Cryog. Eng.* **10**, 468–476.

Danilov, I. B., and Kovatchev, V. T. (1972). "Two-Stage Expansion Engine with Differential Piston." *Proc. Fourth Int. Cryog. Eng. Conf.*, pp. 87–89, IPC Sci. and Tech. Press Ltd., Guildford, U.K.

Daunt, J. G. (1956). *The Production of Low Temperatures down to Hydrogen Temperatures. Handbuch der Physik* Vol. **IV**, Springer-Verlag, Berlin.

Davies, M. (1949). *The Physical Principles of Gas Liquefaction and Low Temperature Rectification.* Longmans Green and Co. Ltd., London.

Dobrov, V. M., and Orlov, A. V. (1974). "The Valveless Piston Expansion Engine with Six-Phase Gas Distribution Cycle." *Proc. Fifth Cryog. Eng. Conf., Kyoto,* IPC Sci. and Tech. Press Ltd., Guildford, U.K.

Doll, R., and Eder, F. X. (1964). "Gas Lubricated Low Temperature Piston Expansion Engine without Control Valves." *Adv. Cryog. Eng.* **9**, 561–564.

Furihara, A., Kikuchi, K., and Nagano, H. (1968). "Closed Cycle Helium Refrigerator for Laser Applications." *Proc. First Int. Cryog. Eng. Conf., Tokyo*, pp. 213–214, Heywood-Temple Ind. Pub., London.

Gardner, J. B., and Smith, K. C. (1960). "Power Consumption and Thermodynamic Reversibility in Low Temperature Refrigeration and Separation Processes." *Adv. Cryog. Eng.* **5**, 32–46.

Gessner, R. L., and Colyer, D. B. (1968). "Miniature Claude and Reverse Brayton Cycle Turbomachinery Refrigerators." *Adv. Cryog. Eng.* **13**, 474–484. See also: Colyer, D. B., and Gessner, R. L. (1968). "Miniature Cryogenic Refrigerator Turbomachinery." *Adv. Cryog. Eng.* **13**, 485–493.

Hood, C. B., Vogelhuber, W. W., and Barnes, C. B. (1964). "Helium Refrigerators for Operation in the 10–30 K Range." *Adv. Cryog. Eng.* **9**, 496–506.

Johnson, R. W., Collins, S. C., and Smith, J. L. (1971). "Hydraulically Operated Two-Phase Helium Expansion Engine." *Adv. Cryog. Eng.* **16**, 171–177.

Kapitza, P. (1934). "The Liquefaction of Helium by an Adiabatic Method." *Proc. R. Soc. London Ser. A* **147**, 189.

Kapitza, P. L., and Danilov, I. B. (1968). "Cascade Helium Liquefiers with Piston Type Engines." *Proc. First Int. Cryog. Eng. Conf., Tokyo*, pp. 228–231, Heywood-Temple Ind. Pub., London.

Kneuer, R., and Turnwald, E. (1968). "Measurements and Experience with Valveless Piston Expansion Engines." *Proc. Second Inst. Cryog. Eng. Conf., Brighton, U.K.*, pp. 12–15, Iliffe Sci. and Tech. Pub. Ltd., Guildford, U.K.

Land, M. L. (1960). "Expansion Turbines and Engines for Low Temperature Processing." *Adv. Cryog. Eng.* **5**, 250–260.

Leger, P., Thomas, P., and Zemati, C. (1971). "Design of Valveless-Type Piston Expanders with Variable Clearance for a 1.5 W, 4.5 K Helium Refrigerator." *Proc. Third Int. Cryog. Eng. Conf., Berlin*, pp. 259–263, Iliffe Sci. and Tech. Pub., Guildford, U.K.

Long, H. M., and Simon, F. E. (1954). *Appl. Sci. Res.* **4**, 237.

Maddocks, F. E. (1968). "Applications of Turbomachinery to Small-Capacity Closed Cycle Cryogenic Systems." *Adv. Cryog. Eng.* **13**, 463–473.

Malkov, M. P., and Danilov, I. B. (1968). "Cryogenic Equipment of the Institute for Physical Problems." *Proc. First Int. Cryog. Eng. Conf., Tokyo*, pp. 225–228, Heywood-Temple Ind. Pub., London.

Matsubara, Y., Ishizaki, Y., and Oshima, K. (1968). "New Type Expansion Engine for Refrigeration." *Proc. First Cryog. Eng. Conf., Tokyo*, pp. 210–211, Heywood-Temple Ind. Pub., London.

Meier, R. H., and Currie, R. B. (1968). "A 4 K Single Engine Cycle Helium Refrigerator." *Adv. Cryog. Eng.* **13**, 441–449.

Morain, W. A., and Homes, J. W. (1963). "An Analysis of the Performance of Large Reciprocating Expansion Engines with the Aid of a Computer and Laboratory Prototype." *Adv. Cryog. Eng.* **8**, 228–235.

Parulekar, B. B., and Narayankhedkar, K. G. (1972). "Small-Capacity Valveless Piston Expansion Engine Type Configuration." *Proc. Fourth Int. Cryog. Eng. Conf., Eindhoven*, pp. 90–92, IPC Sci. and Tech. Press Ltd., Guildford, U.K.

Parulekar, B. B., and Narayankhedkar, K. G. (1976). "Performance of the Valveless Low Temperature Heat Pump." *Proc. Sixth Int. Cryog. Eng. Conf.*, pp. 77–79, IPC Sci. and Tech. Press, Guildford, U.K.

Pastuhov, A. (1960). "Helium Refrigeration." *Adv. Cryog. Eng.* **1**, 41–43.

Robinson, G. Y. (1965). "Large Capacity Helium Liquefier." *Adv. Cryog. Eng.* **10**, 22–26.

Robinson, G. Y., and Hogan, W. H. (1968). "Large Scale Helium Refrigeration System for the Range 2–20 K." *Proc. First Cryog. Eng. Conf., Tokyo*, pp. 219–220, Heywood-Temple Ind. Pubs., London.

Sato, T. (1968). "Some Experiments on the Reciprocating Expander for Large Helium Liquefiers." *Proc. First Cryog. Eng. Conf., Tokyo*, pp. 217–218, Heywood-Temple Ind. Pubs., London.

Schulte, C. A., Fowle, A. A., Huechling, T. P., and Kronauer, R. E. (1965). "A Cryogenic Refrigerator for Long-Life Applications in Satellites." *Adv. Cryog. Eng.* **10**, 477–485.

Smith, J. L. (1967). "A Metal Bellows Expansion Engine." *Adv. Cryog. Eng.* **12**, 595–601.

Terbot, J. W. (1976). "A New Helium Refrigerator for Superconducting Cable Systems." *Adv. Cryog. Eng.* **21**, 190–196.

Walker, G. (1980). *Stirling Engines.* Oxford University Press, Oxford.

Weinstock, H. (1968). "A Review of Cryogenic History and Principles of Refrigeration." *Appl. Cryog. Technol.* **1**.

Winters, A. R., and Snow, W. A. (1966). "Capacity and Economic Performance of a Large 5 K Helium Refrigerator." *Adv. Cryog. Eng.* **11**, 116–125.

Witter, C. E. (1966). "Design of a Closed Cycle Helium Temperature Liquefier." *Adv. Cryog. Eng.* **11**, 107–115.

Zeitz, K., Geist, J. M., and Lashmet, P. K. (1965). "Methods for Producing Low Temperature Refrigeration." U.S. Patent 3,199,304.

Zeitz, K., and Woolfenden, B. K. (1963). "A Closed Cycle Helium Refrigerator for 2.5 K." *Adv. Cryog. Eng.* **8**, 206–212.

Name Index

Subject Index